Books Published in Industrial Biotechnology Series

1. Silkworm Biofactory: Silk to Biology
   Enoch Y Park, Katsumi Maenaka (eds.), 2019
2. Probiotics, the Natural Microbiota in Living Organisms: Fundamentals and Applications
   Hesham El Enshasy, Shang Tian Yang (eds.), 2021
3. Microbial Surfactants—Volume 1: Production and Applications
   Riyazali Zafarali Sayyed, Hesham Ali El Enshasy, Hameeda Bee (eds.), 2021
4. Microbial Surfactants—Volume 2: Applications in Food and Agriculture
   R.Z. Sayyed and Hesham Ali El-Enshasy (eds.), In Press

Series: Industrial Biotechnology

MICROBIAL SURFACTANTS
Volume 2: Applications in Food and Agriculture

Editors

R.Z. Sayyed

Head, Department of Microbiology
PSGVP Mandal's Arts Science & Commerce College
Shahada, India

Hesham Ali El-Enshasy

Institute of Bioproducts Development
University of Technology Malaysia
Johor Bahru, Malaysia

CRC Press
Taylor & Francis Group
Boca Raton London New York

CRC Press is an imprint of the
Taylor & Francis Group, an **informa** business
A SCIENCE PUBLISHERS BOOK

First edition published 2022
by CRC Press
6000 Broken Sound Parkway NW, Suite 300, Boca Raton, FL 33487-2742

and by CRC Press
2 Park Square, Milton Park, Abingdon, Oxon, OX14 4RN

CRC Press is an imprint of Taylor & Francis Group, LLC

Library of Congress Cataloging-in-Publication Data (applied for)

ISBN: 978-1-032-16247-8 (hbk)
ISBN: 978-1-032-16248-5 (pbk)
ISBN: 978-1-003-24773-9 (ebk)

DOI: 10.1201/9781003247739

Typeset in Times New Roman
by Radiant Productions

Preface to the Series

Industrial biotechnology has a deep impact on our lives, and is the focus of attention of academia, industry and governmental agencies and become one of the main pillars of knowledge based economy. The enormous growth of biotechnology industries has been driven by our increased knowledge and developments in physics, chemistry, biology, and engineering. Therefore, the growth of this industry in any part of the world can be directly related to the overall development in that region.

The interdisciplinary ***Industrial Biotechnology*** book series will comprise a number of edited volumes that review the recent trends in research and emerging technologies in the field. Each volume will covers specific class of bioproduct or particular biofactory in modern industrial biotechnology and will be written by internationally recognized experts of high reputation.

The main objective of this work is to provide up to date knowledge of the recent developments in this field based on the published works or technology developed in recent years. This book series is designed to serve as comprehensive reference and to be one of the main sources of information about cutting-edge technologies in the field of industrial biotechnology. Therefore, this series can serve as one of the major professional references for students, researchers, lecturers, and policy makers. I am grateful to all readers and we hope they will benefit from reading this new book series.

Series Editor

Prof. Dr. rer. Nat. Hesham A El Enshasy
Johor Bahru, Malaysia

Preface

Biosurfactants or green surfactants are the surface-active biomolecules produced by a wide variety of microorganisms. Different types of biosurfactants are produced by microorganisms. Their enormous diversity makes them an interesting molecule for applications in many fields such as agriculture and food. They promote seed germination, inhibit phytopathogens and induce systemic resistance in plants. Biosurfactants play vital role in motility, signaling and biofilm formation. They increase plant-microbe interaction and are the best and sustainable alternatives to chemical surfactants.

Biosurfactants are commercially important molecules due to their unique and useful features, such as high surface activity, low toxicity, stability over a wide range of pH values, temperature and ionic concentrations, biodegradability, excellent emulsifying ability and antimicrobial activity.

The prime aim of this book is to highlight over various aspects including basics to advances on concept, classification, production and applications of biosurfactants in various fields such as food and agriculture. It also provides excellent literature on fermentation, recovery, genomics and metagenomics of biosurfactant production. This volume is presented in an easy-to-understand manner, with well-illustrated diagrams, protocols, figures, and recent data on the industrial demand—Market and Economics and the production of biosurfactants from novel substrates are particularly worthwhile additions. As such, this book will prove useful for students, researchers, teachers, and entrepreneurs in the area of biosurfactants and its allied fields.

Contents

1

Microbial Biosurfactants

Sources, Classification, Properties and Mechanism of Interaction

P Saranraj,[1,*] *R Z Sayyed,*[2] *P Sivasakthivelan,*[3] *M Durga Devi,*[4] *Abdel Rahman Mohammad Al–Tawaha*[5] and *S Sivasakthi*[6]

1. Introduction

Surfactants are among the most versatile products of the industry. Surfactant is an abridgment of the term surface-active chemical compounds. A surfactant is a substance present in a system, having the property of adsorbing onto the surface or interface of the system and alters the interfacial surface energy to an exceptional extent. The term interface denotes the difference between the two immiscible phases, and the term surface denotes the interface where one phase is gas which is generally air. The interfacial energy is the minimum amount of work required to create that interface. In order to work out interfacial surface tension between two phases, interfacial energy per unit area is to be measured. It is the minimum amount of work required to create a unit area of the interface or to expand it by

[1] Department of Microbiology, Sacred Heart College (Autonomous), Tirupattur – 635 601, Tamil Nadu, India.

[2] Department of Microbiology, PSGVP Mandal's Arts, Science and Commerce College, Shahada – 425 409, Maharashtra, India.

[3] Department of Agricultural Microbiology, Faculty of Agriculture, Annamalai University, Annamalai Nagar – 608 002, Tamil Nadu, India.

[4] Department of Biochemistry, Sacred Heart College (Autonomous), Tirupattur – 635 601, Tamil Nadu, India.

[5] Department of Biological Sciences, Al - Hussein Bin Talal University, Maan, Jordan.

[6] Department of Microbiology, Shanmuga Industries Arts and Science College, Tiruvannamalai, Tamil Nadu, India.

* Corresponding author: microsaranraj@gmail.com

unit area. The physical phenomenon is additionally a measure of the difference in the nature of two phases meeting at the surface. The interfacial energy per unit area required to make the extra amount of that interface is the product of interfacial surface tension and increase in the area of the interface. A surfactant is, therefore, a substance which at a low concentration adsorbs at some or all interfaces in the system, and significantly changes the amount of work required to expand those interfaces (Rosen and Kunjappu 2012). Surfactants are surface-active agents with wide-ranging properties including the lowering of surface and interfacial tensions of liquids (Christofi and Ivshina 2002). Biosurfactants are biological compounds that exhibit high surface-active properties (Georgiou et al. 1992, Kuiper et al. 2004). Surfactants are characteristically organic compounds containing both hydrophobic groups which form an integral part of tails and therefore, the heads are composed of hydrophilic groups. The surfactant molecule contains both a water-insoluble which is an oil-soluble component and a water-soluble component. They are surface-active amphiphilic agents possessing both hydrophilic as well as hydrophobic moieties that reduce the surface and interfacial tensions by accumulating at the interface between two fluids such as oil and water, signifying that surfactants moreover assist the solubility of polar compounds in organic solvents (Fig. 1). They are of synthetic or biological origin. Due to their interesting properties like lower toxicity level, a higher degree of biodegradability, high foaming potential, and optimal activity at extreme conditions of temperature, pH level, and salinity, they have been attracting the attention of scientific and industrial communities. Increasing public awareness about environmental pollution influences the research and development of novel technologies that help in the clean-up of organic and inorganic contaminants like hydrocarbons and metals accumulated in society. An alternative and eco-friendly method of bioremediation technology for contaminated sites with these pollutants is the use of biosurfactants and biosurfactant-producing microorganisms. The diversity of biosurfactants makes them a vital group of compounds for potential use during a big variety of commercials and biotechnological applications (Pacwa Płociniczak et al. 2011).

The special properties of biosurfactants are that they allow their use and partial replacement of artificially or chemically synthesized surfactants in industrial operations. Chemically synthesized biosurfactants are mostly derived from oil and are widely utilized in cosmetics. Biosurfactants in aqueous solutions and hydrocarbon mixtures reduce surface tension, Critical Micelle Concentration (CMC), and interfacial tension (Rahman et al. 2002). Currently, Microbial biosurfactants have gained considerable interest in recent years due to their low toxicity, biodegradable nature, and diversity. Their range of potential industrial applications include enhanced

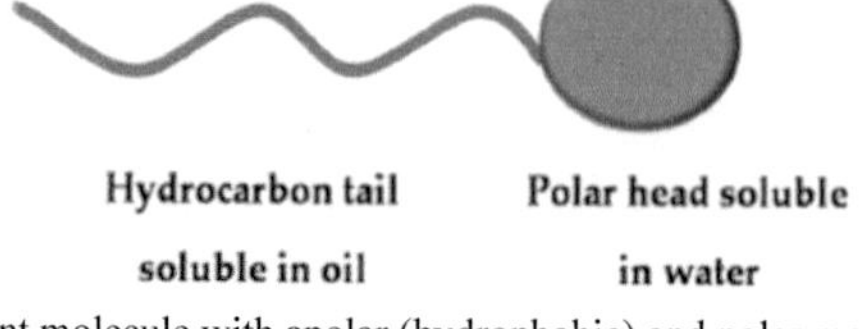

Figure 1. Surfactant molecule with apolar (hydrophobic) and polar moieties (hydrophilic) (Source: Danyelle Khadydja Santos et al. 2016).

oil recovery, crude oil drilling, lubricants, surfactant-aided bioremediation of water-insoluble pollutants, health care, and food processing (Sullivan 1998). Other areas that tap and use the potentials of biosurfactants are cosmetics and formulations of soaps, foods, and both dermal and transdermal drug delivery systems.

2. Microbial Biosurfactant

Microorganisms produce a wide range of secondary metabolites namely surfactants, generally called Biosurfactants. Microbial Biosurfactants are low molecular weight surface-active compounds widely produced by bacteria, fungi, and yeast. The term "Surfactants" was first coined by Antara products in 1950 which covered all the products having surface activity, such as, wetting agents, detergents, foaming agents, emulsifiers, and dispersants. Microbial surfactants are produced by a variety of microbes that possess different structures chemically and physically (Amaral et al. 2010). Microbial surface-active compounds are diverse molecules produced by different microorganisms in various structures and are mainly classified by their chemical structure and their microbial origin. They are made up of a hydrophilic moiety, comprising an acid, peptide cations, or anions, monosaccharides, disaccharides or polysaccharides, and a hydrophobic moiety of unsaturated or saturated hydrocarbon chains of fatty acids. The hydrophilic (polar) part of the biosurfactants is usually mentioned as 'head' and therefore the hydrophobic part (non-polar) is understood as 'tail' (Karanth et al. 1999). These structures confer a good range of properties, including the power to lower surface and interfacial surface tension of liquids and to make micelles and microemulsions between two different phases (Smyth et al. 2010).

Microorganisms utilize a spread of organic compounds because of the source of carbon and energy for their growth. When the carbon source is an insoluble substrate such as a hydrocarbon, microbes facilitate their diffusion into the cell by producing a spread of drugs, the biosurfactants. Some bacteria excrete ionic surfactant, which emulsifies hydrocarbon substrates in the growth medium (Guerra Santos et al. 1986). Some microbes are capable of adapting their cell wall, by synthesizing lipopolysaccharides or non-ionic surfactants in their cell wall. The exact reason why some microorganisms produce surfactant is unclear (Deziel et al. 2000).

According to Danyelle Khadydja Santos et al. (2016), microorganisms use a group of carbon sources and energy for growth. The mix of carbon sources with insoluble substrates favors the intracellular diffusion and production of various substances. Microorganisms like bacteria, filamentous fungi, and yeasts can produce biosurfactants with different molecular structures and surface activities (Campos et al. 2013). In recent decades, there has been a rise in scientific interest regarding the isolation of microorganisms that produce tensioactive molecules with good surfactant characteristics, like a coffee Critical Micelle Concentration (CMC), low toxicity, and high emulsifying activity (Silva et al. 2014).

The literature describes the genera *Pseudomonas* and *Bacillus* as great biosurfactant producers (Silva et al. 2014). However, most biosurfactants of a bacterial origin are inadequate for use in the food industry due to their possible pathogenic nature (Shepherd et al. 1995). *Candida bombicola* and *Candida lipolytica* are among

Table 1. Microorganisms involved in production of major Biosurfactants.

Biosurfactant class	**Microorganisms**
Glycolipids	
Rhamnolipids	*Pseudomonas aeruginosa*
Trehalose lipids or Trehalolipids	*Mycobacterium* sp., *Nocardia* sp., *Corynebacterium* sp., *Rhodococcus erythropolis*, *Rhodococcus ruber* and *Arthrobacter* sp.
Sophorolipids	*Torulopsis bombicola*, *Torulopsis apicola*, *Torulopsis petrophilum*, *Wickerhamiella domericqiae* and *Candida bogoriensis*
Mannosylerythritol lipids	*Candida antarctica*, *Psdozyma antarctica* and *Psdozyma rugulosa*
Cellobiolipids	*Ustilago zeae* and *Ustilago maydis*
Lipopeptides and Lipoproteins	
Surfactin	*Bacillus subtilis*
Subitisin	*Bacillus subtilis*
Lichenysin	*Bacillus licheniformis*
Iturnin	*Bacillus subtilis*
Fengycin	*Bacillus subtilis*
Viscosin	*Pseudomonas fluorescens*
Arthrofactin	*Pseudomonas* sp.
Amphisin	*Pseudomonas syringae*
Putisolvin	*Pseudomonas putida*
Serrawettin	*Serratia marcescens, Serratia liquefaciens* and *Serratia rubidaea*
Gramicidin	*Brevibacterium brevis*
Polymyxin	*Bacillus polymyxa*
Peptide lipid	*Bacillus licheniformis*
Fatty acids, Phospholipids and Neutral lipids	
Fatty acids	*Corynebacterium lepus*
Corynomicolic acids	*Corynebacterium insidibasseosum*
Phospholipids	*Rhodococcus erythropolis*, *Thiobacillus thiooxidans, Corynebacterium lepus*, *Micrococcus* sp., *Acinetobacter* sp. and *Candida* sp.
Neutral lipids	*Thiobacillus thiooxidans*
Ornithine lipids	*Pseudomonas* sp., *Thiobacillus* sp., *Gluconobacter* sp. and *Micrococcus* sp.
Polymeric Biosurfactants	
Emulsan	*Acinetobacter calcoaceticus*
Apoemulsan	*Acinetobacter venetianus*
Alasan	*Acinetobacter calcoaceticus* and *Acinetobacter radioresistens*
Liposan	*Candida lipolytica*
Biodispersan	*Acinetobacter calcoaceticus*
Lipomanan	*Candida tropicalis*
Carbohydrate – Protein - Lipid	*Pseudomonas fluorescens* and *Debaryomyces polymorphis*
Protein PA	*Pseudomonas aeruginosa*
Aminoacid – Lipids	*Bacillus* sp.
Particulate Biosurfactants	
Vesicles and Fimbriae	*Acinetobacter calcoaceticus*
Whole cells	Variety of bacteria

Source: Jitendra Desai and Ibrahim Banat (1997), Krishnaswamy Muthusamy et al. (2008)

the foremost commonly studied yeasts for the assembly of biosurfactants. A key advantage of using yeasts, like *Yarrowia lipolytica*, baker's yeast, and *Kluyveromyces lactis*, resides in their "Generally considered safe" (GRAS) status. Organisms with GRAS status do not offer the risks of toxicity or pathogenicity, which allows their use in the food and pharmaceutical industries (Campos et al. 2013). Biosurfactants have many environmental applications like bioremediation and dispersion of oil spills, enhanced oil recovery, and transfer of petroleum. Biosurfactants are potentially employed in the areas related to food, cosmetics, health care industries, and cleaning toxic chemicals of industrial and agricultural origin. Biosurfactants produced by various microorganisms together with their properties are listed in Table 1.

3. Classification of Biosurfactants

Biosurfactants were categorized according to their microbial origin and chemical structure. Rosenberg and Ron (1999) classified biosurfactants into two categories based on their Molecular weight. They are (i) High molecular weight molecules (high mass) and (ii) Low molecular weight molecules (Low mass). Low mass biosurfactants include Glycolipids, Lipopeptides, Lipoproteins, Fatty acids, and Phospholipids, whereas High mass surfactants include Polymeric and Particulate surfactants. Bacteria are considered the major producers of biosurfactants among microbes. The production of biosurfactants from microbes depends on the conditions in which they are fermented, such as, environmental factors, and availability of nutrients. The High molecular weight microbial surfactants are generally polyanionic heteropolysaccharides containing both polysaccharides and proteins, the Low molecular weight microbial surfactants are often Glycolipids. The production of microbially mediated biosurfactants differs from the environment of the growing microorganism based on its nutritional status. Microbial cells with high cell surface hydrophobicity are also surfactants and they themselves play a natural role in the growth of microbial cells on hydrocarbons, and sulfur which are water-insoluble substrates. Cell adhesion, cell aggregation, emulsification, flocculation, dispersion, and desorption phenomena were mediated by the exocellular surfactants (Karanth et al. 1999).

Polar grouping was the base in classifying the Artificially synthesized surfactants, whereas the biosurfactants were categorized by their chemical composition and their microbial origin. In general, their structure includes a hydrophilic moiety consisting of amino acids or peptides anions or cations; monosaccharides, disaccharides, or polysaccharides; and a hydrophobic moiety consisting of unsaturated, saturated, or fatty acids. The different groups of biosurfactants include Glycolipids, lipoproteins and Lipopeptides, Phospholipids and Fatty acids, Polymeric surfactants, and Particulate surfactants. A number of reports on the synthesis of biosurfactants by hydrocarbon-degrading microorganisms were recorded along with the data stating that water-soluble compounds such as Glucose, Sucrose, Glycerol, or Ethanol were also capable of producing the same (Cooper and Goldenberg 1987, Hommel et al. 1994, Passeri 1992). The biosurfactant-producing microbes are distributed among a wide group of genera.

3.1 Glycolipids

Glycolipids are carbohydrates like monosaccharides, disaccharides, trisaccharides and tetrasaccharides that include glucose, mannose, galactose, glucuronic acid, rhamnose and galactose sulphate combined with long-chain aliphatic acids or hydroxy aliphatic acids. Popular biosurfactants are Glycolipids which are carbohydrates in combination with long-chain aliphatic acids or hydroxy aliphatic acids which are linked by ether or an ester group. The best-known Glycolipids are Rhamnolipids, Trehalolipids and Sophorolipids. Other types of Glycolipids have been reported in the literature such as Glycoglycerolipid (Nakata 2000), Sugar based bioemulsifiers (Kim et al. 2000), Cellobiolipids, Mannosylerythritol lipid A (Van Hoogmoed et al. 2000) and many different Hexose lipids (Golyshin et al. 1999).

3.1.1 Rhamnolipids

Certain species of Pseudomonas are characterized to produce large amounts of biosurfactants containing one or two molecules of rhamnose linked to one or two molecules of hydroxydecanoic acid (Benincasa et al. 2004, Nitschke et al. 2005, Monteiro et al. 2007, Pornsunthorntawee et al. 2008). In Rhamnolipids, the molecules of Rhamnose are linked to one or two molecules of β-hydroxydecanoic acid, and are the best-studied Glycolipids. The first report on the Production of rhamnose containing glycolipids was reported in *Pseudomonas aeruginosa* by Jarvis and Johnson (1949). L-Rhamnosyl-L-rhamnosyl-β-hydroxydecanoyl-β-hydroxydecanoate (Fig. 2) and L-rhamnosyl-β-hydroxydecanoyl-β-hydroxydecanoate, denoted as Rhamnolipid 1 and 2, respectively, are the principal Glycolipids produced by *Pseudomonas aeruginosa* (Edward and Hayashi 1965, Hisatsuka et al. 1977). The formation of rhamnolipid types 3 and 4 containing one b-hydroxydecanoic acid with one and two rhamnose units, respectively (Syldatk et al. 1985), methyl ester derivatives of Rhamnolipids 1 and 2 (Hirayamo and Kato 1982), and Rhamnolipids with alternative fatty acid chains have also been reported (Parra et al. 1989, Rendell et al. 1990). Edward and Hayashi (1965) have reported formation of Glycolipid, type R-1 containing two rhamnose and two hydroxydecanoic units by *Pseudomonas aeruginosa*. A second kind of rhamnolipid (R-2) containing one rhamnose unit was reported by Rendell et al. (1990). These mutants develop when the growth medium is supplemented with rhamnolipid. Rhamnolipids from *Pseudomonas* spp. have been demonstrated to lower the interfacial tension against n-hexadecane to 1 mN/m and the surface tension to 25 to 30 mN/m (Lang and Wagner 1987, Parra et al. 1989). They also emulsify alkanes and stimulate the growth of *Pseudomonas aeruginosa* on

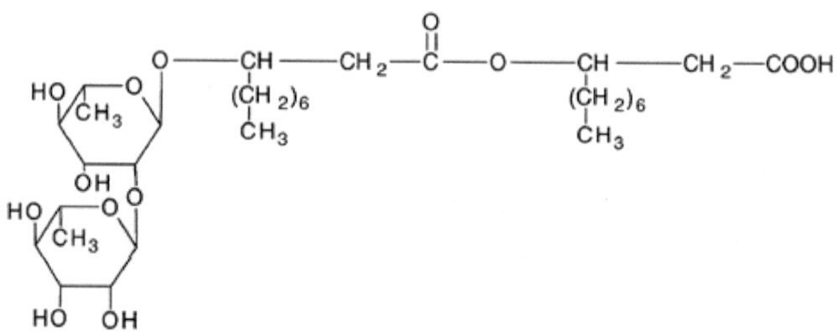

Figure 2. Structure of Rhamnolipid.

hexadecane (Hisatsuka et al. 1977). Gas chromatographic analysis of hydroxyl fatty acids rhamnolipid produced by *Pseudomonas aeruginosa* showed that positions of the fatty acids in the lipid moiety were variable (Monteiro et al. 2007).

3.1.2 Trehalose Lipids or Trehalolipids

Several structural types of microbial trehalose lipid biosurfactants have been reported and the structure of Trehalolipids or Trehalose lipid is given in Fig. 3. Disaccharide trehalose linked at C-6 and C-6' to mycolic acids is associated with most species of *Mycobacterium* sp., *Nocardia* sp. and *Corynebacterium* sp. (Lang and Wagner 1987, Cooper et al. 1989). Mycolic acids are long-chain, α-branched-β-hydroxy fatty acids. Microorganisms produce Trehalolipids in different sizes and structures of mycolic acid and they also differ in the number of carbon atoms and degrees of unsaturation (Lang and Wagner 1987, Desai and Banat 1997). Trehalose dimycolate produced by *Rhodococcus erythropolis* and *Arthrobacter* sp. has been extensively studied (Kretschmer et al. 1982). *Rhodococcus erythropolis* also synthesizes a novel anionic trehalose lipid (Ristau and Wagner 1993). Trehalose lipids from *Rhodococcus erythropolis* and *Arthrobacter* sp. reduced the surface and interfacial tensions in the culture broth from 25 to 40 and 1 to 5 mN/m, respectively (Lang and Wagner 1987, Li et al. 1984). Philp (2002) reported the production of trehalose lipids from alkanotrophic *Rhodococcus ruber* on gaseous alkanes propane and butane.

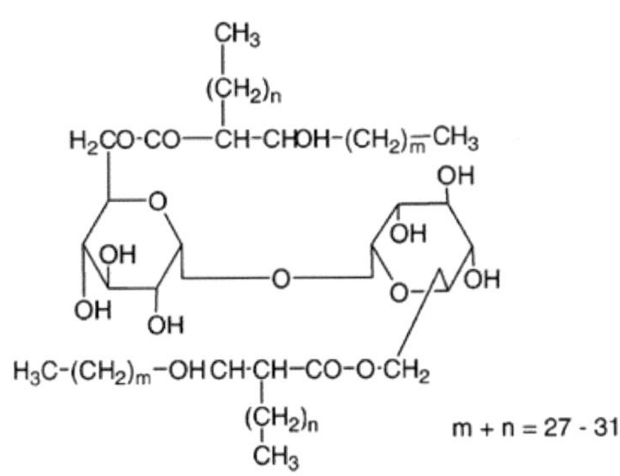

Figure 3. Trehalose lipids or Trehalolipids.

3.1.3 Sophorolipids

Sophorolipids consist of a dimeric carbohydrate sophorose attached with a long-chain hydroxyl fatty acid and are mainly produced by yeasts such as *Torulopsis bombicola*, *Torulopsis apicola* (Tullock et al. 1967) and *Wickerhamiella domericqiae* (Chen et al. 2006) and consist of a dimeric carbohydrate sophorose linked to a long-chain hydroxyl fatty acid by glycosidic linkage. Biosurfactants are made up of a mixture of six to nine various hydrophobic sophorosides. Similar mixtures of water-soluble sophorolipids from several yeasts have also been reported (Hommel et al. 1987). Cutler and Light (1979) showed that *Candida bogoriensis* produces glycolipids in which sophorose is linked to docosanoic acid diacetate. Sophorolipids have the capacity to lower the surface tension of water from 72.8 mN/m to 40–30 mN/m, with a CMC of 40 to 100 mg/L (Van Bogaert et al. 2007). Sophorolipids were produced by *Torulopsis petrophilum* in water-insoluble substrates like alkanes and vegetable oils (Cooper and Paddock 1983). These sophorolipids, which were chemically identical to those produced by *Torulopsis bombicola,* did not emulsify alkanes or vegetable oils. Sophorolipids were not produced, when *Torulopsis petrophilum* was grown on a glucose yeast extract medium, instead, an effective protein-containing alkane emulsifying agent was formed (Cooper and Paddock 1983). These findings seem to contradict the conventional statement that surfactants as well as microbial emulsifiers are produced to enhance the uptake of water-insoluble substrates. Sophorolipids lower surface and interfacial tension but they are not effective emulsifying agents (Cooper and Paddock 1984). Lactonic and acidic sophorolipids lowered the interfacial tension between n-hexadecane and water from 40 to 5 mN/m and exhibited stability towards changes in pH and temperature (Lang et al. 1989). Mostly Sophorolipids occur as a mixture of macrolactones and free acid form. It is reported that the lactone form of the Sophorolipid is required, or preferable, for various applications (Hu and Ju 2001). These biosurfactants are made up of a mixture of about six to nine different hydrophobic Sophorolipids (Fig. 4).

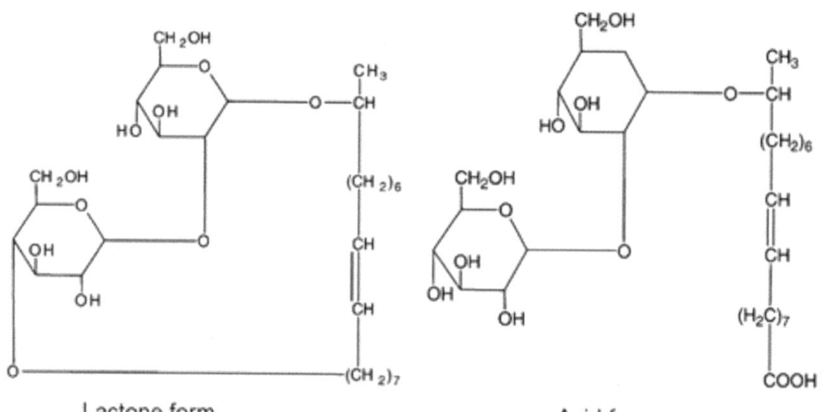

Figure 4. Structure of Lactonized and Free-acid forms of Sophorolipids.

3.1.4 *Mannosylerythritollipids*

Mannosylerythritollipids are the glycolipid biosurfactants consisting of a sugar called mannosylerythritol and are synthesized by yeast like *Candida antarctica* (Crich et al. 2002) and *Candida* sp. (Kim et al. 1999). The fatty acid component of biosurfactant was determined to be hexanoic, dodecanoic, tetradecanoic or tetradecenoic acids (Kim et al. 1999). Mannosylerythritol lipids (Fig. 5) synthesized by *Candida* sp. lowered the surface tension of water to 29 dyne/cm at CMC of 10 mg/L and the minimum Interfacial tension was 0.1 dyne/cm against Kerosene (Kim et al. 1999). Fukuoka et al. (2007) have characterized the surface-active properties of a new glycolipid biosurfactant, mono acylated mannosylerythritol lipid produced by *Psdozyma antarctica* and *Psdozyma rugulosa.*

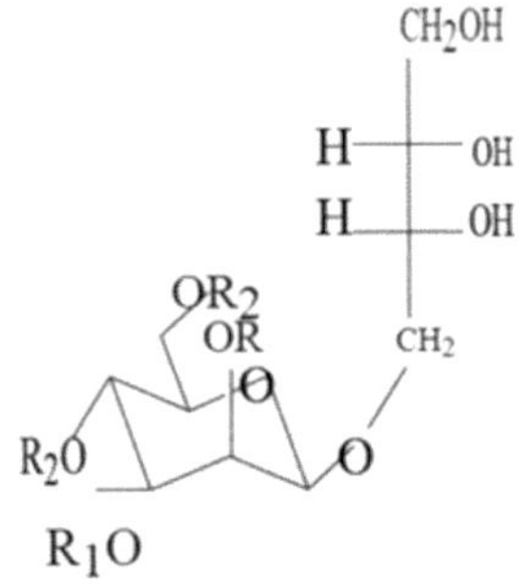

Figure 5. Mannosylerythritollipids.

3.2 Lipopeptides and Lipoproteins

Cyclic lipopeptides including decapeptide antibiotics (gramicidins) and lipopeptide antibiotics (polymyxins) are produced by *Bacillus brevis* (Marahiel et al. 1977) and *Bacillus polymyxa* (Suzuki et al. 1965), respectively, which possess surface-active properties. Ornithine-containing lipids from *Pterostylis rubescens* (Yamane 1987) and *Thiobacillus thiooxidans* (Knoche and Shiveley 1972), cerilipin, an ornithine- and taurine containing lipid from *Gluconobacter cerinus* (Tahara et al. 1976), and lysine-containing lipids from *Agrobacterium tumefaciens* (Tahara et al. 1976) also exhibit excellent biosurfactant activity. An aminolipid biosurfactant called serratamolide has been isolated from *Serratia marcescens* (Mutsuyama et al. 1985). Biosurfactant increased cell hydrophilicity by blocking the hydrophobic sites on the surface of the cells when negatively mutated by serratamolide (Bar Ness et al. 1988).

3.2.1 Surfactin

Surfactin, a cyclic lipopeptide is one of the most effective powerful biosurfactants known so far, which was first reported in *Bacillus subtilis* (Fig. 6) (Kappeli et al. 1979). Because of its exceptional surfactant activity, it is named as Surfactin (Rosenberg et al. 1979). Surfactins can lower the surface tension from 72 mN/m to 27.9 mN/m (Zosim et al. 1982) and have a CMC of 0.017 g/L (Cirigliano and Carman 1984). The surfactin groups of compounds are shown to be a cyclic lipoheptapeptides which contain a β-hydroxy fatty acid in its side chain (Cooper et al. 1981). Recent studies indicate that surfactins show potent antiviral, antimycoplasma, antitumoral, anticoagulant activities as well as inhibitors of enzymes (Hisatsuka et al. 1971, Zosim et al. 1982). Although, such properties of surfactins qualify them for potential applications in medicine or biotechnology, they have not been exploited extensively till date.

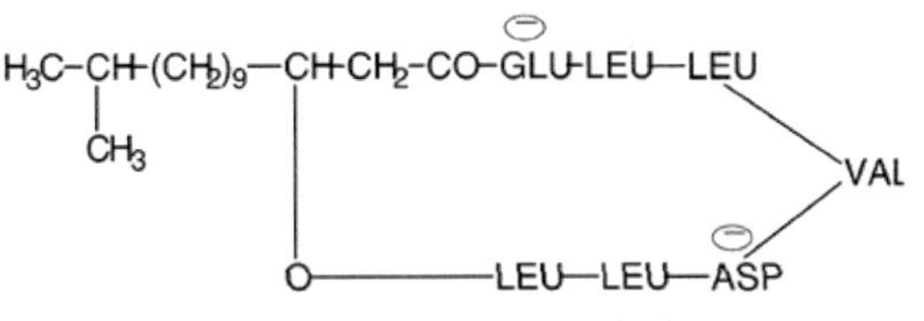

Figure 6. Structure of Surfactin.

3.2.2 *Lichenysin*

Lichenysin produced by *Bacillus licheniformis* exhibits similar structure and physiochemical properties to that of surfactin (McInerney et al. 1990). *Bacillus licheniformis* also produce several other surface-active agents that act synergistically and exhibit excellent temperature, pH and salt stability (McInerney et al. 1990). Lichenysin A produced by *Bacillus licheniformis* strain is characterized to contain a long-chain β-hydroxy fatty acid molecule (Yakimov et al. 1996). The Lichenysin produced by *Bacillus licheniformis* is capable of lowering the surface tension of water to 27 mN/m and the interfacial tension between water and n-hexadecane to 0.36 mN/m. Lichenysin is stable even under a wide range of temperature fluctuations, pH variations and NaCl concentrations which promote the dispersion of colloidal 3-silicon carbide and aluminum nitride slurries more efficiently compared to other chemical agents (Horowitz and Currie 1990). It has also been reported that lichenysin is a more efficient cation chelator compared to Surfactin (Grangemard et al. 2001).

3.2.3 *Iturin*

Iturin A, the first discovered compound of the Iturin group is the best-known member isolated from a *Bacillus subtilis* strain taken from the soil (Peypoux et al. 1978). The subsequent isolation from other strains of *Bacillus subtilis* of five other Lipopeptides such as Iturin AL, Mycosubtilin, Bacillomycin L, D, F, and LC (or Bacillopeptin), all having a common pattern of chemical constitution, led to the adoption of the generic name of "Iturins" for this group of Lipopeptides (Kajimura et al. 1995). The compounds of Iturin groups are cyclic lipoheptapeptides which contain a β-amino fatty acid in the side chain. Lipopeptids belonging to the iturin family are potent antifungal agents that can also be used as biopesticides for plant protection (Vater et al. 2002, Romero et al. 2007, Mizumoto et al. 2007).

3.2.4 *Fengycin*

Fengycin is a lipodecapeptide containing a β-hydroxy fatty acid in its side chain which comprises of C15 to C17 variants which have a characteristic Ala-Val dimorphy at its 6th position in its peptide ring (Vater et al. 2002). Wang et al. (2004) have demonstrated the identification of Fengycin homologues produced by *Bacillus subtilis* by using Electrospray Ionization Mass Spectrometry (ESI-MS) technique.

3.2.5 *Viscosin*

Viscosin is majorly produced by *Pseudomonas fluorescens*. The bacterium can adhere to the broccoli heads as a wetting agent and cause decay of the wounded and unwounded florets of broccoli. Mutants which are deficient in viscosin obtained by transposon mutagenesis affect the wounded broccoli florets but they are devoid of the ability to decay unwounded ones unlike the wild type bacterium. Mating of these mutants with their triparent corresponding wild type clones and the helper *Escherichia coli* (with the mobilizable plasmid pPK2013) yielded transconjugants. Their linkage maps indicated that a 25 kb chromosomal DNA after transcription and translation forms three proteins as a synthetase complex and it is required to produce Viscosin. A probe was made from the DNA region hybridized with DNA fragments of other plant-pathogenic *Pseudomonas* to varying degrees (Braun et al. 2001).

3.2.6 Arthrofactin

Arthrofactin produced by *Pseudomonas* sp. MIS38, is the most potent cyclic lipopeptide-type biosurfactant ever reported. Three genes termed *Arf*A, *Arf*B and *Arf*C form the Arthrofactin synthetase gene cluster and encode *ArfA*, *ArfB* and *ArfC* which assemble to form a unique structure. *ArfA*, *ArfB* and *ArfC* contain two, four, and five functional modules, respectively. A module is defined as the unit that catalyzes the incorporation of a specific amino acid into the peptide product. The arrangement of the modules of a peptide synthetase is usually colinear with the amino acid sequence of the peptide. The modules can be further subdivided into different domains that are characterized by a set of short conserved sequence motifs). Each module bears a condensation domain [C] (responsible for formation of peptide bond between two consecutively bound amino acids), adenylation domain [A] (responsible for amino acid recognition and adenylation at the expense of ATP) and thiolation domain [T] (serves as an attachment site of 4-phosphopantetheine cofactor and a carrier of thioesterified amino acid intermediates). However, none of the 11 modules possess the epimerization domain [E] responsible for the conversion of amino acid residues from L to D form. Moreover, two thioesterase domains are tandemly located at the C-terminal end of *ArfC*. *Arf*B is the gene absolutely essential for arthrofactin production as its disruption impairs this act (Roongsawan et al. 2003).

3.2.7 Amphisin

Amphisin is produced by *Pseudomonas* sp. It has both biosurfactant and antifungal properties and brings about the inhibition of plant pathogenic fungi. The two-component regulatory system *GacA/GacS* (*GacA* is a response regulator and *GacS* is a sensor kinase) controls the amphisin synthetase gene (*amsY*) (Koch et al. 2002). The surface motility of this bacterium requires the production of this biosurfactant as is indicated by the mutants defective in the genes *gacS* and *amsY*. Amphisin synthesis is regulated by *gacS* gene as the *gacS* mutant regains the property of surface motility upon the introduction of a plasmid encoding the heterologous wild-type *gacS* gene from *Pseudomonas syringae* (Andersen et al. 2003).

3.2.8 Putisolvin

Pseudomonas putida produces two surface active cyclic lipopeptides designated as Putisolvins I and II. The ORF (open reading frame) encoding the synthesis of the Putisolvins bears amino acid homology to various lipopeptide synthetases (Kuiper et al. 2004). Putisolvins are produced by a Putisolvin synthetase designated as *psoA*. Three heat shock genes *dnaK, dnaJ* and *grpE* positively regulate the biosynthesis of Putisolvin (Dubern et al. 2005). The *ppuI-rsaL-ppuR* quorum sensing system controls putisolvin biosynthesis. *ppuI* and *ppuR* mutants exhibit decreased Putisolvin production whereas *rsaL* mutants show enhanced Putisolvin production (Dubern et al. 2006).

3.2.9 Serrawettin

A group of Gram negative bacteria such as *Serratia* produces surface active cyclodepsipeptides known as Serrawettin W1, W2 and W3 (Matsuyama et al. 1986, Matsuyama et al. 1989). Varied strains of *Serratia marcescens* produces different

Serrawettins, e.g., Serrawettin W1 was produced by *Serratia marcescens* strains 274 and *Serratia marcescens* ATCC 13880 or *Serratia marcescens* NS 38. Serrawettin W2 was produced by *Serratia marcescens* strain NS 25 and W3 was produced by *Serratia marcescens* strain NS 45. Besides this *Serratia liquefaciens* produces Serrawettin W2. Temperature dependent synthesis of novel lipids—Rubiwettin R1 and RG1 was recorded in *Serratia rubidaea* (Matsuyama et al. 1990).

3.3 *Fatty Acids, Phospholipids and Neutral Lipids*

Large quantities of fatty acid and phospholipid surfactants are produced by bacterial and yeast species when grown on n-alkanes (Robert et al. 1989). Rich vesicles are produced by *Acinetobacter* sp. (Fig. 7), Phosphatidylethanolamine (Kappeli et al. 1979), which form clear microemulsions of alkanes in water. The quantitative production of phospholipids has been recorded in a few species of *Aspergillus* (Kappeli et al. 1979) and *Thiobacillus thiooxidans* (Beeba and Umbreit 1971). *Arthrobacter* strain (Wayman et al. 1984) and *Pseudomonas aeruginosa* (Robert et al. 1989) accumulate up to 40 to 80% (wt/wt) of such lipids when cultivated on hexadecane and olive oil, respectively. Phosphatidylethanolamine produced by *Rhodococcus* sp. Reduction in interfacial tension between water and hexadecane to less than 1 mN/m and a CMC of 30 mg/liter is noticed when *Erythropolis* is grown on n-alkane (Kretschmer et al. 1982).

When grown on alkanes few hydrocarbon-degrading microbes produce extracellular free fatty acids that exhibit good surfactant activity. The fatty acid biosurfactants are saturated fatty acids in the range of C12 to C14 and complex fatty acids containing hydroxyl groups and alkyl branches (Mac Donald et al. 1981, Kretschmer et al. 1982). It was shown that *Arthrobacter* strain (Wayman et al. 1984) and *Pseudomonas aeruginosa* (Robert et al. 1989) accumulated up to 40–80% (w/w) of such lipids when cultivated on Hexadecane and Olive oil respectively.

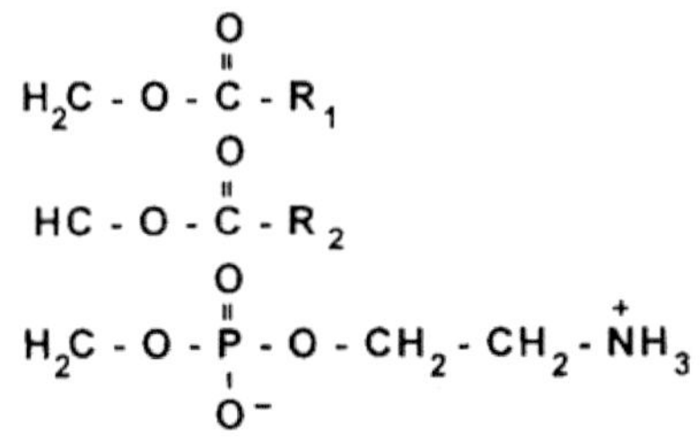

Figure 7. Structure of Phosphatidylethanolamine.

3.4 *Polymeric Biosurfactants*

Polymeric biosurfactants are high molecular weight biopolymers, which show certain properties like high viscosity, tensile strength and resistance to shear. The following are examples of different classes of polymeric biosurfactants.

3.4.1 Emulsan

Acinetobacter calcoaceticus produces a potent extracellular polymeric bioemulsifier called Emulsan (Rosenberg et al. 1979) which is characterized as a polyanionic

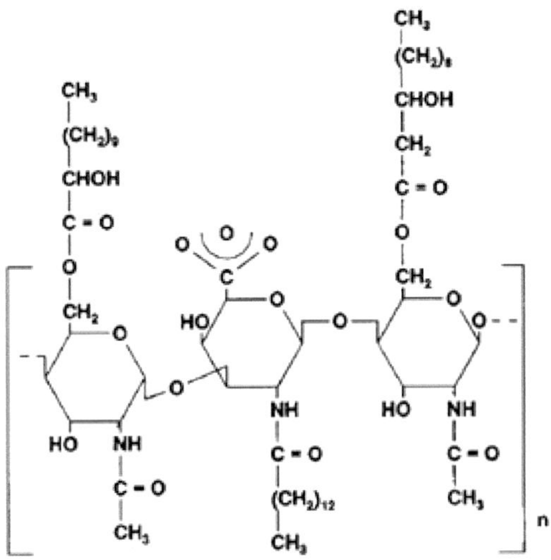

Figure 8. Structure of Emulsan.

amphipathic heteropolysaccharide (Fig. 8). The heteropolysaccharide backbone consists of repeating units of trisaccharide of N-acetyl-d-galactosamine, N-acetylgalactosamineuronic acid and an unidentified N-acetylamino sugar (Zukereberg et al. 1979). Apoemulsan is the product obtained after the removal of a protein fraction which exhibits low emulsifying activity on hydrophobic substrates such as n-hexadecane. Major protein associated with the emulsan complex is a cell surface esterase (Bach et al. 2003).

3.4.2 *Apoemulsan*

Apoemulsan is an extracellular, polymeric lipoheteropolysaccharide produced by *Acinetobacter venetianus.* Deproteinized Emulsan (apoemulsan, 103 kDa) consists of d-galactosamine, l-galactosamine uronic acid (pKa, 3.05) and a diamino, 2-desoxy n-acetylglucosamine (Mac Donald et al. 1981). It retained emulsifying activity towards certain hydrocarbon substrates but was unable to emulsify relatively non-polar, hydrophobic, aliphatic materials (Nitschke and Costa 2007, Singh et al. 2007). It is now known that polymers are synthesized from Wzy pathway. However, there also appears a differing report which claims that the process is based on presence of Polysaccharide-copolymerase (PCP) (Makkar and Cameotra 2002, Van Hamme et al. 2006). Singh et al. (2007) proved that synthesis of this polymer was dependant on Wzy pathway where, PCP protein controlled the length of the polymer. This was proved by inducing defined point mutations in the proline-glycine-rich region of apoemulsan PCP protein (Wzc). Five out of eight mutants produced higher weight BE than that of the wild type while four had modified biological properties. It has been suggested that emulsifying activity and release of polymer is mediated via esterase gene est (34.5 kD). A study carried out by Das and Mukherjee (2005) proved that lipase is responsible for enhanced emulsification properties.

3.4.3 *Alasan*

Alasan is an anionic alanine-containing heteropolysaccharide protein biosurfactant produced by *Acinetobacter radioresistens* and it was reported to solubilize and degrade polyaromatic hydrocarbons (Barkay et al. 1999). Alasan is a surface-active component with 35.77 kD protein called AlnA. This surface active protein AlnA has a high amino acid sequence homology to *Escherichia coli* Outer membrane protein A (OmpA), but however OmpA does not possess any emulsifying activity (Toren et al. 2002). Three of Alasan proteins were purified from *Acinetobacter radioresistens* are having molecular masses of 16, 31 and 45 kD and it was demonstrated that the 45 kD protein had the highest specific emulsifying activity, 11% higher than the intact Alasan complex (Toren et al. 2002). The 16- and 31-kD proteins gave relatively low emulsifying activities, but they were significantly higher than that of apo-alasan (Toren et al. 2002).

3.4.4 *Liposan*

Candida lipolytica produce an extracellular water soluble emulsifier called Liposan which is composed of 83% (v/w) carbohydrate and 17% (w/v) protein. The carbohydrate portion is a heteropolysaccharide consisting of glucose, galactose, galactosamine and galacturonic acid (Singh et al. 2007).

3.4.5 *Biodispersan*

Biodispersan is an extracellular, anionic polysaccharide produced by *Acinetobacter calcoaceticus* which acts as a dispersing agent for water insoluble solids (Das and Mukherjee 2005). It is non-dialyzable, with an average molecular weight of 51,400 and contains four reducing sugars, namely, glucosamine, 6-methylaminohexose, galactosamine uronic acid and an unidentified amino sugar (Singh et al. 2007). Rich protein was also secreted along with the extracellular polysaccharides. Protein defective mutants produced an equal or enhanced biodispersion as compared to the parent strain (Bach et al. 2003).

3.5 *Particulate Biosurfactants*

Partition of hydrocarbons by Extracellular membrane vesicles leads to the formation of a microemulsion which plays a vital role in the uptake of alkanes by microbial cells. *Acinetobacter* sp. strain with a vesicle diameter of 20 to 50 nm and a buoyant density of 1.158 g/cm^3 was made up of phospholipid, protein, and lipopolysaccharide. The membrane vesicles contain about 5 times as much phospholipid and about 350 times as many polysaccharides as does the outer membrane of the same organism. Hydrocarbon-degrading and pathogenic bacteria are attributed to surface activity by several cell surface components, which include different structures as follows, M protein and lipoteichoic acid in Group A *Streptococci*, protein A in *Staphylococcus aureus*, layer A in *Aeromonas salmonicida*, Prodigiosin in *Serratia* spp., Gramicidins in *Bacillus brevis* spores, and thin fimbriae in *Acinetobacter calcoaceticus* (Roggiani and Dubnau 1993).

4. Properties of Biosurfactants

Biosurfactants have increasing demand in commercial use because of the wide spectrum of available sources. There are many advantages of microbial biosurfactants compared to their chemically synthesized counterpart. Microbial surfactants are selected based on varied parameters such as surface movement, resilience to pH, adaptation to varied temperature and ionic quality, biodegradability, low poisonous nature, emulsifying and demulsifying capacity, and antiadhesive activity. The main characteristic features of biosurfactants and a short description of each property are given below.

4.1 *Surface and Interface Activity*

Efficiency and effectiveness are basic characteristics of a better surfactant. Efficiency is measured by the CMC, whereas effectiveness is related to surface and interfacial tensions (Barros et al. 2007). The CMC of biosurfactants lies in between 1 to 2000 mg/L, whereas interfacial (oil/water) and surface tensions are respectively 1 and 30 mN/m. A good surfactant lowers surface tension of water from 72 to 35 mN/m and the interfacial tension of water or hexadecane from 40 to 1 mN/m (Mullian 2005). Surfactin produced by *Bacillus subtilis* reduces the surface tension of water to 25 mN/m and interfacial tension of water or hexadecane to < 1 mN/m (Cooper et al. 1981). Rhamnolipids from *Pseudomonas aeruginosa* decrease the surface tension of water to 26 mN/m and the interfacial tension of water/hexadecane to < 1 mN/m (Hisatsuka et al. 1971). The Sophorolipids from *Torulopsis bombicola* have been reported to reduce the surface tension to 33 mN/m and the interfacial tension to 5 mN/m (Cooper and Cavalero 2003). In general, biosurfactants are more effective and efficient and their CMC is about 10–40 times lower than that of chemical surfactants, i.e., less surfactant is necessary to get a maximum decrease in surface tension (Desai and Banat 1997).

A surfactant helps in decreasing surface strain and the interfacial pressure. Surfactin produced by *Bacillus subtilis* can lessen the surface tension of water to 25 mN m^{-1} and interfacial strain of water or hexadecane to under 1 mN m^{-1} (Cavalero and Cooper 2003). *Pseudomonas aeruginosa* produces Rhamnolipids which diminished surface tension of water to 26 mN m^{-1} and interfacial strain of water or hexadecane to under 1 mN m^{-1} (Chakrabarti 2012). Biosurfactants are more powerful and effective and their CMC is around a few times lower than chemical surfactants, i.e., for maximal decline on surface strain, less surfactant is fundamental (Das and Mukherjee 2007).

4.2 *Temperature, pH and Ionic Strength Tolerance*

Most of the biosurfactants and their surface activities are not influenced by varying external factors such as temperature and pH. McInerney et al. (1990) reported that Lichenysin from *Bacillus licheniformis* and another biosurfactant produced by *Arthrobacter protophormiae* was not affected by temperature (30°C to 100°C), pH

(2.0–12.0) and by Sodium chloride (10% NaCl concentration whereas 2% NaCl is enough to inactivate synthetic surfactants) and Calcium concentrations (up to 25 g/L). A lipopeptide of *Bacillus subtilis* was stable even after autoclaving (121°C/20 min) and after 6 months at –18°C no changes in the surface activity were noticed from pH 5 to 11 and in the concentration of Sodium chloride up to 20% (Nitschke and Pastore 1990). Extremophile biosurfactants have gained importance in recent years for their commercial utilization. Since, industrial procedures include extremes of temperature, pH and weight, it is important to separate novel microbial items which are ready to work under these conditions (Cooper et al. 1981).

4.3 *Biodegradability*

Microbial derived compounds degrade easily compared to the Synthetic surfactants. Biosurfactants are easily degraded by microorganisms in water and soil, making these compounds adequate for bioremediation of toxic substances, oil spills, and waste treatment (Desai and Banat 1997). Synthetic surfactants cause various environmental issues and thus, biodegradable biosurfactants from marine microorganisms are concerned with biosorption of ineffective solvent polycyclic sweet-smelling hydrocarbon, phenanthrene contaminated in aquatic surfaces (Gautam and Tyagi 2006). Marine algae, *Cochlodinium* utilizes the biodegradable biosurfactant Sophorolipids with a removal efficiency of 90% for every 30 minutes of treatment (Gharaei Fathabad 2011).

4.4 *Low Toxicity*

Microbial surfactants are generally considered as low in toxicity or nontoxic products and therefore are appropriate for many industries. A low degree of toxicity allows the use of biosurfactants in foods, cosmetics and pharmaceutical industries. A low level of toxicity is essential for environmental applications. Biosurfactants can be produced from industrial wastes. There was less work recorded with respect to the poisonous nature of biosurfactants. Poremba et al. (1991) showed that the elevated level of toxicity in chemically-derived surfactants displayed an LC50 against *Photobacterium phosphoreum* and was found to be 10 times lower than that of Rhamnolipids. The low toxicity profile of biosurfactants, such as Sophorolipids from *Candida bombicola* made them helpful in nourishment ventures (Hatha et al. 2007).

By analyzing the toxicity of synthetic surfactants and commercial dispersants, it was found that most biosurfactants degraded faster, except synthetic sucrose-stearate which showed structure homology to Glycolipids and degraded quicker than the biogenic Glycolipids. It was also reported that biosurfactants showed higher EC50 (effective concentration to decrease 50% of test population) values than synthetic dispersants (Poremba et al. 1991). A biosurfactant from *Pseudomonas aeruginosa* was tested along with a synthetic surfactant (Marlon A-350) which is mostly used in the industry, for its toxicity and mutagenic properties. The results revealed a higher level of toxicity and mutagenic effect in the chemical-derived surfactant and non-toxic and non-mutagenic in the biosurfactant (Flasz et al. 1998).

4.5 Specificity

Biosurfactants are multiple molecules with specific groups with varied functions such as in the detoxification of different pollutants and de-emulsification of industrial emulsions and in the fields of food, pharmaceuticals, and cosmetics with a specific action.

4.6 Biocompatibility and Digestibility

Properties such as Biological compatibility and digestibility of microbial biosurfactants allow the use of biomolecules in many industries, especially food, pharmaceutical and cosmetics.

4.7 Emulsion Forming and Emulsion Breaking

Stable emulsions can be produced with a life span of months and years. Higher molecular-mass biosurfactants are in general better emulsify than low-molecular-mass biosurfactants. Sophorolipids from *Torulopsis bombicola* have been shown to reduce surface tension, but are not good emulsifiers. By contrast, Liposan does not reduce the surface tension, but has been used successfully to emulsify edible oils. Polymeric surfactants offer additional advantages because they coat droplets of oil, thereby forming a stable emulsion. This property is especially useful for making oil/water emulsions for cosmetics and food.

Biosurfactants act as emulsifiers or de-emulsifiers. An emulsion can be depicted as a heterogeneous framework, comprising of one immiscible fluid scattered in another as beads, whose distance across by and large surpasses 0.1 mm. Emulsions are of two types: oil-in-water or water-in-oil emulsions. They have a minimal stability which might be balanced out by added substances, for example, biosurfactants and can be kept up as steady emulsions for a considerable length of time to years (Hu and Ju 2001).

Liposan, which is a water soluble emulsifier synthesized by *Candida lipolytica* has been used with edible oils to form stable emulsions. Liposans are commonly used in the cosmetic and food industries for producing stable oil/water emulsions (Campos et al. 2013). Polymeric surfactants offer additional advantages because they coat droplets of oil, thereby forming stable emulsions. This property is especially useful for making oil/water emulsions for cosmetics and food.

4.8 Chemical Diversity

The chemical diversity of naturally produced Biosurfactants offers a wide selection of surface-active agents with properties closely related to specific applications.

4.9 Antiadhesive Agent

A biofilm can be depicted as a group of microbes or other organic matter that have aggregated on any surface (Hatha et al. 2007). The initial step on biofilm foundation is bacterial adherence over the surface which is influenced by different components including microorganisms (sort of), hydrophobicity and electrical charges on the surface, prevailing ecological conditions and the capacity of microorganisms to deliver extracellular polymers that assist cells to grapple to surfaces (Jadhav et al. 2011). The biosurfactants can be utilized in changing the hydrophobicity of the surface which thus influences the bonding of microorganisms over the surface. A surfactant from *Streptococcus thermophilus* backs off the colonization of other thermophilic strains of *Streptococcus* on the steel which are in charge of fouling. So also, a biosurfactant from *Pseudomonas fluorescens* hindered the connection of *Listeria monocytogenes* onto the steel surface (Konishi et al. 2008).

5. Mechanism of Interaction of Biosurfactants

Biosurfactants are microbial amphiphilic and polyphilic polymers that tend to interact with the phase boundary between two phases in a heterogeneous system, defined as the interface. For all interfacial systems, it is known that organic molecules from the aqueous phase tend to immobilize at the solid interface. There they eventually form a film known as a conditioning film, which will change the properties (wettability and surface energy) of the original surface (Neu 1996). In an analogy to organic conditioning films, biosurfactants may interact with the interfaces and affect the adhesion and detachment of bacteria. In addition, the substratum surface properties determine the composition and orientation of the molecules conditioning the surface during the first hour of exposure. After about 4 hrs, a certain degree of uniformity is reached and the composition of the adsorbed material becomes substratum independent (Neu 1996).

Owing to the amphiphilic nature of biosurfactants, not only hydrophobic but a range of interactions are involved in the possible adsorption of charged biosurfactants to interfaces. Most natural interfaces have an overall negative or, rarely, positive charge. Thus, the ionic conditions and the pH are important parameters if interactions of ionic biosurfactants with interfaces to be investigated (Craig et al. 1993). Gottenbos et al. (2001) demonstrated that positively charged biomaterial surfaces exert an antimicrobial effect on adhering Gram negative bacteria, but not on Gram positive bacteria. In addition, the molecular structure of a surfactant will influence its behaviour at the interfaces. In describing the surface-active approach, an effort is made to elaborate on the possible theoretical locations and orientations of the biosurfactants. Nevertheless, it must be kept in mind that the situation in natural systems is far more complex and requires the consideration of many additional parameters.

6. Conclusion

Surfactants have long been among the most versatile of process chemicals. Their market is extremely competitive, and manufacturers will have to expand their arsenal to develop products for the 1990s and beyond. In this regard, biosurfactants are promising candidates. Biosurfactants are unique biomolecules with a variety of functions that are fast becoming a more efficient and greener alternative to their predecessor chemical surfactants. The unique properties of biosurfactants allow their use and possible replacement of chemically synthesized surfactants in a great number of industrial operations. During the last 2–3 decades a wide variety of microorganisms have been reported to produce numerous types of biosurfactants. Their biodegradability and lower toxicity gives them an advantage over their chemical counterparts and therefore may make them suitable for replacing chemicals. The emulsifying activity of the biosurfactant is another branch having full scope of their use as emulsion forming agents for hydrocarbons and oils giving stable emulsions. While many types of biosurfactants are in use, no single biosurfactant is suitable for all potential applications. To date, biosurfactants are unable to compete economically with chemically synthesized compounds in the market, mainly due to their high production costs and the lack of comprehensive toxicity testing. Biosurfactants are used by many industries and one could easily say that there is almost no modern industrial operation where properties of surfaces and surface-active agents are not exploited. The potential application of biosurfactants in industries is also a reality.

References

Amaral, P.F., M.A. Coelho, I.M. Marrucho and J.A. Coutinho. 2010. Biosurfactants from yeasts: Characteristics, production and application. Adv. Exp. Med. Bio. 672: 236–249.

Andersen, J.B., B. Koch, T.H. Nielsen, D. Sorensen, M. Hansen et al. 2003. Surface motility in *Pseudomonas* sp. DSS73 is required for efficient biological containment of the root pathogenic microfungi *Rhizoctonia solani* and *Pythium ultimum*. Microbio. 149: 37–46.

Bach, H., Y. Berdichevsky and D. Gutnick. 2003. An exocellular protein from the oil-degrading microbe *Acinetobacter venetianus* RAG-1 enhances the emulsifying activity of the polymeric bioemulsifier emulsan. Appl. Environ. Microbiol. 69: 2608–2615.

Bar Ness, R., N. Avrahamy, T. Matsuyama and M. Rosenberg. 1988. Increased cell surface hydrophobicity of a *Serratia marcescens* NS 38 mutant lacking wetting activity. J. Bacteriol. 170: 4361–4364.

Barkay, T., S. Navon-Venezia and E.Z. Ron. 1999. Enhancement of solubilization and biodegradation of polyaromatic hydrocarbons by the bioemulsifier alasan. Appl. Environ. Microbiol. 65: 2697–2702.

Barros, F.F.C., C.P. Quadros, M.R. Maróstica and G.M. Pastore. 2007. Surfactina: Propriedades químicas, tecnológicas e funcionais para aplicações em alimentos. Quím. Nova. 30: 1–14.

Beeba, J.L. and W.W. Umbreit. 1971. Extracellular lipid of *Thiobacillus thiooxidans*. J. Bacteriol. 108: 612–615.

Benincasa, M., A. Abalos and I. Oliveria. 2004. Chemical structure, surface properties and biological activities of the biosurfactant produced by *Pseudomonas aeruginosa* LBI from soapstock. Antonie van Leeuwenhoek 85: 1–8.

Braun, P.G., P.D. Hildebrand, T.C. Ells and D.Y. Kobayashi. 2001. Evidence and characterization of a gene cluster required for the production of Viscosin, a Lipopeptide biosurfactant, by a strain of *Pseudomonas fluorescens*. Can. J. Micro. 47: 294–301.

Campos, J.M., T.L.M. Stamford, L.A. Sarubbo, J.M. Luna, R.D. Rufino et al. 2013. Microbial biosurfactants as additives for food industries. Biotechnol. Prog. 29: 1097–1108.

Cavalero, D.A. and D.G. Cooper. 2003. The effect of medium composition on the structure and physical state of Sophorolipids produced by *Candida bombicola* ATCC 22214. J. Biotechnol. 103: 31–41.

Chakrabarti, S. 2012. Bacterial biosurfactant: Characterization, antimicrobial and metal remediation properties. Ph.D. Thesis, National Institute of Technology, Mumbai, India.

Chen, J., X. Song and H. Zhang. 2006. Production, structure elucidation and anticancer properties of Sophorolipid from *Wickerhamiella domercqiae*. Enzyme Microb. Technol. 39: 501–506.

Christofi, N. and I.B. Ivshina. 2002. Microbial surfactants and their use in field studies of soil remediation. J. Appl. Micro. 93: 915–929.

Cirigliano, M. and G. Carman. 1984. Isolation of a bioemulsifier from *Candida lipodytica*. Appl. Environ. Microbiol. 48: 747–750.

Cooper, D.G., C.R. MacDonald, S.J.B. Duff and N. Kosaric. 1981. Enhanced production of Surfactin from *Bacillu subtilis* by continuous product removal and metal cation additions. Appl. Environ. Microbiol. 42: 408–412.

Cooper, D.G and D.A. Paddock. 1983. *Torulopsis petrophilum* and surface activity. Appl. Environ. Microbiol. 46: 1426–1429.

Cooper, D.G. and D.A. Paddock. 1984. Production of a biosurfactant from *Torulopsis bombicola*. Appl. Environ. Microbiol. 47: 173–176.

Cooper, D.G. and B.G. Goldenberg. 1987. Surface active agents from two *Bacillus* species. Appl. Environ. Micro. 53: 224–229.

Cooper, D.G. and D.A. Cavalero. 2003. The effect of medium composition on the structure and physical state of Sophorolipids produced by *Candida bombicola* ATCC 22214. J. Biotechnol. 103: 31–41.

Craig, V.S.J., B.W. Ninham and R.M. Pashley. 1993. Effect of electrolytes on bubble coalescence. Nature 364: 317–319.

Crich, D., M.A. De la Mora and R. Cruz. 2002. Synthesis of the Mannosyl erythritol lipid MEL A; confirmation of the configuration of the meso-erythritol moiety. Tetrahedron 58: 35–44.

Cutler, A.J and R.J. Light. 1979. Regulation of hydroxydocosanoic and Sophoroside production in *Candida bogoriensis* by the level of glucose and yeast extract in the growth medium. J. Biol. Chem. 254: 1944–1950.

Danyelle Khadydja, F. Santos, Raquel D. Rufino, Juliana M. Luna, A. Valdemir et al. 2016. Biosurfactants: Multifunctional Biomolecules of the 21st Century. Int. J. Mol. Sci. 17(401): 1–31.

Das, K. and A.K. Mukherjee. 2005. Characterization of biochemical properties and biological activities of biosurfactants produced by *Pseudomonas aeruginosa* mucoid and non-mucoid strains. Appl. Microbiol. Biotechnol. 69: 192–199.

Das, K. and A.K. Mukherjee. 2007. Crude petroleum-oil biodegradation efficiency of *Bacillus subtilis* and *Pseudomonas aeruginosa* strains isolated from petroleum oil contaminated soil from north-east India. Biores. Tech. 98: 1339–1345.

Desai, J.D. and I.M. Banat. 1997. Microbial production of surfactants and their commercial potential. Microbiol. Mol. Biol. Rev. 61: 47–64.

Deziel, E., F. Lepine, S. Milot and R. Villemur. 2000. Mass spectrometry monitoring of Rhamnolipids from a growing culture of *Pseudomonas aeruginosa* Strain 57RP. Biochimica et Biophysica Acta 1485: 145–152.

Dubern, J.F., E.L. Lagendijk, B.J.J. Lugt enberg and G.V. Bloemberg. 2005. The heat shock genes *dna*K, *dna*J, and *grp*E are involved in regulation of Putisolvin biosynthesis in *Pseudomonas putida* PCL1445. J. Bacteriol. 187: 5967–5976.

Dubern, J.F., B.J.J. Lugt enberg and G.V. Bloemberg. 2006. The ppul-rsaL-ppuR quorum sensing system regulates biofilm formation of *Pseudomonas putida* PCL 1445 by controlling biosynthesis of the cyclic lipopeptides Putisolvins I and II. Journal of Bacteriology 188(8): 2898–2906.

Edward, J.R. and J.A. Hayashi. 1965. Structure of a Rhamnolipid from *Pseudomonas aeruginosa*. Arch. Biochem. Biophys. 111: 415–421.

Flasz, A., C.A. Rocha, B. Mosquera and C. Sajo. 1998. A comparative study of the toxicity of a synthetic surfactant and one produced by *Pseudomonas aeruginosa* ATCC 55925. Med. Sci. Res. 26: 181–185.

Fukuoka, T., T. Morita and T. Konishi. 2007. Characterization of new Glycolipid biosurfactants, tri-acylated mannosylerythritol lipids, produced by pseudozyma yeasts. Biotechnol. Lett. 29: 1111–1118.

Gautam, K.K. and V.K. Tyagi. 2006. Microbial surfactants: A Review. J. Oleo Sci. 55: 155–166.

Georgiou, G., S.C. Lin and M.M. Sharma. 1992. Surface active compounds from microorganisms. Biotech. 10: 60–65.

Gharaei Fathabad, E. 2011. Biosurfactants in pharmaceutical industry. Ameri. J. Drug Disco. Develop. 1(1): 58–69.

Golyshin, P.M., H.L. Fredrickson and L. Giuliano. 1999. Effect of novel biosurfactants on biodegradation of polychlorinated biphenyls by pure and mixed bacterial cultures. Microbiologica. 22: 257–267.

Gottenbos, B., D. Grijpma and C. Vander Mei. 2001. Antimicrobial effects of positively charged surfaces on adhering Gram positive and Gram negative bacteria. J. Antimicrob. Chemother. 48: 7–13.

Grangemard, I., J. Wallach and R. Maget Dana. 2001. Lichenysin: a more efficient cation cheNlator than Surfactin. Appl. Biochem. Biotechnol. 90: 199–210.

Guerra Santos, L., O. Kappeli and A. Fiechter. 1986. Dependence of *Pseudomonas aeruginosa* continuous culture biosurfactant production on nutritional and environmental factors. Appl. Microbio. Biotech. 24: 443–448.

Hatha, A.A.M., G. Edward and K.S.M.P. Rahman. 2007. Microbial biosurfactants—Review. J. Mar. Atmos. Res. 3: 1–17.

Hirayama, T. and I. Kato. 1982. Novel Rhamnolipids from *Pseudomonas aeruginosa*. FEBS Lett. 139: 81–85.

Hisatsuka, K., T. Nakahara, N. Sano and K. Yamada. 1971. Formation of rhamnolipid by *Pseudomonas aeruginosa*: Its function in hydrocarbon fermentations. Agric. Biol. Chem. 35: 686–692.

Hisatsuka, K., T. Nakahara, Y. Minoda and K. Yamada. 1977. Formation of protein like activator for n-alkane oxidation and its properties. Agric. Biol. Chem. 41: 445–450.

Hommel, R.K., O. Stuwer, W. Stuber, D. Haferburg, H.P. Kleber et al. 1987. Production of water soluble surface active exolipids by *Torulopsis apicola*. Appl. Microbiol. Biotechnol. 26: 199–205.

Hommel, R.K., L. Weber, A. Weiss, U. Himelreich, O. Rilke et al. 1994. Production of Sophorose lipid by *Candida apicola* grown on glucose. J. Biotechnol. 33: 147–155.

Horowitz, S. and J.K. Currie. 1990. Novel dispersants of silicon carbide and aluminium nitrate. J. Dispersion Sci. Technol. 11: 637–659.

Hu, Y. and J.K. Ju. 2001. Purification of lactonic sophorolipids by crystallization. J. Biotechnol. 87: 263–272.

Jadhav, M., S. Kalme, D. Tamboli and S. Govindwar. 2011. Rhamnolipid from *Pseudomonas desmolyticum* NCIM-2112 and its role in the degradation of Brown 3REL. J. Basic Microbiol. 51: 385–396.

Jarvis, F.G. and M.J. Johnson. 1949. A glycolipid produced by *Pseudomonas aeruginosa*. J. Am. Chem. Soc. 71: 4124–4126.

Jitendra Desai and Ibrahim Banat. 1997. Microbial production of surfactants and their commercial potential. Microbiol. Mol. Bio. Rev. 61(1): 47–64.

Kajimura, Y., M. Sugiyama and M. Kaneda. 1995. Bacillopeptins, new cyclic lipopeptide antibiotics from *Bacillus subtilis* FR-2. J. Antibiot. (Tokyo). 48: 1095–1103.

Kappeli, O., R. Shah and W.R. Finnerty. 1979. Partition of alkane by an extracellular vesicle derived from hexadecane grown *Acinetobacter.* J. Bacteriol. 140: 707–712.

Karanth, N.G.K., P.G. Deo and N.K. Veenanadig. 1999. Microbial biosurfactant and their importance. Cur. Sci. 77: 116–126.

Kim, H.S., B.D. Yoon and D.H. Choung. 1999. Characterization of a biosurfactant, mannosylerythritol lipid produced from *Candida* sp. SY16. Appl. Microbiol. Biotechnol. 52: 713–721.

Kim, H.S., E.J. Lim and S.O. Lee. 2000. Purification and characterization of biosurfactants from *Nocardia* sp. L-417. Biotechnol. Appl. Biochem. 31: 249–253.

Knoche, H.W. and J.M. Shiveley. 1972. The structure of an Ornithine containing lipid from *Thiobacillus thiooxidans*. J. Biol. Chem. 247: 170–178.

Koch, B., T.H. Nielsen, D. Sorensen, J.B. Andersen and C. Christophersen. 2002. Lipopeptide production in *Pseudomonas* sp. DSS73 is regulated by components of sugar beet seed exudate *via* the Gac two-component regulatory system. Appl. Environ. Microbio. 68: N4509–4516.

Konishi, M., T. Fukuoka, T. Morita, T. Imura and D. Kitamoto. 2008. Production of new types of Sophorolipids by *Candida batistae*. J. Oleo. Sci. 57: 359–369.

Kretschmer, A., H. Bock and F. Wagner. 1982. Chemical and physical characterization of interfacial-active lipids from *Rhodococcus erythropolis* grown on *n*-alkane. Appl. Environ. Microbiol. 44: 864–870.

Krishnaswamy Muthusamy, Subbuchettiar Gopalakrishnan, Thiengungal Kochupappy Ravi, Panchaksharam Sivachidambaram et al. 2008. Biosurfactants: Properties, commercial production and application. Cur. Sci. 94(25): 739–747.

Kuiper, I., L. Ellen, R.P. Lagendijk, P.D. Jeremy, E.M.L. Gerda et al. 2004. Characterization of two *Pseudomonas putida* lipopeptide biosurfactants, Putisolvin I and II, which inhibit biofilm formation and break down existing biofilms. Mol. Microbio. 51(1): 97–113.

Lang, S. and F. Wagner. 1987. Structure and properties of biosurfactants, p. 21–47. *In*: Kosaric, N., W.L. Cairns and N.C.C. Gray (eds.). Biosurfactants and Biotechnology. Marcel Dekker, Inc., New York, N.Y.

Lang, S., E. Katsiwela and F. Wagner. 1989. Antimicrobial effects of biosurfactants. Fat. Sci. Technol. 91: 363–366.

Li, Z.Y., S. Lang, F. Wagner, L. Witte and V. Wray et al. 1984. Formation and identification of interfacial-active glycolipids from resting microbial cells of *Arthrobacter* sp. and potential use in tertiary oil recovery. Appl. Environ. Microbiol. 48: 610–617.

Mac Donald, C.R., D.G. Cooper and J.E. Zajic. 1981. Surface-active lipids from *Nocardia erythropolis* grown on hydrocarbons. Appl. Environ. Microbiol. 41: 117–123.

Makkar, R. and S.S. Cameotra. 2002. An update on the use of unconventional substrates for biosurfactant production and their application. Appl. Microbiol. Biotechnol. 58: 428–434.

Marahiel, M., W. Denders, M. Krause and H. Kleinkauf. 1977. Biological role of gramicidin S in spore functions. Studies on gramicidin-S negative mutants of *Bacillus brevis* 9999. Eur. J. Biochem. 99: 49–52.

Matsuyama, T., K. Kameda and I. Yano. 1985. Two kinds of bacterial wetting agents: aminolipid and glycolipid. Proc. Japan Soc. Mass Spec. 11: 125–128.

Matsuyama, T., M. Sogawa and Y. Nakagawa. 1989. Fractal spreading growth of *Serratia marcescens* which produces surface-active exolipids. FEMS Microbio. Lett. 61: 243–246.

Matsuyama, T., K. Keneda, I. Ishizuka, T. Toida, I. Yano et al. 1990. Surface active novel glycolipid and linked 3-hydroxy fatty acids produced by *Serratia rubidaea*. J. Bact. 172(6): 3015–3022.

McInerney, M.J., M. Javaheri and D.P. Nagle. 1990. Properties of the biosurfactant produced by *Bacillus liqueniformis* strain JF-2. J. Microbiol. Biotechnol. 5: 95–102.

Mizumoto, S., M. Hirai and M. Shoda. 2007. Enhanced Iturin a production by *Bacillus subtilis* and its effect on suppression of the plant pathogen *Rhizoctonia solani*. Appl. Microbiol. Biotechnol. 75: 1267–1274.

Monteiro, S.A., G.L. Sassaki and L.M. De Souza. 2007. Molecular and structural characterization of the biosurfactant produced by *Pseudomonas aeruginosa* DAUPE 614. Chem. Phys. Lip. 147: 1–13.

Mulligan, C.N. 2005. Environmental applications for biosurfactants. Environ. Pollut. 133(2): 183–198.

Mutsuyama, T., M. Fujita and I. Yano. 1985. Wetting agent produced by *Serratia marcescens*. FEMS Microbiol. Lett. 28: 125–129.

Nakata, K. 2000. Two glycolipids increase in the bioremediation of halogenated aromatic compounds. J. Biosci. Bioeng. 89: 577–581.

Neu, T. 1996. Significance of bacterial surface active compounds in interaction of bacteria with interfaces. Microbiol. Rev. 60: 151–166.

Nitschke, M. and G.M. Pastore. 1990. Production and properties of a surfactant obtained from *Bacillus subtilis* grown on cassava wastewater. Bioresour. Technol. 97: 336–341.

Nitschke, M., S.G. Costa and J. Contiero. 2005. Rhamnolipid surfactants: an update on the general aspects of these remarkable biomolecules. Biotechnol. Prog. 21: 1593–1600.

Nitschke, M. and S.G.V.A.O. Costa. 2007. Biosurfactants in food industry. Trends Food Sci. Technol. 18: 252–259.

Pacwa Płociniczak, M., G.A. Płaza, Z. Piotrowska Seget and S.S. Cameotra. 2011. Environmental applications of biosurfactants: Recent advances. Int. J. Mol. Sci. 1: 633–654.

Parra, J.L., J. Guinea, M.A. Manresa, M. Robert, M.E. Mercade, F. Comelles and M. P. Bosch. 1989. Chemical characterization and physicochemical behaviour of biosurfactants. J. Am. Oil Chem. Soc. 66: 141–145.

Passeri, A. 1992. Marine biosurfactants—Production, characterization and biosynthesis of anionic glucose lipid from marine bacterial strain MM1. Appl. Microbiol. Biotechnol. 37: 281–286.

Peypoux, F., F. Besson and G. Michel. 1978. Structure de Iturine C de *Bacillus subtilis*. Tetrahedron 38: 1147–1152.

Philp, J.C., M.S. Kuyukina and I.B. Ivshina. 2002. *Alkanotripic rhodococcus* ruber as a biosurfactant producer. Appl. Microbiol. Biotechnol. 59: 318–324.

Poremba, K., W. Gunkel, S. Lang and F. Wagner. 1991. Toxicity testing of synthetic and biogenic surfactants on marine microorganisms. Environ. Toxicol. Water Qual. 6: 157–163.

Pornsunthorntawee, O., P. Wongpanit and S. Chavadej. 2008. Structural and physicochemical characterization of crude biosurfactant produced by *Pseudomonas aeruginosa* SP4 isolated from petroleum-contaminated soil. Bioresour. Technol. 99: 1589–1595.

Rahman, K.S.M., I.M. Banat, T.J. Rahman, T. Thayumanavan, P. Lakshmanaperumalsamy et al. 2002. Bioremediation of gasoline contaminated soil by bacterial consortium amended with poultry litter, coir pith and rhamnolipid biosurfactant. Biores. Tech. 81: 25–32.

Rendell, N.B., G.W. Taylor, M. Somerville, H. Todd, R. Wilson et al. 1990. Characterization of *Pseudomonas* rhamnolipids. Biochim. Biophys. Acta. 1045: 189–193.

Ristau, E. and F. Wagner. 1993. Formation of novel anionic trehalosetetraesters from *Rhodococcus erythropolis* under growth limiting conditions. Biotechnol. Lett. 5: 95–100.

Robert, M., M.E. Mercade, M.P. Bosch, J.L. Parra, M.J. Espuny et al. 1989. Effect of the carbon source on biosurfactant production by *Pseudomonas aeruginosa* 44T. Biotechnol. Lett. 11: 871–874.

Roggiani, M. and D. Dubnau. 1993. ComA, a phosphorylated response regulator protein of *Bacillus subtilis*, binds to the promoter region of *srfA*. J. Bacteriol. 175: 3182–3187.

Romero, D., A. De Vicente and J.L. Olmos. 2007. Effect of lipopeptides of antagonistic strains of *Bacillus subtilis* on the morphology and ultrastructure of the cucurbit fungal pathogen *Podosphaera fusca*. J. Appl. Microbiol. 103: 969–976.

Roongsawang, N., K. Hase, M. Haruki, T. Imanaka, M. Morikawa et al. 2003. Cloning and characterization of the gene cluster encoding arthrofactin synthetase from *Pseudomonas* sp. MIS38. Chem. Biol. 10: 869–880.

Rosen, M.J. and J.T. Kunjappu. 2012. Characteristic features of surfactants. *In*: Milton, J.T. and J. Rosen (eds.). Surfactants and Interfacial Phenomena (Chapter 1). Hoboken, NJ, USA. John Wiley & Sons, Inc.

Rosenberg, E., A. Zuckerberg, C. Rubinovitz and D.L. Gutinck. 1979. Emulsifier *Arthrobacter* RAG-1: Isolation and emulsifying properties. Appl. Environ. Microbiol. 37: 402–408.

Rosenberg, E. and E.Z. Ron. 1999. High and low molecular mass microbial surfactants. Appl. Microbio. Biotech. 52(2): 154–162.

Shepherd, R., J. Rockey, I.W. Shutherland and S. Roller. 1995. Novel bioemulsifier from microorganisms for use in foods. J. Biotech. 40: 207–217.

Silva, R.C.F.S., A.G. Almeida, J.M. Luna, R.D. Rufino, V.A. Santos et al. 2014. Applications of biosurfactants in the petroleum industry and the remediation of oil spills. Int. J. Mol. Sci. 15: 12523–12542.

Singh, A., J.D. Van Hamme and O.P. Ward. 2007. Surfactants in microbiology and biotechnology. Biotechnol. Adv. 25: 99–121.

Smyth, T.J., A. Perfumo, R. Marchant, I.M. Banat, M. Chen et al. 2010. Directed microbial biosynthesis of deuterated biosurfactants and potential future application to other bioactive molecules. Appl. Microbio. Biotech. 87(4): 1347–1354.

Sullivan, E.R. 1998. Molecular genetics of biosurfactant production. Curr. Opin. Biotechnol. 9: 263–269.

Suzuki, T., K. Hayashi, K. Fujikawa and K. Tsukamoto. 1965. The chemical structure of Polymyxin E. The identies of polymyxin E1 with colistin A and polymyxin E2 with colistin B. J. Biol. Chem. 57: 226–227.

Syldatk, C., S. Lang, F. Wagner, V. Wray and L. Witte. 1985. Chemical and physical characterization of four interfacial-active rhamnolipids from *Pseudomonas* sp. DSM 2874 grown on n-alkanes. ZeitschriftNaturforschung. 40: 51–60.

Tahara, Y., Y. Yamada and K. Kondo. 1976. A new lipid; the ornithine and taurine-containing 'cerilipin'. Agric. Biol. Chem. 40: 243–244.

Toren, A., G. Segal and E.Z. Ron. 2002. Structure—function studies of the recombinant protein Bioemulsifier Aln A. Environ. Microbiol. 4: 257–261.

Tullock, P., A. Hill and J.F.T. Spencer. 1967. A new type of marocyclic lactone from *Torulopsis apicola*. J. Chem. Soc. Chem. Commun. 21: 584–586.

Van Bogaert, I.N., K. Saerens and C. De Muynck. 2007. Microbial production and application of sophorolipids. Appl. Microbiol. Biotechnol. 76: 23–34.

Van Hamme, J.D., A. Singh and O.P. Ward. 2006. Physiological aspect. Part-1 in a series of papers devoted to surfactants in microbiology and biotechnology. Biotechnol. Adv. 24: 604–620.

Van Hoogmoed, C.G., M. Van der Kuijl Booij and H.C. Van der Mei. 2000. Inhibition of *Streptococcus mutans* NS adhesion to glass with and without a salivary conditioning film by biosurfactant-releasing *Streptococcus mitis* strains. Appl. Environ. Microbiol. 66: 659–663.

Vater, J., B. Kablitz and C. Wilde. 2002. Matrix assisted laser desorption ionization-time of flight mass spectrometry of lipopeptide biosurfactants in whole cells and culture filtrates of *Bacillus subtilis* C-1 isolated from petroleum sludge. Appl. Environ. Microbiol. 68: 6210–6219.

Wang, J., J. Liu and X. Wang. 2004. Application of electrospray ionization mass spectrometry in rapid typing of Fengycin homologues produced by *Bacillus subtilis*. Lett. Appl. Microbiol. 39: 98–102.

Wayman, M., A.D. Jenkins and A.G. Kormady. 1984. Biotechnology for oil and fat industry. J. Am. Oil Chem. Soc. 61: 129–131.

Yakimov, M.M., H.L. Fredrickson and K.N. Timmis. 1996. Effect of heterogeneity of hydrophobic moieties on surface activity of Lichenysin A, a lipopeptide biosurfactant from *Bacillus* licheniformis BAS50. Biotechnol. Appl. Biochem. 23: 13–18.

Yamane, T. 1987. Enzyme technology for the lipid industry: an engineering overview. J. Ameri. Oil Chem. Soci. 64(12): 1657–1662.

Zosim, Z., D.L. Guntick and E. Rosenberg. 1982. Properties of hydrocarbon in water emulsion. Biotechnol. Bioeng. 24: 281–292.

Zukerberg, A., A. Diver and Z. Peeri. 1979. Emulsifier of *Arthrobacter* RAG-1: chemical and physical properties. Appl. Environ. Microbiol. 37: 414–420.

2

Microbial Fermentation Technology for Biosurfactants Production

P Saranraj,[1,*] *P Sivasakthivelan,*[2]
Karrar Jasim Hamzah,[3] *Mustafa Salah Hasan*[4] and
Abdel Rahman Mohammad Al–Tawaha[5]

1. Introduction

The target market is of fundamental importance to the implementation of an industrial biosurfactant production project (Gudina et al. 2016). For cosmetic, medicinal and food products, production is only viable on a small-scale, as the column chromatography methods required to separate molecules are not economical on a large scale. Thus, the use of crude fermentation broths could be a viable solution, especially if the application is in an environmental context, as biosurfactants in such cases do not need to be pure and can be synthesized using a blend of inexpensive carbon sources, which would allow the creation of an economically and environmentally viable technology for bioremediation processes (Kumar et al. 2016). Although, improvements in biosurfactant technology have enabled a 10 to 20 fold increase in the production of these biomolecules, it is likely that further, significant advances (even if of a smaller magnitude) are needed to make this technology commercially viable (Geetha et al. 2018).

[1] Department of Microbiology, Sacred Heart College (Autonomous), Tirupattur – 635 601, Tamil Nadu, India.
[2] Department of Agricultural Microbiology, Faculty of Agriculture, Annamalai University, Annamalai Nagar – 608 002, Tamil Nadu, India.
[3] Department of Internal and Preventive Veterinary Medicine, College of Veterinary Medicine, AL-Qasim Green University, Babylon, Iraq.
[4] University of Fallujah, College of Veterinary Medicine, Department of Internal and Preventive Medicine, Fallujah, Iraq.
[5] Department of Biological Sciences, Al - Hussein Bin Talal University, Maan, Jordan.
* Corresponding author: microsaranraj@gmail.com

Diverse microorganisms are known to produce a number of surface-active agents primarily in order to adapt and grow on a variety of substrates among other natural functions. These biosurfactants are produced under various growth and environmental conditions and are reported to be mainly involved in increasing the solubility and availability of various water immiscible substrates. Members of *Pseudomonas*, *Bacillus*, *Rhodococcus* and *Candida* genera are the most widely implicated in the production of different types of biosurfactants. The biosurfactant industry has demonstrated remarkable growth in recent decades, although the large-scale production of these biomolecules remains a challenge from the economic stand point. This is mainly due to the enormous difference between the financial investment required and viable industrial production (Luna et al. 2011). Thus, the following are the main criteria to be considered for biosurfactant production to become truly viable: (a) type of raw materials; (b) continuous provision of the same composition of ingredients; (c) types of microorganisms; (d) the adequate design of industrial fermentors; (e) financial investments; (f) the target market; (g) purification processes; (h) biosurfactant properties; (i) production conditions, especially the time required for fermentation; (j) adequate production yields; and (k) the processing of recycled products (minimal or able to sell for more than the drop in value) (Santos et al. 2013).

2. Fermentation Technology for Biosurfactant Production

Fermentation may be defined as a set of chemical reactions that cause the degradation of complex organic molecules into simpler compounds. This bioconversion of complex compounds into simpler ones may be induced by living organisms like bacteria and fungi. Apart from major products like ethanol, carbon dioxide, etc., additional compounds or by products are also produced during the fermentation process. These by products are also referred to as secondary metabolites. These secondary metabolites are bioactive compounds such as antibiotics, enzymes, growth factors, peptides biosurfactants, etc. (Subramaniyam and Vimala 2012). Fermentation process can be divided into two major types: (i) Submerged or Liquid state fermentation and (ii) Surface or Solid state fermentation. Among the two types of fermentation, Solid-state fermentation is emerging as a promising strategy for biosurfactant production.

Currently, three different types of fermenter operation processes which are frequently used for culturing bacteria. These are Batch, Fedbatch and Continuous fermentation processes. Batch fermentation is the process of culturing with all of the required nutrients provided at the start of the fermentation process and the process is run until all of the nutrients are exhausted and the broth is then harvested, all of the ingredients required for fermentation are added to the fermenter before inoculation with the seed culture. Batch fermentation has the advantage of being simple and having low risk of external contamination as no further additions are required except for pH stabilizers. The process is best for fermentations of cultures with high yield and for substances that can tolerate high initial nutrient conditions (Anderson 2009).

Fedbatch fermentation is similar to batch fermentation but only starts with some of the required nutrients at the inoculation stage in order to prevent inhibition of product production at high concentrations of substrate, further nutrients are added as the fermentation progresses in order to maintain substrate concentration for the production of the desired product (Chang et al. 2016). The advantages of Fedbatch fermentation include reduction of substrate and product inhibition and can decrease overall fermentation time, this then allows higher concentration of product without being inhibited by high levels of nutrients in the broth. Fedbatch fermentation however, carries the risk of potential contamination due to the addition of nutrients through a sterilizer, and the increased costs for specialized sterilization equipment. Batch and fed batch fermentations can be repeated using the same fermenter system after harvesting the culture by leaving a small amount of the previous batch in the fermenter as inoculum, this adds the risk of contamination, and degradation of the culture limits the number of repeat batches to about 2 or 3 before the fermenter must be cleaned and sterilized (Chang et al. 2012).

Continuous fermentation processes start with the medium and inoculum in the fermenter, after the culture has grown, the broth is withdrawn at the same rate as the fermenter is fed nutrients in order to maintain a constant volume of broth in the fermenter. Under ideal conditions the dilution rate will be the same as the culture growth rate, when this balance is maintained for long enough, there are no changes in the conditions within the reactor; this is called steady state operation (Brethauer and Wyman 2010). Compared to batch fermentation processes, continuous fermentation reduces down time for cleaning and sterilization between batches, although continuous fermentation cannot be run indefinitely, fermentations of several hundred hours can be completed under aseptic conditions. Continuous fermentation has better control at steady state operation which in turn reduces costs (Brethauer and Wyman 2010), but contamination from adapted cultures is difficult to avoid as they can grow back through the continuous harvest line (Anderson 2009). Khopade et al. (2012) completed their investigations using shake flasks in batch fermentation. When designing and optimizing a fermentation process, the optimum growth conditions of the isolated microorganism need to be identified, this is most effectively achieved at small scale using shake flasks by measuring optical density of the culture medium throughout the culture time to produce a growth curve for each of the variables such as temperature, salinity and medium composition. Parameters such as pH, O_2 content and O_2 uptake and other environmental factors cannot be as easily monitored and controlled at small scale (Smith 2009). Following this, optimum conditions can be established for culture of the microorganism. It should be noted that optimum conditions for growth of the microorganism may not be the optimum conditions for the production of the desired product. Following the optimization in shake flasks, the process can be scaled up to larger volumes for further optimization and development for potential use at an industrial scale.

The development of a suitable growth medium depends on the nutritional requirements of the microorganism to be cultured. In order to ensure that the production of biosurfactants is economical, low cost substrates with sufficient nutritional value need to be used as this can account for 10–30% of the overall costs (Silva et al. 2010). Khopade et al. (2012) chose to optimize the carbon and nitrogen sources available for utilization in order to obtain higher productivity of the biosurfactant. This was done using several carbon sources whilst keeping the nitrogen source constant, then using the optimum carbon source, varying nitrogen sources were compared and the optimums were chosen. The optimal growth conditions required for high cell density is not the same as the optimum conditions for biosurfactant production as previously noted, and in the case of *Pseudomonas aeruginosa*, when producing Rhamnolipid, fed batch fermentation with the carbon source in the feed produces a very low dry cell weight concentration (g/L) whereas the Rhamnolipid concentration is at its highest producing over 3.5 g/L (Ghomi Avili et al. 2012).

2.1 *Submerged or Liquid State Fermentation for Biosurfactant Production*

Liquid state fermentation of Submerged Fermentation (SmF) is the type of fermentation in which microorganisms are able to grow in the medium present in the form of a solution. This type of fermentation process is known for utilizing the substrates present in free flowing liquid, e.g., broth and others. During this process, the secondary metabolites or bioactive compounds such as biosurfactants, antibiotics, peptides among others are secreted into the fermentation broth. SmF or SSF is predominantly used on industrial scale. It is best suited for the microorganisms such as bacteria and some fungal strains which require moisture for the optimum growth and production of secondary metabolites.

The major advantage of SmF is that the purification of the bioactive compounds produced during the process is much easier as compared to surface fermentation (Subramaniyam and Vimala 2012). On the other hand, the purification of biosurfactants produced by solid state fermentation is difficult and more complicated. During the extraction of product after fermentation, some of the other water soluble compounds apart from the desired product may leach out, which makes the purification process quite difficult (Cameotra 2011). Not much research has been done on comparative study of the two techniques for the production of bioactive compounds. Tabaraie et al. (2012) compared use of both the techniques, i.e., solid state fermentation and liquid state for the production of a bioactive compound (cephalosporin-C). According to this study it was reported that solid state fermentation was better than liquid state fermentation for the production of antibiotic compounds by filamentous fungi. This conclusion was based on better control of the operating conditions and low costs involved during this process. However, other comparative studies show that for certain strains SmF is better and vice versa. Thus, implying that the fermentation technique should be chosen based on the microorganism used for production (Subramaniyam and Vimala 2012).

2.2 *Solid State Fermentation for Biosurfactant Production*

Surface or Solid state fermentation (SSF) is a type of fermentation that occurs in the absence or near absence of aqueous media. The substrates employed during this type of fermentation are usually cost free renewable wastes that are rich in carbon and protein content (Cameotra 2011). A few examples of solid substrates used for SSF are banana peel, wheat bran, tapioca peel (Vijayaraghavan et al. 2011), cassava dregs (Hong et al. 2001), rice husk, sugarcane, cassava bagasse, oil cakes such as palm kernel cake, coffee husk (Cameotra 2011). Solid state fermentation is a simple process and is considered to be effective because it produces concentrated products. Generally, for this type of fermentation the microorganisms involved are those which require less moisture content for growth, e.g., fungi and others. However, for bacteria, which require high water activity for growth, substrate fermentation is not preferred very often (Subramaniyam and Vimala 2012). Solid state fermentation has a number of advantages over the liquid state fermentation process. In this process less not much effluent is released because of the scarce amount of water used. This reduces pollution to a large extent as otherwise caused by the discharge of liquid state fermentation. It is also cost effective as it uses low volume equipments. The aeration process during solid state fermentation is easier compared to during the liquid state process, which is essential for the growth of aerobic microorganisms (Cameotra 2011).

Solid state fermentation is emerging as a promising strategy for biosurfactant production especially for overcoming the problem of foam production encountered in the more widely followed Submerged fermentations (SmF) (Camilios Neto et al. 2011). A medium based on okara with the addition of sugarcane bagasse as a bulking agent was employed in a SSF set up for the production of Surfactin by *Bacillus pumilus*. Under optimized conditions, 809 mg/L of Surfactin was produced upon cultivation in Column bioreactors with forced aeration which was at competitive levels to those reported in the literature for production by SmF. Extraction of Surfactin was simple with the whole content being mixed in water and filtered to get a cell-free preparation from which the biosurfactant was precipitated out and then extracted using Chloroform: Methanol (4:1 v/v) mixture (Slivinski et al. 2012). However, heat production in the system needs to be critically monitored as a variation in temperature would vary the Surfactin homologues formed. In a work on the use of agroindustrial by-products for bioprocesses, SSF of soybean flour and rice straw as substrate were studied for the production of an antibacterial lipopeptide by *Bacillus amyloliquefaciens* (Zhu et al. 2012).

In another work on Lipopeptides production by SSF, *Bacillus subtilis* was grown on a mixture of olive leaf residue flour and olive cake flour. A total of 30–67 mg of lipopeptides per gram of solid substrate was produced signifying the feasibility in the use of SSF for lipopeptides production, thus saving on operational costs associated with SmF (Zouari et al. 2014). In a comparative study on the production of biosurfactant by a fungus, *Pleurotus ostreatus* in SmF and SSF, with and without shaking, sunflower seed shells were used for SSF while sunflower oil was incorporated in the media for SmF. Although a good emulsification index was observed in SmF, SSF without shaking emerged as the best and most competitive option for the cultivation of the biosurfactant producing organism owing to the low cost of the process set up. The biosurfactant produced was found to be a carbohydrate–peptide–lipid complex (Velioglu and Urek 2015). SSF hence holds much promise for economizing biosurfactant production but much optimization needs to be done to make this approach feasible and market friendly.

Not much work has been done on analyzing the competitiveness of these fermentation processes for the production of Glycolipids. In a recent study, Mahua oil cake was used as a substrate for the production of biosurfactant with *Serratia rubidaea*. Rhamnolipid type biosurfactant with good antifungal activity was found to be produced by the organism under SSF in media optimized using Randomized Surface Methodology (RSM) (Nalini and Parthasarathi 2014). Although, work has been done on low-cost production of Rhamnolipids, much needs to be done with respect to Rhamnolipid production using SSF especially using agroindustrial by-products such as coconut waste, sesame waste, soybean waste among others, as substrates. Current biosurfactant production, however, is mostly limited to batch or fed-batch production processes due to limitations on sustained nutrient feeding strategies, and biomass growth and heat and mass transfer reactions which limit the process efficiency (Winterburn and Martin 2012). Hence, more research is needed to devise an integrated bioprocess for continuous biosurfactant production and recovery using low-cost waste biomass in a SSF setup.

3. Yield Enhancement in Biosurfactant Production by Media Modulation

Media constituents play a vital role in the type and quantity of the biosurfactant produced. Use of a carrier in the growth medium proved to be a novel approach for enhanced biosurfactant production but not much work has been done to fully exploit this strategy. The addition of activated charcoal in the fermentation medium increased Surfactin yield approximately 36 times as compared to a medium without solid support giving a biosurfactant yield of 3–6 g/L (Yeh et al. 2005). Enhancement in the growth of the producing organism was stipulated to be due to the presence of activated carbon barriers which partly were being used for Biofilm associated growth of cells. Use of growth enhancers like lactones is another promising strategy for enhanced yield. The addition of endogenous homoserine lactones and recycling of 20% of the spent medium was found to stimulate the Rhamnolipid production under Fed-batch conditions.

Lactones served as inducers for Rhamnolipid production and an almost 100% increase in Rhamnolipid production was observed when the spent medium was recycled in the reactor (Santos et al. 2016). This assumes importance as Lactonic SLs have been reported to have better surface tension lowering activity (Van Bogaert et al. 2007). Thus, it seems promising to use lactones for enhanced biosurfactant production, especially for Glycolipids. Lactones can also lead to the production of derivatized or a specific biosurfactant congener, more suitable for a particular application, thus making the overall process more profitable. Effect of lactones in production media of other biosurfactant types, however, needs to be further explored. Use of a specific producer strain could also lead to either production of a specific product or a derivatized or crystallized product which would make recovery easier and economical. A notable development in this is the specific production of SLs from *Starmerella bombicola* which are produced mostly as a mixture of 23 homologues containing acidic and Lactonic SLs. Acidic SLs find major application in wound healing while Lactonic SLs are mostly used as anticancer and antimicrobial agents for application in cosmetic industry. Exclusive synthesis of Lactonic or acidic SLs was successfully done by engineering *Starmerella bombicola* strains which generated less foam as well as yielded a selective 100% Lactone product (Roelants et al. 2016).

Use of Nanoparticles (NP) is another upcoming approach for enhanced biosurfactant production. Biosurfactant production has been confirmed to be significantly affected by many metal salts especially Iron (Fe). Hence, an upcoming potential strategy for enhanced biosurfactant production is the use of low concentrations of Fe-NPs. An increase of around 80% biosurfactant production by *Nocardiopsis* was observed in the presence of 10 mg/L. Fe-NP as compared to the control (Kiran et al. 2014). This is critical considering the phenomenon of Fe limitation during biosurfactant production. In another study, low Fe-NP concentration (1 mg/L) was found to increase the production of biosurfactant by 63% as compared to the control in fermentation by *Serratia* species. However, the growth of the producing organism was inhibited at higher concentrations of Fe-NP confirming the requirement for limiting Fe ions in the media for optimum Glycolipids biosurfactant production (Liu et al. 2013). In a recent study, 1 mg/L of Fe-silica NP (Fe-Si-NP) with 6 hrs addition time was found to increase rhamnolipid production by *Pseudomonas aeruginosa* strain by 57 % as compared to a medium free of NPs (Sahebnazar et al. 2018). This increase was attributed to an increase in growth and subsequent cell lysis at higher NP concentrations thus releasing the biosurfactant in the culture medium and increasing the yield.

Another aspect, partly related to the Downstream process, which could contribute to production economics is the aesthetic value of the recovered product, especially if the product is intended for food or medical applications. Some substrates used in the media for biosurfactant production have been known to give a coloured biosurfactant with currently available downstream processes which might make them aesthetically unacceptable. Distillery wastewater is one such substrate. Adsorption—desorption process using wood-based activated carbon was tested when distillery wastewater was used as a substrate for biosurfactant production by *Pseudomonas* strain. Not only was the biosurfactant free of any colour, a fivefold higher biosurfactant was recovered from collapsed foam than from fermented distilled wastewater (Dubey et al. 2005). Standardization of such studies on media composition could open a whole new approach for biosurfactant production enhancement as well as a low production cost.

4. Factors Affecting Biosurfactant Production

4.1 Carbon source

Large quantities of carbon sources have been utilized by many investigations for the creation of biosurfactants. Diesel, Mustard oil, Raw petroleum, Ethanol, Glucose, Rhamnose, Sucrose, Mannitol and Glycerol have been accounted for as a decent wellspring of carbon substrates for biosurfactant production. It is clear that the significance of the carbon substrate plays a noteworthy part in biosurfactant union, however its significance is life form needy as amid a production with *Pseudomonas* sp., diverse carbon sources in the medium influenced the synthesis of the biosurfactant creation yet a substrate with various chain lengths displayed no impact on the chain length of unsaturated fat moieties in Glycolipid (Desai and Banat 1997).

Production levels of biosurfactants fall when using carbon sources which are immiscible in water such as olive oil and alkanes. Also, it has been shown that the various sources of carbon used by microorganisms modify the composition of biosurfactants in its polar fraction while keeping the glycolipid chain's length in the non-polar fraction unchanged by fatty acids. Nonetheless, a qualitative variation exists among biosurfactant production by *Acinetobacter* sp. as is evidenced by the number of alkane carbons (Desai and Banat 1997). *Bacillus subtilis* was used as a carbon source with potato (60 g/L) for the production of Surfactin (Fox and Bala 2000). Linhardt et al. (1989) observed that the *Pseudomonas aeruginosa* strain used corn oil as a carbon source and produced Rhamnolipids,and the highest production of Rhamnose (5.4 g/L) was obtained when the concentration in the culture medium was 40 g/L of corn oil. *Cellulomonas cellulans* produced Glycolipids (8.9 g/L, expressed as glycerol per litre) when it grew in a liquid medium with 30 g glycerol/L (Arino et al. 1998).

Pseudomonas aeruginosa used hydrocarbons with carbon chains upto C12 and olive oil as carbon sources to produce Rhamnolipids. However, in the presence of fructose, production is inhibited (Robert et al. 1989). Benincasa et al. (2002) proved that *Pseudomonas aeruginosa* uses soapstock (3 g/L) as a carbon source to produce Rhamnolipids (12 g/L). *Torulopsis magnolie* employs long-chain fatty acids, hydrocarbons and glycerin to produce Sophorolipids (Cameotra and Makkar 1998). When *Arthrobacter paraffineus* grows in D-glucose and when hexadecane is added during its latent growth phase, the biosurfactant performance has a significant peak. Patel and Desai (1997) reported that *Pseudomonas aeruginosa* increased their biosurfactant production with molasses (48% w/v of carbohydrates) present in the medium with carbon source at 7% concentration, which generated 0.24 g/L of Rhamnose after 96 hrs of incubation.

Corynebacterium lepusc, grows in glucose but if hexadecane is added it is able to synthesize large amounts of biosurfactants. The way *Torulopsis bombicola* handles its carbon sources allows us to observe an increase in production of Glycolipids. Tuelva et al. (2002) reported that *Pseudomonas putida* was able to produce Rhamnolipids utilizing hexadecane carbon as its only source. Martinez Toledo et al. (2006) reported that *Pseudomonas putida* was able to increase production of Rhamnolipids when Phenanthrene (200 mg/L) was added to the culture. Prabhu and Phale (2003) reported that *Pseudomonas aeruginosa* produces biosurfactants when palm oil is used as a carbon source. Other studies have shown that *Pseudomonas* sp. could produce biosurfactants growing on either Phenanthrene or dextrose as carbon sources, but Phenanthrene showed the higher production levels compared to dextrose. De Lima et al. (2009) observed the rhamnolipid production due to *Pseudomonas aeruginosa* growth on different waste frying soybean oils, where the maximum Rhamnose production (4.3 g/L) was obtained using soybean oil. *Pseudomonas aeruginosa* used palm oil as a carbon source and produced biosurfactant at oil loading rates of 2 kg/m^3 in 24 hrs, the surface tension of the liquid medium was reduced by 58%.

4.2 Nitrogen source

A Nitrogen source is vital for biosurfactant creation. A medium containing nitrogen plays a fundamental part in the microbial protein development and chemical blends rely on it. Distinctive nitrogen sources have been utilized for the creation of biosurfactants like yeast separate, ammonium sulfate, ammonium nitrate, sodium nitrate, meat concentrate and malt extricates. Nitrate sources are utilized to the extreme for surfactant creation in *Pseudomonas aeruginosa*. Ammonium salts and urea are favored for *Arthrobacter paraffineus*. In spite of the fact that yeast separate is the most utilized nitrogen hotspot for biosurfactant creation, its use concerning focus is creature and culture medium ward. The creation of surface-dynamic mixes frequently happens when the nitrogen source is exhausted in the way of life medium, amid the stationary period of cell development.

The Nitrogen source is the second most important supplement for the production of biosurfactants by microorganisms. In fermentative processes, the C/N ratio affects the buildup of metabolites. High C/N ratios (i.e., low nitrogen levels) limit bacterial growth, favouring cell metabolism towards the production of metabolites. In contrast, excessive nitrogen leads to the synthesis of cellular material and limits the buildup of products (Robert et al. 1989). Different organic and inorganic nitrogen sources have been used in the production of biosurfactants. Santa Anna et al. (2002) describe the importance of nitrogen for the production of a biosurfactant by *Pseudomonas aeruginosa* cultivated in a mineral medium containing 3% glycerol. As $NaNO_3$ proved more effective than $(NH_4)_2SO_4$, nutritional limitations clearly guide the cell metabolism to the formation of the product. Mulligan and Gibbs (1989) report that *Pseudomonas aeruginosa* uses nitrates, ammonium and amino acids as nitrogen sources. Nitrates are first reduced to nitrite and then ammonium forms. The Ammonium form is assimilated either by glutamate dehydrogenase to form glutamate or glutamine synthetase to form glutamine. Glutamine and α-ketoglutarate are then converted to glutamine by L-glutamine 2-oxoglutarate aminotransferase. However, lipid formation rather than sugar is the rate-determining factor in the biosynthesis of Rhamnolipids and nitrogen limitation can lead to the accumulation of lipids. In comparison to the ammonium form, the assimilation of the nitrate form is slower and simulates nitrogen limitation, which is favourable for the production of Rhamnolipids. High yields of Sophorose lipids, which are biosurfactants produced by the fungi *Torulopsis bombicola* and *Candida bombicola* have been achieved using yeast extract and urea as the nitrogen source (Deshpande and Daniels 1995). Moreover, high yields of mannosylerythritol lipid by *Candida* sp., *Candida lipolytica* and *Candida glabrata* have been achieved with ammonium nitrate and yeast extract (Sarrubo et al. 2007, Rufino et al. 2008).

4.3 Phosphorous source

There is evidence of an existing connection between rhamnolipid synthesis and glutamine synthetase activity in *Pseudomonas aeruginosa*. This enzyme delivers maximum activity levels when its exponential growth phase ends, and at that time its production of biosurfactants begins. Furthermore, Mulligan et al. (1999) reported that the shift in metabolism to phosphates coincides with the production of biosurfactants. Phosphorous limitation in the culture media was more effective for biosurfactant production by *Pseudomonas aeruginosa* than Nitrogen limitation, only when hexadecane or palmitic acid were used as carbon sources in anaerobic conditions (Chayabutra et al. 2000). In studies done by Martinez Toledo and Rodriguez Vazquez

(2010) it was also noted that *Pseudomonas putida* is able to produce Rhamnolipids with variations in its polar fraction due to the relationship changes among C:N:P, where glucose and Phenanthrene were used as carbon sources in the culture medium.

4.4 *Micronutrients*

Several papers indicated the great importance of micronutrients in the culture media for bacteria. Guerra Santos et al. (1986) proved that under limited ion conditions Magnesium, Calcium, Potassium, Sodium, Iron and other trace elements, a high rate of biosurfactant production by *Pseudomonas aeruginosa* is delivered. Various authors have shown that an overproduction of biosurfactants by *Pseudomonas* spp. is obtained when it reaches its latency growth stage, under limited nitrogen and iron conditions.

Phosphorus, iron, magnesium and sodium are all elements of importance to the production of biosurfactants by *Rhodococcus* sp. but potassium and calcium are the elements of greatest importance. Iron also has an important impact on the production of Rhamnolipids by *Pseudomonas aeruginosa*. The concentration of manganese and iron affects the production of Surfactin by *Bacillus* subtilis. A report discusses how *Bacillus subtilis* has an active transportation system for both ions, manganese may act as a cofactor to many enzymes involved in the metabolism of nitrogen, even in reactions from glutamate and ammonium catalyzed by the glutamine synthetase enzyme, this being its main assimilation mechanism of inorganic nitrogen; it was observed that with a concentration of 0.5 mg/L of $FeSO_4$ $7H_2O$ (C/Fe ratio of 72 : 400) in the culture medium Rhamnolipids production by *Pseudomonas aeruginosa* increased (500 mg/L). *Bacillus subtilis* increases its production of Surfactin in the presence of 1.7 mM of iron sulfate in the culture medium (Wei and Chu 1998). The iron concentration of 50–100 μg/L was the optimal concentration in the liquid medium for rhamnolipid production by *Pseudomonas aeruginosa* strain (Fernandes et al. 2016). Amezcua Vega et al. (2004) observed an increase in the production of Rhamnolipids by *Pseudomonas putida* at a low C/Fe ratio (26000) in the culture medium, where the carbon source was corn oil at 2% (w/v). Similar results were observed with *Rhodococcus* sp. that produced Glycolipids (3 g/L) at a low C/Fe ratio of 12000 (Robert et al. 1989).

4.5 *Growth Conditions and Environmental Factors*

Environmental factors are extremely important in the yield and characteristics of the biosurfactant produced. In order to obtain large quantities of biosurfactant it is necessary to optimize the process conditions because the production of a biosurfactant is affected by variables such as pH, temperature, aeration and agitation speed.

4.5.1 *pH*

The effect of pH in the biosurfactant production by *Candida antarctica* has been investigated using phosphate buffer with pH values varying from 4 to 8. All conditions used resulted in a reduction of biosurfactant yield when compared to distilled water (Kitamoto et al. 2001). Zinjarde and Pant (2002) studied the influence of initial pH in the production of a biosurfactants by *Candida lipolytica*. The best production of biosurfactants occurred when the pH was 8.0, which is the natural pH of sea water. The acidity of the production medium was the parameter studied in the synthesis of Glycolipids by *Candida antarctica* and *Candida apicola*. When pH is maintained at 5.5, the production of Glycolipids reaches a maximum. The synthesis of the biosurfactants decreased without pH control indicating the importance of maintaining it throughout the fermentation process (Bednarski et al. 2004).

4.5.2 *Temperature*

Most of the biosurfactant productions reported so far have been performed in a temperature range of 25 to 30°C. Casas and Ochoa (1999) noticed that the amount of Sophorolipids obtained in the culture medium of *C. bombicola* at a temperature of 25°C or 30°C is similar. Nevertheless, fermentation performed at 25°C presents a lower biomass growth and a higher glucose consumption rate in comparison to the fermentation performed at 30°C. Desphande and Daniels (1995) observed that the growth of *Candida bombicola* reaches a maximum at a temperature of 30°C while 27°C is the best temperature for the production of Sophorolipids. In the culture of *Candida antarctica*, temperature causes variations in the biosurfactant production. The highest mannosylerythritol lipids production was observed at 25°C for that conducted with both growing and resting cells (Satpute et al. 2010).

4.5.3 *Metal ion Concentration*

Metal ion concentration plays a very important role in the production of some biosurfactants as they form important cofactors of many enzymes. The overproduction of Surfactin biosurfactant occurs in presence of Fe_2^+ in mineral salt medium. The properties of Surfactin are modified in the presence of inorganic cations such as Potassium ions and Sodium ions overproduction (Thimon et al. 1992).

4.5.4 *Natural Elements*

Environmental elements are critical in the yield of the biosurfactant delivered. To acquire huge amounts of biosurfactants, it is constantly important to upgrade the bioprocess as the item might be influenced by changes in temperature, pH, air circulation or unsettling speed. Most biosurfactant preparations are accounted for to be performed in a temperature scope of 25–300°C, be that as it may, in *Acinetobacter paraffineus* and *Pseudomonas* sp. strain, this temperature extent caused modification in the organization of biosurfactant created. The impact of pH on biosurfactant delivered was considered by Zinjarde and Pant (2002) who announced that the best creation occurred when the pH was 8.0, the regular pH of ocean water which is *Candida lipolytica's* regular natural environment. With *Pseudomonas* sp., Rhamnolipids creation was at its highest at a pH from 6 to 6.5 and diminished drastically at a pH above 7.

4.5.5 *Aeration and Agitation*

Aeration and Agitation are imperative factors that impact the production of biosurfactants as both encourage the oxygen exchange from the gas to the fluid stage. It might likewise be connected to the physiological capacity of the microbial emulsifier, and it has been recommended that the creation of bioemulsifiers can improve the solubilization of water insoluble substrates and subsequently encourages supplement transportation to microorganisms. It was noted that the highest yield estimate of the surfactant (45.5 g/L) was attained when the wind current rate was 1 vvm and the broken up oxygen focus was kept up at half of immersion (Krishnaswamy et al. 2008).

4.5.6 Salinity

Salinity of a specific medium additionally assumes a critical part in the biosurfactant production. In any case, opposite perceptions were seen for some biosurfactant items which were not influenced by fixations up to 10% (weight/volume) albeit slight diminishments in the CMC were distinguished (Kugler et al. 2015).

5. Microbial Production of Biosurfactants

Bushnell Haas Broth is used as a production medium and inoculated with a 24–48 hrs old bacterial culture prepared in a Nutrient broth medium or 144–168 hrs old fungal culture prepared in Potato Dextrose Broth medium placed at room temperature in a Shaking condition. The inoculated culture is allowed to grow under optimum condition for 7–10 days. The culture broth is centrifuged at 10000 rpm for 15 min to remove the cells in order to obtain clear sterile supernatant (Abouseoud et al. 2007, Vandana and Peter 2014).

Optimization of the production of biosurfactant can be achieved by testing the biosurfactant production throughout the fermentation process whilst changing the variables accordingly. Using a Tensiometer in order to monitor any changes in surface tension is a good indicator of biosurfactant production, foaming in shake flasks during culturing is also a good indicator of the presence of biosurfactants in the media and this can be analyzed further by testing the emulsification index of the biosurfactant produced as described in Shavandi et al. (2011). It can be noted that Fed-batch fermentation is more effective than Batch fermentation processes in order to produce higher concentrations of Rhamnolipid by *Pseudomonas aeruginosa*, when the carbon source is limited by the feed process (Ghomi Avili et al. 2012). A higher concentration does not mean that the biosurfactant produced cannot be isolated and studied at lower concentrations when using the batch fermentation process however, which may be more cost effective during the initial screening for suitable biosurfactants.

6. Recovery of Biosurfactants

6.1 Cold Acetone Precipitation Method

In the Cold acetone precipitation method, three volumes of chilled acetone were added to the crude biosurfactant solution and allowed to stand for 10 hrs at 4°C. Precipitates were collected by centrifugation at 10000 rpm for 20 min and the resulting pellet was served as partially purified biosurfactant which was further evaporated to dryness to remove residual acetone after its dissolution in sterile water (Ilori et al. 2005, Vandana and Peter 2014).

6.2 Acid Precipitation Method

A biosurfactant can also be precipitated by adjusting the pH of the cell-free broth culture to 2.0 using 6 N HCl and maintaining a temperature of 4°C overnight. Pellets thus precipitated are collected by centrifugation (8000 rpm for 15 min at 20°C) and dissolved in sterile distilled water. After that, the pH is adjusted at 8.0 by using 1 N NaOH for further use (Abouseoud et al. 2008).

6.3 Chloroform: Methanol Precipitation Method

The cell free broth was acidified to pH 2.0 with HCl. There after biosurfactant was extracted twice using equal volumes of Chloroform: Methanol (2:1) solutions in a Separatory funnel (Parhi et al. 2016).

6.4 Ammonium Sulphate Precipitation

Ammonium sulphate precipitation is used for precipitation of high-molecular weight biosurfactants such as Emulsions and Biodispersions (protein rich compounds). As per the type of biosurfactant, a different concentration of ammonium sulphate is used. In case of Ammonium sulphate precipitation, the Rhamnolipid is precipitated by a salting out process and the product is further purified by a Dialysis procedure and Lyophilized (Rosenberg et al. 1979).

6.5 Ethanol Precipitation

Like other solvents, ethanol is also used for obtaining a crude extract of the biosurfactant from the supernatant culture of microbes. Broth culture is centrifuged at 11000 rpm for 20 min at 4°C and the biosurfactant is precipitated from the supernatant by using cold Ethanol. Phetrong et al. (2008) found that the precipitation of the emulsifier from *Acinetobacter calcoaceticus* sub sp. *anitratus* with Ethanol was the most efficient method when compared with other precipitation methods.

7. Purification of Biosurfactants

7.1 Dialysis

Dialysis is easy and cost effective for the purification of the biosurfactant. Dialysis and Ultra filtration techniques are widely exploited to enhance the purity of the Biosurfactant by using seamless Cellulose dialysis bags. The collected precipitate samples containing biosurfactant dissolved in 5–10 ml of sterile distilled water and dialyzed against double distilled water for 48 hrs at 10°C. The dialysate was stored at 4°C in an air tight container for further use. Kaplan and Rosenberg (1982) reported the production of the biosurfactant from *Acinetobacter calcoaceticus* which after Ammonium sulphate precipitation was dissolved in deionized water and Dialyzed in cold distilled water.

7.2 Thin Layer Chromatography (TLC)

Thin Layer Chromatography (TLC) is used for the preliminary characterization of the biosurfactant. A portion of the crude biosurfactant is separated on a Silica gel plate using Chloroform: Methanol water (10: 10: 0.5 v/v/v). The type of biosurfactant is characterized by using a developing solvent system with a different colour developing reagent. Ninhydrin reagent is used to detect Lipopeptides biosurfactant as a red spot produced by the Biosurfactant (Maheswari and Parveen 2012).

8. Immobilization as a Tool for Biosurfactant Production

A major hindrance in continuous culturing and downstream processing of biosurfactants is the washout of cells from the reactor as well as the effect of changing reactor conditions and unwanted metabolites on cell growth apart from the foam-inducing property of free cells. An alternative to using foam-inducing cells for the production process is to explore the option of immobilized cells. This is especially advantageous in the case of a bioprocess using resting microbial cells, as growth and product formation stages would be separate making the product separation easier along with maintaining the continuity of the bioprocess. Alginate immobilized cells of *Pseudomonas fluorescens* were used for rhamnolipid production. Although pH of the media varied considerably in the presence of immobilized cells as compared to free cells, the Rhamnolipid produced had good and stable emulsifying activity and fewer by-products interfered with biosurfactant production. Also, immobilized organisms were easy to separate from the broth during downstream processing (Abouseoud et al. 2008).

In another work, cells of *Bacillus subtilis* immobilized onto Fe-enriched light polymer particles, which were found to produce 2.09–4.3 times more biosurfactant than planktonic cells in a batch reaction process. *Acinetobacter threephase* inverse fluidized bed biofilm reactor with cells of *Bacillus subtilis* immobilized on Fe-polypropylene foam particles and was used for the study. At the same time, direct addition of Fe-polypropylene pellets in a culture medium selectively enhanced the production of Fengycin as compared to Surfactin. An immobilized cell carrier system together with varying Fe concentrations hence provided an interesting tool for steering the biosurfactant production process towards a selective biosurfactant yield (Gancel et al. 2009). In a similar but modified strategy, *Pseudomonas aeruginosa* cells immobilized on Ca-alginate beads were used for Rhamnolipid production. Use of a high-density magnetic gradient facilitated its retention from the foam and flushing back into the reactor during foam fractionation and recovery. This facilitated a continuous rhamnolipid production in a 10^{-1} bioreactor yielding a final RL amount of 70 g after four production cycles (Heyd et al. 2011). The efficacy of this process was also tested by immobilization of *Pseudomonas nitroreducens* using Ca-alginate beads under resting cell conditions. Palm oil and diesel were used as carbon sources and 5.1 g/L Rhamnolipid yield was obtained (Oliveria et al. 2015). Combinations of novel immobilization techniques and reactor conditions hence offer greener pastures of research.

Other strategies which could hold strategic importance in studying and enhancing the large-scale yield of biosurfactants include the use of Microbioreactors for optimization studies, in situ product removal by automated surface enrichment, employment of a novel oxygenation process and use of novel techniques like pertraction, and others (Chtioui et al. 2012). Biosurfactants have a variety of applications, each differing in the associated purity required as well as a specific structure of the compound used. Hence, utilization of a crude product without any costly purification processes would contribute immensely to lowering the overall production cost. This would especially be extremely profitable in case of applications like MEOR, bioremediation, wastewater treatment, metal bioremediation, and others where the application of a crude product would be equally effective (Singh and Cameotra 2004).

9. Conclusion

The fermentation process holds the key to improving the overall process economics in biosurfactant production. A number of attempts have been made to increase biosurfactant productivity by manipulating physiological conditions and medium composition. Microorganisms such as yeast, bacteria or fungi can produce Biosurfactants surface active compounds using different substrates such as oils, glycerol, alkanes, sugars and wastes. Biosurfactants are biodegradable, making them an attractive alternative to chemically synthesized surfactants which are normally petroleum-based and environmentally hazardous. Biosurfactant production can be a costly process which can be made less so by varying production modalities and parameters. Production can be either by batch, fed-batch or continuous fermentation methods when down streaming and growth controlling factors especially the nutrient can be altered to optimize production. Biosurfactants can therefore be produced in adequate quantities using bioreactors and cheap feedstock as nutrient sources. Knowledge of biosurfactants, their characteristics and uses is expanding and more invaluable research is being conducted into optimizing productivity and reducing costs. The benefits of biologically produced surfactants cannot be denied and they surpass conventional chemical surfactants in many ways but there are major limitations still facing their industrial application. Their low yield and high cost when compared to chemical surfactants has started to receive more biotechnological research in order to successfully overcome these limitations. If momentum is maintained, we will start to see commercially available biosurfactant products being utilized by industries such as oil recovery, fuel extraction and medicine within the next decade.

References

Abouseoud, M., R. Maachi, A. Amrane, S. Boudergua, A. Nabia et al. 2007. Evaluation of different carbon and nitrogen sources in production of biosurfactant by *Pseudomonas fluorescens*. J. Microbio. Biotech. 223: 143–151.

Abouseoud, M., A. Yataghene, A. Amrane and R. Maachi. 2008. Biosurfactant production by free and alginate entrapped cells of *Pseudomonas fluorescens*. J. Ind. Microbio. Biotech. 35: 1303–1308.

AmezcuaVega, C.A., P.H.M. Varaldo and F. Garcia. 2004. Effect of culture conditions on fatty acids composition of a biosurfactant produced by *Candida ingens* and changes of surface tension of culture media. Biores. Tech. 98: 237–240.

Anderson, T.M. 2009. Industrial Fermentation Processes. Encyclopedia of Microbiology (Third Edition). Oxford: Academic Press, pp. 349–361.

Arino, S., R. Marchal and J.P. Vandecasteele. 1998. Production of new extracellular Glycolipids by a strain of *Cellulomonas cellulans* (*Oreskoviax anthineolytica*) and their structural characterization. Can. J. Micro. Biol. 44: 238–243.

Bednarski, W., M. Adamczak and J. Tomasik. 2004. Application of oil refinery waste in the biosynthesis of Glycolipids by yeast. Biores. Techn. 95: 15–18.

Benincasa, M., J. Contiero, M.A. Manresa and J.O. Moraes. 2002. Rhamnolipid productions by *Pseudomonas aeruginosa* LBI growing on soap stock as the sole carbon source. J. Food En. 54: 283–288.

Brethauer, S. and C.E. Wyman. 2010. Continuous hydrolysis and fermentation for cellulosic ethanol production. Biores. Tech. 101(13): 4862–4874.

Cameotra, S.S. and R.S. Makkar. 1998. Synthesis of biosurfactants in extreme conditions. Appl. Microbiol. Biotechnol. 50: 520–529.

Cameotra, S.S. 2011. Environmental applications of biosurfactants: Recent advances. Int. J. Mol. Sci. 1: 633–654.

Camilios Neto, D., C. Bugay, A.P. de Santana Filho, T. Joslin, L.M. de Souza et al. 2011. Production of Rhamnolipids in solid state cultivation using a mixture of sugarcane bagasse and corn bran supplemented with glycerol and soybean oil. Appl. Microbiol. Biotechnol. 89: 1395–1403.

Casas, J.A. and F. Garcia Ochoa. 1999. Sophorolipid production by *Candida bombicola* medium composition and culture methods. J. Biosci. Bioeng. 88: 488–494.

Chang, H., W. Xing, T. Xia, R. Fu, L. Tao et al. 2016. Biological characteristics of biosurfactant producing Petroleum degrader bacterium *Bacillus* BS-8. Agri. Sci. Tech. 17(1): 1–3.

Chang, Y., K. Chang, C. Huang, C. Hsu, H. Jang et al. 2012. Comparison of batch and fed-batch fermentations using corncob hydrolysate for bioethanol production. Fuel. 97: 166–173.

Chayabutra, C., J. Wu and L.K. Ju. 2000. Rhamnolipid production by *Pseudomonas aeruginosa* under denitrification: effects of limiting nutrients and carbon substrates. Biotechnol. Bioeng. 72(1): 25–33.

Chtioui, O., K. Dimitrov, F. Gancel, P. Dhulster, I. Nikov et al. 2012. Rotating discs bioreactor, a new tool for lipopeptides production. Process Biochem. 47: 2020–2024.

De Lima, C.J.B., E.J. Ribeiro, E.F.C. Sérvulo, M.M. Resende, V.L. Cardoso et al. 2009. Biosurfactant production by *Pseudomonas aeruginosa* grown in residual Soybean Oil. Appl. Biochem. Biotechnol. 152: 156–168.

Desai, J.D. and I.M. Banat. 1997. Microbial production of surfactants and their commercial potential. Microbiol. Mol. Bio. Rev. 61: 47–64.

Deshpande, M. and L. Daniels. 1995. Evaluation of sophorolipid biosurfactant production by *Candida bombicola* using animal fat. Biores. Tech. 54: 143–150.

Dubey, K.V., A.A. Juwarkar and S.K. Singh. 2005. Adsorption—Desorption process using wood based activated carbon for recovery of biosurfactant from fermented distillery wastewater. Biotechnol. Prog. 21: 860–867.

Fernandes, P.L., E.M. Rodrigues, F.R. Paiva, B.A.L. Ayupe, M.J. McInerney et al. 2016. Biosurfactan, solvent and polymer production by *Bacillus subtilis* RI4914 and their application for enhanced oil recovery. Fuel. 180: 551–557.

Fox, S.L. and G.A. Bala. 2000. Production of surfactant from *Bacillus subtilis* ATCC 21332 using potato substrates. Biores. Tech. 75: 325–240.

Gancel, F., L. Montastruc, T. Liu, L. Zhao, I. Nikov et al. 2009. Lipopeptide overproduction by cell immobilization on iron-enriched light polymer particles. Process Biochem. 44: 975–978.

Geetha, S.J., I.M. Banat and S.J. Joshi. 2018. Biosurfactants: production and potential applications in Microbial Enhanced Oil Recovery (MEOR). Biocatal. Agri. Biotechnol. 14: 23–32.

Ghomi Avili, M., M. Hasan Fazaelipoor, S. Ali Jafari, S. Ahmad Ataei et al. 2012. Comparison between batch and fed-batch production of rhamnolipid by *Pseudomonas aeruginosa*. Irani. J. Biotech. 10(4): 263–269.

Gudina, E.J., A.I. Rodrigues, V. de Freitas, Z. Azevedo, J.A. Teixeira et al. 2016. Valorization of agro-industrial wastes towards the production of Rhamnolipids. Biores. Tech. 212: 144–150.

Guerra Santos, L.H., O. Kappeli and A. Flechter. 1986. Dependence of *Pseudomonas aeruginosa* continuous culture biosurfactant production on nutritional and environmental factors. Appl. Microbiol. Biotechnol. 24: 443–448.

Heyd, M., M. Franzreb and S. Berensmeier. 2011. Continuous rhamnolipid production with integrated product removal by foam fractionation and magnetic separation of immobilized *Pseudomonas aeruginosa*. Biotechnol. Prog. 27: 706–716.

Hong, K., Y. Ma and M. Li. 2001. Solid state fermentation of phytase from cassava dregs. Appl. Biochem. Biotech. 93(1-9): 777–785.

Ilori, M.O., C.J. Amobi and A.C. Odocha. 2005. Factors affecting Biosurfactant production by oil degrading *Aeromonas* spp. isolated from a tropical environment. Chemosph. 6: 110–116.

Kaplan, N. and E. Rosenberg. 1982. Exopolysaccharide distribution of and bioemulsifier production by *Acinetobacter calcoaceticus* BD4 and BD413. Appl. Environ. Microbiol. 44: 1335–1341.

Khopade, A., B. Ren, X. Liu, K. Mahadik, L. Zhang et al. 2012. Production and characterization of biosurfactant from marine *Streptomyces* species. J. Collo. Interf. Sci. 367(1): 311–318.

Kiran, G.S., B. Sabarathnam, N. Thajuddin and J. Selvin. 2014. Production of glycolipid biosurfactant from sponge-associated marine Actinobacterium *Brachybacterium paraconglomeratum* MSA21. J. Surfactant Deterg. 17: 531–542.

Kitamoto, D., T. Ikegami and G.T. Suzuki. 2001. Microbial conversion of n-alkanes into glycolipid biosurfactants, mannosylerythritol lipids, by *Pseudozyma* (*Candida antarctica*). Biotech. Let. 23: 1709–1714.

Krishnaswamy, M., G. Subbuchettiar, T.K. Ravi and S. Panchaksharam. 2008. Biosurfactants properties, commercial production and application. Cur. Sci. 94: 736–747.

Kugler, J.H., M. Le Roes Hill, C. Syldatk and R. Hausmann. 2015. Surfactants tailored by the class Actinobacteria. Front Microbiol. 6: 212.

Kumar, A.P., A. Janardhan, B. Viswanath, K. Monika, J.Y. Jung et al. 2016. Evaluation of orange peel for biosurfactant production by *Bacillus licheniformis* and their ability to degrade naphthalene and crude oil. Biotech. 6: 1–10.

Linhardt, R.J., R. Bakhit and L. Daniels. 1989. Microbially produced rhamnolipid as a source of rhamnose. Biotechnol. Bioeng. 33: 365–368.

Liu, J., C. Vipulanandan, T.F. Cooper and G. Vipulanandan. 2013. Effects of iron nanoparticles on bacterial growth and biosurfactant production. J. Nanopart. Res. 15: 1–13.

Luna, J.M., R.D. Rufino, C.D.C. Albuquerque, L.A. Sarubbo, G.M. Campos Takaki et al. 2011. Economic optimized medium for tenso-active agent production by *Candida sphaerica* UCP 0995 and application in the removal of hydrophobic contaminant from sand. Int. J. Mol. Sci. 12: 2463–2476.

Maheswari, N.U. and I.F. Parveen. 2012. Comparative study of biosurfactant by using *Bacillus licheniformis* and *Trichoderma viride* from paper waste contaminated soil. Int. J. Chem. Sci. 10(3): 1687–1697.

Martinez Toledo, A., E. Ríos Leal, F. Vázquez Duhalt, M. González Chávez, C. Del et al. 2006. Role of phenanthrene in rhamnolipid production by *Pseudomonas putida* in different media. Environ. Technol. 27: 137–142.

Mulligan, C.N. and B.F. Gibbs. 1989. Correlation of nitrogen metabolism with biosurfactant production by *Pseudomonas aeruginosa*. Appl. Environ. Microbiol. 55: 3016–3019.

Mulligan, C.N., T.Y.K. Chow and B.F. Gibbs. 1989. Enhanced biosurfactant production by a mutant *Bacillus subtilis* strain. Appl. Microbiol. Biotechnol. 31: 486–489.

Mulligan, C.N., R.N. Young, B.F. Gibbs, S. James and H.P.J. Bennett. 1999. Metal removal from contaminated soil and sediments by the biosurfactant Surfactin. Environ. Sci. Tech. 32: 3812–3820.

Nalini, S. and R. Parthasarathi. 2014. Production and characterization of Rhamnolipids produced by *Serratia rubidaea* SNAU02 under solid-state fermentation and its application as biocontrol agent. Biores. Techn. 173: 231–238.

Oliveira, M.R., A. Magri, C. Baldo, D. Camilios Neto, T. Minucelli et al. 2015. Review: Sophorolipids a promising biosurfactant and its applications. Int. J. Adv. Biotech. Res. 6: 161–174.

Parhi, P., V.V. Jadhav and R. Bhadekar. 2016. Increase in production of biosurfactant from *Oceanobacillus* sp. BRI 10 using low cost substrates. Songklanakarin J. Sci. Tech. 38(2): 207–211.

Phetrong, K., H. Kittikun and A.S. Maneerat. 2008. Production and characterization of bioemulsifier from a marine bacterium, *Acinetobacter calcoaceticus* sub sp. *Anitratus* SM7. Songklanakarin J. Sci. Tech. 30(3): 297–305.

Prabhu, Y. and P.S. Phale. 2003. Biodegradation of phenanthrene by *Pseudomonas* sp. Strain PP2: novel metabolic pathway, role of biosurfactant and cell surface hydrophobicity in hydrocarbons assimilation. Appl. Microbiol. Biotechnol. 61: 342–351.

Robert, M., M.E. Mercade, M.P. Bosch, J.L. Parra, M.J. Espuny et al. 1989. Effect of the carbon source on biosurfactant production by *Pseudomonas aeruginosa* 44T. Biotech. Lett. 11: 871–874.

Roelants, S.L., K. Ciesielska, S.L. De Maeseneire, H. Moens, B. Everaert et al. 2016. Towards the industrialization of new biosurfactants: biotechnological opportunities for the lactone esterase gene from Starmerella bombicola. Biotechnol. Bioeng. 113: 550–559.

Rosenberg, E., A. Zuckerberg, C. Rubinovitz and D.L. Gutnick. 1979. Emulsifier of *Arthrobacter* RAG-1: isolation and emulsifying properties. Appl. Environ. Microbiol. 37: 402–408.

Rufino, R.D., L.A. Sarubbo, B.N. Benicio and G.M. Campos Takaki. 2008. Experimental design for the production of tensio-active agent by *Candida lipolytica*. J. Ind. Microbiol. Biotechnol. 35: 907–914.

Sahebnazar, Z., D. Mowla and G. Karimi. 2018. Enhancement of *Pseudomonas aeruginosa* growth and rhamnolipid production using iron-silica nanoparticles in low cost medium. J. Nanostructures 8: 1–10.

Santa Anna, I.M., G.V. Sebastian, N. Pereira, T.L.M. Alves, E.P. Menezes et al. 2002. Production of biosurfactant from a new and promising strain of *Pseudomonas aeruginosa* PA1. Appl. Biochem. Biotechnol. 91: 459–467.

Santos, D.K.F., R.D. Rufino, J.M. Luna, V.A. Santos, A.A. Salgueiro et al. 2013. Synthesis and evaluation of biosurfactant produced by *Candida lipolytica* using animal fat and corn steep liquor. J. Pet. Sci. Eng. 105: 43–50.

Santos, D.K.F., R.D. Rufino, J.M. Luna, V.A. Santos, L.A. Sarubbo et al. 2016. Biosurfactants: multifunctional biomolecules of the 21st century. Int. J. Mol. Sci. 17: 401.

Sarrubo, L.A., C.B.B. Farias and G.M. Campos Takaki. 2007. Co-utilization of canola oil and glucose on the production of a surfactant by *Candida lipolytica*. Curr. Microbiol. 54: 68–73.

Satpute, S.K., A.G. Banpurkar, P.K. Dhakephalkar, I.M. Banat, B.A. Chopade et al. 2010. Methods for investigating biosurfactants and bioemulsifiers: A review. Crit. Rev. Biotech. 8: 1–18.

Shavandi, M., G. Mohebali, A. Haddadi, H. Shakarami and A. Nuhi. 2011. Emulsification potential of a newly isolated biosurfactant-producing bacterium, *Rhodococcus* sp. strain TA6. Colloi. Surf. Biointer. 82(2): 477–482.

Silva, S.N.R.L., C.B.B. Farias, R.D. Rufino, J.M. Luna and L.A. Sarubbo. 2010. Glycerol as substrate for the production of biosurfactant by *Pseudomonas aeruginosa* UCP0992. Colloi. Surf. Biointer. 79(1): 174–183.

Singh, P. and S.S. Cameotra. 2004. Potential applications of microbial surfactants in biomedical sciences. Trends Biotech. 22: 142–146.

Slivinski, C.T., E. Mallmann, J.M. de Araujo, D.A. Mitchell and N. Krieger. 2012. Production of Surfactin by *Bacillus pumilus* UFPEDA 448 in Solid state fermentation using a medium based on okara with sugarcane bagasse as a bulking agent. Process Biochem. 47: 1848–1855.

Smith, J.E. 2009. Biotechnology. 5th edition. Cambridge: Cambridge University Press.

Subramaniyam, R. and R. Vimala. 2012. Solid state and submerged fermentation for the production of bioactive substances: A comparative study. Int. J. Sci. Nat. 3(3): 480–486.

Tabaraie, B., E. Ghasemian, T. Tabaraie, E. Parvizi, M. Rezazarandi et al. 2012. Comparative Evaluation of Cephalosporin-C production in Solid state fermentation and Submerged liquid culture. J. Microbiol. Biotech. Food Sci. 2(1): 83–94.

Thimon, L., F. Peypoux and G. Michel. 1992. Interaction of Surfactin, a biosurfactant from *Bacillus subtilis*, with inorganic cations. Biotech. Lett. 14: 713–718.

Tuelva, K.T., R. George, I. Cristovaa and N.E. Christovaa. 2002. Biosurfactant production by a new *Pseudomonas putida* Strain. Z. Naturforsch. 57(3-4): 356–360.

Van Bogaert, I.N., K. Saerens, C. De Muynck, D. Develter, W. Soetaert et al. 2007. Microbial production and application of Sophorolipids. Appl. Microbiol. Biotechnol. 76: 23–34.

Vandana, P. and J.K. Peter. 2014. Production, partial purification and characterization of biosurfactant from *Pseudomonas fluorescens*. Int. J. Adv. Tech. Engi. Sci. 2(7): 258–264.

Velioglu, Z. and R.O. Urek. 2015. Biosurfactant production by *Pleurotus ostreatus* in submerged and solid-state fermentation systems. Turkish J. Biol. 39: 160–166.

Vijayaraghavan, P., C.S. Remya and S.G. Vincent. 2011. Production of α-Amylase by *Rhizopus microsporus* using Agricultural by-products in Solid State Fermentation. Res. J. Microbiol. 6: 366–375.

Wei, Y.H. and I.M. Chu. 1998. Enhancement of Surfactin production in iron-enriched media by *Bacillus subtilis* ATCC 21332. Enzyme Microb. Tech. 22(8): 724–728.

Winterburn, J.B. and P.J. Martin. 2012. Foam mitigation and exploitation in biosurfactant production. Biotech. Lett. 34: 187–195.

Yeh, M.S., Y.H. Wei and J.S. Chang. 2005. Enhanced production of Surfactin from *Bacillus subtilis* by addition of solid carriers. Biotech. Prog. 21: 1329–1334.

Zhu, Z., G. Zhang, Y. Luo, W. Ran and Q. Shen. 2012. Production of lipopeptides by *Bacillus amyloliquefaciens* XZ-173 in solid state fermentation using soybean flour and rice straw as the substrate. Biores. Tech. 112: 254–260.

Zinjarde, S.S. and A. Pant. 2002. Emulsifier from tropical marine yeast, *Yarrowia lipolytica* NCIM 3589. J. Basic Microbiol. 42: 67–73.

Zouari, R., S. Ellouze Chaabouni and D. Ghribi-Aydi. 2014. Optimization of *Bacillus subtilis* SPB1 biosurfactant production under solid-state fermentation using byproducts of a traditional olive mill factory. Achiev. Life Sci. 8: 162–169.

3

Biosurfactants

Industrial Demand (Market and Economy)

Tesni Collins,[1] *Mark Barber*[2] and *Pattanathu K S M Rahman*[3,*]

1. Introduction

1.1 What are Biosurfactants?

Surfactants refer to amphiphilic molecules as they contain both hydrophobic and hydrophilic domains. For this reason, they are found at liquid-liquid or liquid-gas interfaces between polar and non-polar media, or can form micelles and microemulsions (Mudhoo et al. 2014). Reduction of the repulsive forces at these interphases can reduce the tension between the two phases allowing for better mixing between them (Soberón-Chávez 2011). Biosurfactants refer to surfactants that have been manufactured by plants or microorganisms such as bacteria, fungi or yeast. One suggestion as to why biosurfactants are produced is to make it easier for microorganisms to take in insoluble substrates (Rahman and Gakpe 2008).

Classical surfactants are synthesised from a petroleum feedstock which is detrimental for the environment (Sharma et al. 2014). Biosurfactants have advantages over classical surfactants as they are biocompatible and digestible. They can be produced from relatively low-cost materials, and they can be made to be very specific to their function as well as being biodegradable, less toxic and stable at high pH values and temperatures. These tension-active molecules can be classified according to their chemical composition, molecular weight and antimicrobial source of origin (Soberón-Chávez 2011, Cortes-Sanchez et al. 2013). A small summary of

[1] School of Biological Sciences, University of Portsmouth, Portsmouth, United Kingdom.
[2] Commercialisation Department, University of Portsmouth, Portsmouth, United Kingdom.
[3] TARA Biologics, Woking and TeeGene Biotech, London, United Kingdom.
* Corresponding author: rahman@teegene.co.uk

common biosurfactants is set out below, along with their most prevalent applications in the industry.

1.1.1 Glycolipids

Glycolipids are the largest group of known biosurfactants (Rahman and Gakpe 2008). They are carbohydrates containing a sugar construct such as glucose, mannose, galactose, glucuronic acid, rhamnose and galactose sulphite. They also have long chain aliphatic acids or hydroxy aliphatic acids (Sen 2010, Cortes-Sanchez et al. 2013). Some examples of glycolipids are described below.

Rhamnolipids were first discovered in 1946 by Bergström et al. and are now one of the most researched type of biosurfactants (Sekhon Randhawa and Rahman 2014) as they are produced by well-known organisms (Soberón-Chávez 2011), the *Pseudomonas* genus being the most recognised (Rahman and Gakpe 2008, Soberon-Chavez et al. 2020), and can be produced in high yields after a relatively short incubation period. Rhamnolipids are also one of the factors contributing to the virulence of the microorganism that produces it, this could be because the pathway of production of rhamnolipids appears to be controlled by an elaborate regulatory mechanism controlled by intercellular communication via quorum sensing pathways (Soberón-Chávez 2011).

As rhamnolipids are part of the Glycolipid family they contain a glycon region, this is formed with a rhamnose moiety. Mono-rhamnolipids have one rhamnose whilst di-rhamnolipids have two connected via an α-1,2-glycosidic linkage. The non-glycon region is one, two, or, three (normally saturated) β-hydroxy fatty acids chains (Soberón-Chávez 2011). There are two types of Rhamnolipid; type R-2 contain two rhamnose and two β-hydroxy fatty acids chains, type R-1 contains only one rhamnose unit (Sen 2010).

Rhamnolipids have a wide variety of possible applications (Sekhon Randhawa and Rahman 2014). Some of these include hydrocarbon and heavy metal bioremediation from the environment or an antimicrobial via the disruption of biofilms. In addition, it is possible rhamnolipids may be able to assist in the production techniques of nanoparticles (Kiran et al. 2016, Gudiña et al. 2020).

1.1.2 Sophorolipids

Sophorolipids contain a disaccharide sophorose which is a dimeric carbohydrate (diglucose) bound by a β-1,2 bond (Soberón-Chávez 2011, Sen 2010). This is linked via a β-glycosidic link to a 16 or 18 carbon long hydroxylated fatty acid. This chain can vary, for example, by the number of saturated or unsaturated bonds in the fatty acid tails (Soberón-Chávez 2011). These variations are caused when the biosurfactants are produced by different yeasts for example, *Candida bombicola, Torulopsis magnolia, Torulopsis gropengiesseri, Candida bogoriensis and Rhodobaca bogoriensis* (Soberon Chavez 2011, Soberon Chavez et al. 2020).

Sophorolipids fall into two categories, the first being lactonized where the hydroxyl fatty acid forms a macrocyclic lactone ring with the sophorose. Secondly, they can be acidic where the hydroxyl fatty acid has a free carboxylic acid functional group (Rahman and Gakpe 2008). These groups have different properties offering different applications; lactonized Sophorolipids are better at lowering surface tension and acidic Sophorolipids are better at forming foam (Soberón-Chávez 2011).

Sophorolipids have been found to have a solubilisation ratio comparable to synthetic surfactants and so have good prospects in industrial applications (Sen 2010). Sophorolipids are used in many areas including cleaning products such as hard surface cleaners, dishwasher liquid and laundry detergents. They can also be used in secondary oil recovery, *in situ* bioremediation and degradation of hydrocarbons in soils and water tables. In the food industry they are used in the production of flour and in air conditioning units to prevent ice particle formation, in transportation vehicles and as cleaning agents for fruits and vegetables (Soberón-Chávez 2011). They are also used as ingredients in cosmetics giving antimicrobial properties, as well as possibly improving the elasticity, increasing collagen and depigmentation of the skin (Soberón-Chávez 2011). Chen et al. (2006a) have shown that Sophorolipids have the ability to cause apoptosis of cancer cells. Another paper later in 2006 showed that Sophorolipids could cause human H7402 liver cancer cells to begin apoptosis by blocking the cell cycle in the G1 phase and partly in the S phase. This suggests possible uses in treating liver cancer cells (Chen et al. 2006a, Chen et al. 2006b).

1.1.3 Trehalolipids

Trehalolipids are generally produced by bacteria. Examples include, but are not limited to, *Rhodoccocus, Mycobacterium and Micrococcus* which come from the order Actinomycetales (Soberón-Chávez 2011). These bacteria are normally gram-positive, aerobic bacteria that exhibit filamentous growth (Pepper et al. 2015). Compared to Rhamnolipids and Sophorolipids their hydrophobic regions are more diverse, they can include aliphatic acids or varying lengths of mycolic acids joined in numbers forming mono-, di- and tetraesters (Soberón-Chávez 2011). Trehalose is a disaccharide formed when two glucose molecules bind via an α-1,1 bond. This bond makes it resistant to low pH values (Soberón-Chávez 2011).

1.1.4 Surfactin

Surfactin (a group of approximately 20 lipopeptides) is produced by *Bacillus subtilis* (Sen 2010). Structurally all but esperin are composed of a heptapeptide linked with β-hydroxy fatty acids (Soberón-Chávez 2011). Surfactin is also one of the most effective biosurfactants in terms of reducing surface tension (Rahman and Gakpe 2008).

Different surfactins have different applications including: Antiviral and antibacterial properties in pharmaceuticals, oil extraction in the petroleum industry, bioremediation in the environment and as an emulsifier in cosmetics (Soberón-Chávez 2011).

1.1.5 Polymeric Biosurfactants

The main examples of polymeric biosurfactants are emulsan, liposan, mannoprotein and polysaccharide-protein complexes (Rahman and Gakpe 2008, Sen 2010). Polymeric biosurfactants tend to have high molecular weight and show high viscosity and tensile strength (Sen 2010). Emulsan is a very good emulsifier of hydrocarbons in water and is known to be a powerful emulsification stabilizer (Rahman and Gakpe 2008).

1.2 *What is the Relevance of Intellectual Property?*

Intellectual property (IP) provides a competitive advantage to applicants in industry (Yali 2008). There are four types of Intellectual Property: Patents, trademarks, registered designs and copyright, and also unregistered designs and trade secrets. Patents provide applicants with a limited monopoly right of an invention (Bently et al. 2018). In return for this right the applicant must disclose details of the technical information of the invention, or process, to the public. At some point the information in the patent must be made available for public use (Singh et al. 2019a). Provided patent renewal fees are paid, they usually last for 20 years once they have been granted, although this can depend on the jurisdiction within which it has been granted (Alpin and Davis 2017). Although patents do not provide protection of the work for periods as long as copyrights or trademarks, they do provide significantly more extensive protection (Bently et al. 2018). Trademarks are words, symbols or phrases that are associated with a specific company. In many cases they can be the most valuable asset a company can own (Alpin and Davis 2017). Trademarks prevent others from trading based on another company's reputation (Yali 2008). Copyright protects 'cultural creations'. This includes works such as literature, music and films, but not the ideas behind these works (Yali 2008, Alpin and Davis 2017). Trade secrets protect information from the public domain and there is no expiry date, however, they do not protect against reverse engineering of the invention by another individual (Yali 2008).

A primitive patent system can be traced to the 14th Century in the United Kingdom (Alpin and Davis 2017). However, modern patents have been developing since the 16th Century. These once rudimentary documents are now complex and structured (Bently et al. 2018). All patents now contain an Abstract, Description, and Claims section, although these may be called by alternative titles in different jurisdictions. Abstracts provide a brief summary of what is being patented and is positioned at the start of the document. The claims are laid out to show the scope of the protection and are used to describe the technicalities of what is being patented. The description section is an important part of the document. Its purpose is to provide the public with the knowledge to use the patented works. Biotechnology patents can differ slightly in this area. In some cases, it is not possible to describe a living cell or tissue. In such cases the applicant must also provide a sample of the biological material to a recognised institution. This is according to the European Patent Convention (EPC) in 2000 and the Biotechnology Directive in 1998 (Bently et al. 2018).

There are strict rules of what is patentable in law, however these vary between jurisdictions. Therefore, if a patent is required it must be applied for in every jurisdiction where the protection is desired (Yali 2008). In order for a patent to be granted it must comply with criteria. To be patented the invention or process must be 'patentable subject matter'. This means that the work must have an industrial application. Some matter cannot currently be patented, for example medical and veterinary treatments (Bently et al. 2018). In addition to this, in the biotechnology sector there has been much debate as to whether living organisms should be patentable

(Yali 2008). According to paragraph 3(f) of the schedule A2 to the Patents Act 1977 and Article 53(b) of the EPC 2000 "a patent shall not be granted for any variety of animal or plant or any essentially biological process or other technical process for the product of such a process". In addition to this, 'immoral inventions' are also prohibited from being patented. These include human cloning, and modification of a human germ line. Further to these requirements the invention must also be novel and show an 'inventive step'. When patents are examined by someone practised in the state of the art the invention should not form part of the current knowledge or prior art and should be non-obvious (Bently et al. 2018, Yali 2008).

Increasingly the laws around biotechnology patents are becoming more and more meticulous (Sechley and Schroeder 2002). Debate over patent legislation has been present for century's (Alpin and Davis 2017). One relevant piece of legislation comes from the Convention on Biological Diversity (CBD). This international treaty provides many developing countries a voice to dispute patents granted over their traditional knowledge. An example of this is the 37 patents granted in Europe and the USA over the Neem tree which has been used in traditional Indian medicine for generations. As such the Nagoya Protocol, which was passed in 2012 and enforced from 12th October 2014, states that benefits gleaned from indigenous plants or traditional knowledge should be 'equitably distributed' between the innovator and the people from which it is taken (Bently et al. 2018). Innovation is paramount to the progression and success of companies. In order for innovation to have monetary value, intellectual property must be considered vital (Bogers et al. 2012). Patents provide an incentive for innovation and the improvement of methods and techniques within biotechnology (Françoise and Glen 2017) as without them there would be no way to gain returns from investment as others could copy the works of competitors without the need for overhead costs into the research and production (Yali 2008). Furthermore, in the past, patents have been shown to reflect public attitudes towards different technologies (Bently et al. 2018). It is established that companies that use innovative trends and forecasting, via patent analysis, to their advantage have a competitive edge over their competition (Kim and Bae 2017).

The overarching aim of this investigation is to map and forecast trends in research and development of biosurfactants using the analysis of patent documents. It is anticipated that this study will highlight the main sectors in which innovation is actively taking place and sectors where it is lagging. This study also aims to show what types of biosurfactants are the most popular in research and for applications in industry. These findings will be linked to leading jurisdictions and investigation into commercial literature will aim to forecast where innovation is likely to take place in the near future. Therefore, this systematic review will hopefully link economic growth, leading jurisdictions, types of biosurfactants and major sectors in which patents relating to biosurfactants are currently being granted and will be granted in the future.

2. Investigative Methods

In order to establish the patents to analyse in this review, the patent search engine Espacenet (https://worldwide.espacenet.com/patent/) was utilised. Espacenet allows the search and analysis of 110 million patent documents from around the world (European Patent Office 2020). The patent documents within the database were interrogated using keyword queries. Various data mining techniques were used as described by Tseng et al. (2007). The key terms used were 'biosurfactant', 'biological surfactant' and 'biological surface-active agent'. This was to ensure that patents using synonymous wording were included in the data collection. Granted patents were identified by kind codes which vary in each jurisdiction. The documents were then reviewed, and it was established whether they significantly related to biosurfactants. To minimise subjectivity some guidelines were used to distinguish between related and unrelated documents as shown in Fig. 1.

Guidelines

The biosurfactant should be mentioned in the claims as a main 'ingredient' or 'component' in a method or formulation. If it is included in a list of possible surfactant ingredients it must be included in a list of 3 or less components. For example, cationic, non-ionic or a biosurfactant.

Alternatively, a patent can be described in detail in the description section. It can be described as being of great importance to the method or invention.

Also included were patents where the use of biosurfactants was implied by the description of 'biosurfactant producing bacteria'.

Furthermore, patents where the biosurfactant was a substitution for a synthetic surfactant were only deemed relevant to the investigation if examples were given, or, the biosurfactant alternative was mentioned in the claims of the patent. Patents that simply mention this as a passing notion were not included.

Figure 1. Guidelines used to reduce subjectivity in the inclusion or exclusion of patents process.

However, the subject matter of patent documents is very diverse and therefore there will always be an element of subjectivity. As this investigation will be concluded before the end of the year, patents from 2020 were not investigated. This is summarised in the first three steps of Fig. 2.

A total of 22 jurisdictions were explored in turn. The patent documents were grouped into categories according to the subject of the patent. The categories were not known at the start of data collection, therefore, once all the data was compiled, iterative linking was used to ensure suitable categorisation. The finalised categories used in this study were: 'development and production of biosurfactants', 'oil or gas industry', the 'healthcare or pharmaceutical industry', the 'cleaning and detergent industry', the 'cosmetic industry', 'bioremediation industry', 'antimicrobial, antifungal and biofilm prevention industry', the 'agricultural industry', 'use as foaming or emulsifying agents for the formulations industry', the 'wastewater treatment industry', 'sustainability industry' and 'uncategorised'. The 'uncategorised' patents had a subject matter that was particularly niche and did not fit in the other categories. The year that the patents were granted was also recorded.

The patent documents were categorised into types of biosurfactants. The type of biosurfactant being considered in patent documents can be very specific or vague.

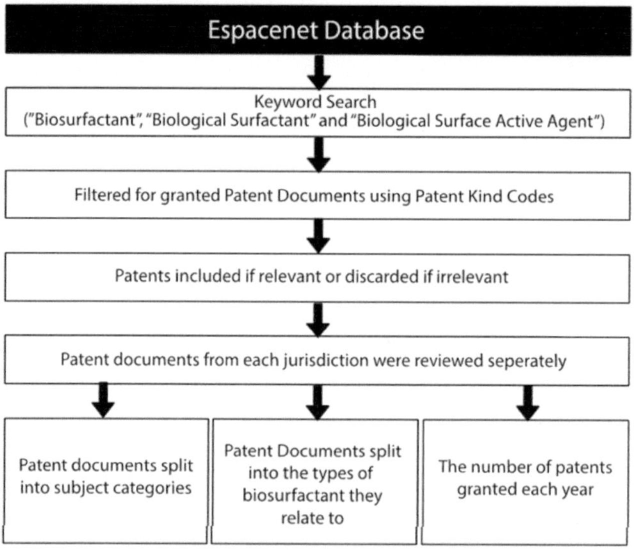

Figure 2. Shows a flow diagram explaining the methodology used in this investigation.

To account for this, patents were tallied under more than one classification where appropriate. This is summarised in the last two steps of Fig. 2.

3. Information and Evaluation

The key terms 'biosurfactant', 'biological surfactant' and 'biological surface-active agent' retrieved a preliminary number of patents of 3204, 546 and 59 respectively. These patents were found across 25 jurisdictions. Only granted patents were investigated further, this reduced the total number of patent documents to 559. This meant that the patents found in France, the United Kingdom, Singapore, Mexico, New Zealand, Romania, Brazil, the European Patent office (EPO) and the World Intellectual Property Organisation (WIPO) were all only applications and therefore not included in this investigation. These 559 documents were reviewed, and 240 relevant patents were found from between 1981 and 2019. For ease of reference, these patents shall be referred to as 'biosurfactant patents' or simply 'patents'. Below are the results of the analysis of these patent documents.

3.1 Biosurfactant Patents Granted in the Research Period

In 1981 the first biosurfactant patent was granted. Between 1981 and 1998 there is a slow increase in the number of patents being granted. In 1998 there were 24 total patents. There is then a more significant increase between 1998 and 2007 from 24 to 113 cumulative patents. However, between 2007 and 2011 there was a lull in the number of granted patents with only two to six patents granted each year (Figs. 3a and 3b).

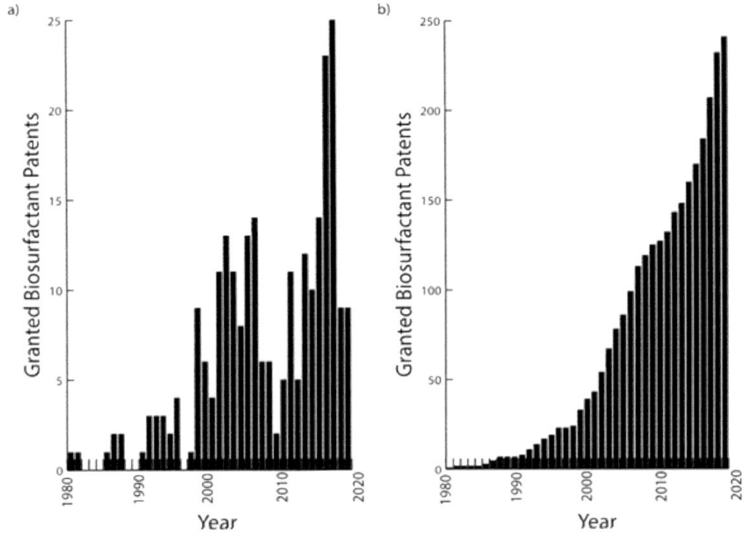

Figure 3(a). Shows the number of biosurfactant patents granted each year. (b) shows the cumulative number of biosurfactant patents granted each year. Both according to Espacenet (https://worldwide.espacenet.com/patent/).

3.2 *Biosurfactant Patents Granted in each Jurisdiction*

South Korea holds the most granted patents of all the jurisdictions investigated at 97. The United States of America (USA) have the second highest at 31, less than half that of South Korea. China, Japan and the Russian Federation all have very similar numbers of patents of between 26 and 28. Australia, Germany, Poland, Chinese Taipei, India and the Eurasian Patent Organisation (EAPO) all hold very few patents of three or less. Furthermore, Chinese Taipei, India and the EAPO all hold the fewest granted patents with only one patent each. This is shown in Fig. 4.

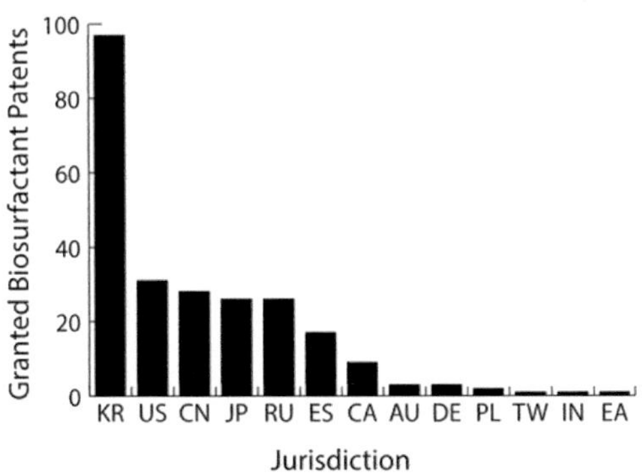

Figure 4. Shows the number of biosurfactant patents granted in each jurisdiction according to Espacenet (https://worldwide.espacenet.com/patent/). Initials: KR = South Korea, US = the USA, CN = China, JP = Japan, RU = the Russian Federation, ES = Spain, CA = Canada, AU = Australia, DE = Germany, PL = Poland, TW = Chinese Taipei, IN = India, EA = Eurasian Patent Organisation (EAPO) (European Patent Office, 2019).

3.3 The Areas in which Biosurfactant Patents are being Granted

The areas of industry and interest where biosurfactant patents are enforced is very wide. Therefore, the category 'Uncategorised' was used for any patents that had a very particular or niche use that did not appropriately fit in another category.

Development and production of biosurfactants had the most relevant patents (72). These relate to specific strains, methods to maximise yield in production, novel biosurfactants and more. This was followed by oil and gas recovery techniques which had 37 related patents and the healthcare and pharmaceutical industry which had 18 patents. The cleaning and detergent industry, the cosmetic industry and bioremediation all had 16 relevant patents. The antimicrobial, antifungal and biofilm prevention industry and the agricultural industry both had a similar number of related patents with nine and eight patents respectively. Use as a foaming or emulsifying agent for the formulations industry and the wastewater treatment industry both have 4 related patents. Sustainability has the least related patents at only three. This is shown in Fig. 5.

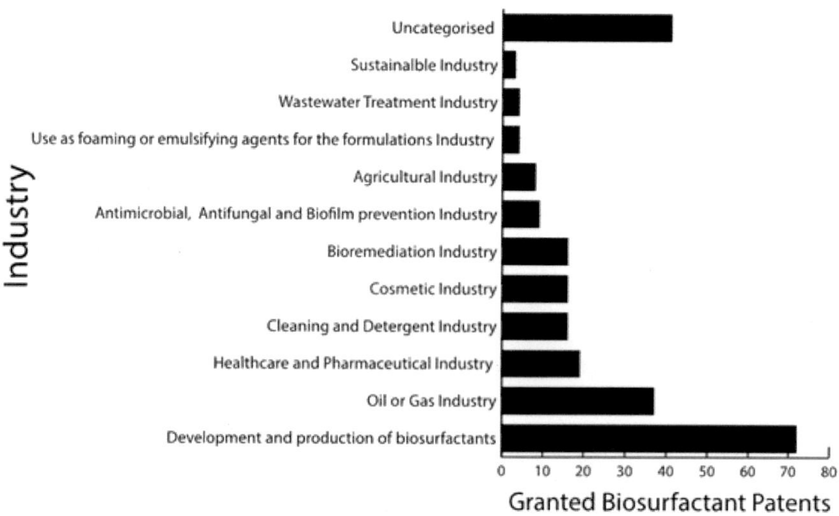

Figure 5. Shows the number of Patents that relate to each category.

There are nine jurisdictions in total that did not have any relevant patents. In Fig. 6 the 'uncategorised' classification was not included as this constitutes patents that relate to very niche areas. For example, most of China's patents were classified in this way. The WIPO, EPO and EAPO are not directly on the map as they are not associated with a particular country but instead provide protection in a 'bundle' of jurisdictions (Bently et al. 2018).

The two most common categories shown in Fig. 6 are oil and gas recovery techniques and the development and production of biosurfactants. In China, India, Russia and Canada the most common category for a patent to relate to was oil and gas

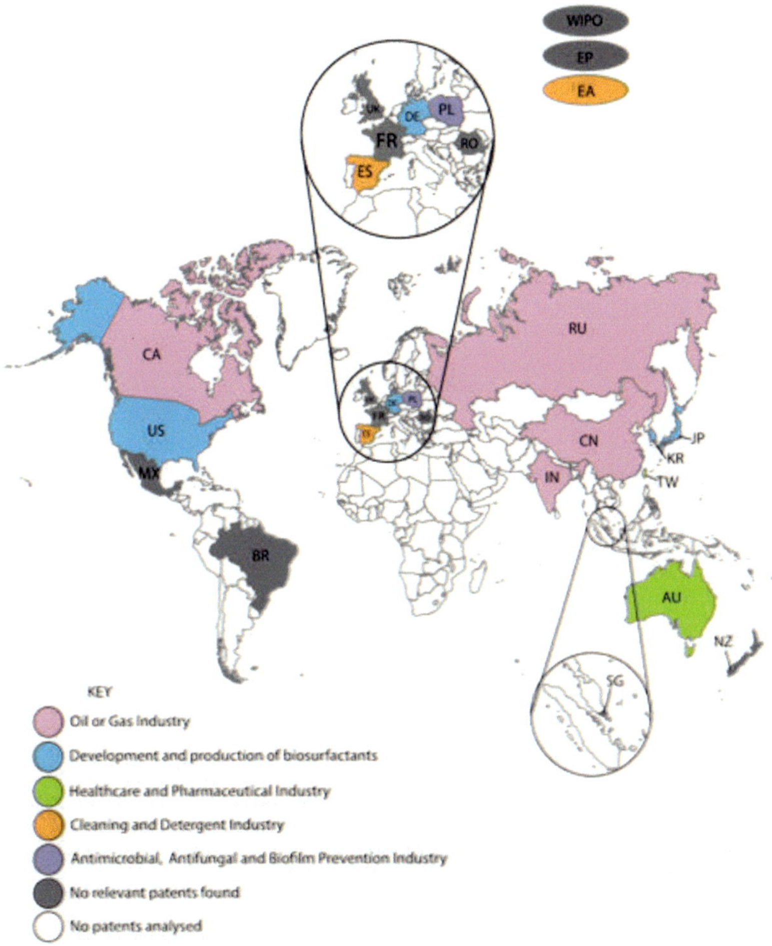

Figure 6. This map shows the category with the most patents granted within each jurisdiction according to Espacenet (https://worldwide.espacenet.com/patent/). KR = South Korea, US = the USA, CN = China, JP = Japan, RU = the Russian Federation, ES = Spain, CA = Canada, AU = Australia, DE = Germany, PL = Poland, TW = Chinese Taipei, IN = India, EAPO = Eurasian Patent Organisation, FR = France, GB = United Kingdom, SG = Singapore, MX = Mexico, NZ = New Zealand, RO = Romania, BR = Brazil, EPO = European Patent Office, WIPO World Intellectual Property Organisation (European Patent Office 2019).

recovery techniques. In the USA, Germany, South Korea and Japan the most common category was the development and production of biosurfactants. The cleaning and detergent industry and the healthcare and pharmaceutical industry were the most common category with two countries each. Comparatively, the antimicrobial, antifungal and biofilm prevention industry was only the most common in Spain.

All the other categories used in this investigation were not found to be the most common in any of the jurisdictions analysed.

3.4 *The types of Biosurfactants the Patents Relate to*

The patents analysed were found to relate to a wide range of biosurfactants. This has been displayed in the graph below, Fig. 7. The most commonly specified type of biosurfactant are Rhamnolipids with 47 related patents. This is followed with 33 patents relating to Sophorolipids. Mannosylerythritol Lipid (MEL) and Surfactin having 19 and 23 patents related to them respectively. Tween, Sphingosine-1-phosphate, Fengyeins, Hydrophobin, Emulsan and Viscosin all had a very similar number of patents relating to them of between two and four. The biosurfactants between Sodium glycocholate and Polybasic acid in Fig. 7 were all only referred to once in the analysed patents.

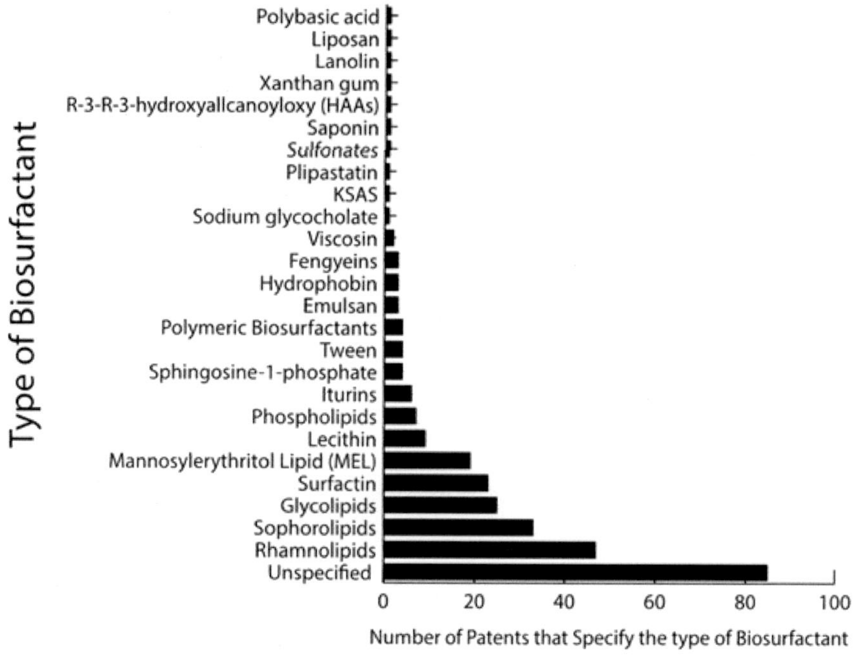

Figure 7. Shows the number of patents that relate to each type of biosurfactant.

3.5 *Patent Categories In 2019*

In 2019 there was 27 patents granted relating to biosurfactants. In 2019 30% of the patents granted were uncategorised in this investigation. The second largest category was the development and production of biosurfactants (26%). Oil and gas recovery techniques and the cosmetic industry both claim 11% of the patents granted in 2019. Antimicrobial compounds and the cleaning and detergent industry claim 7% each and bioremediation and the agricultural industry claim 4% each. These are shown in Fig. 8.

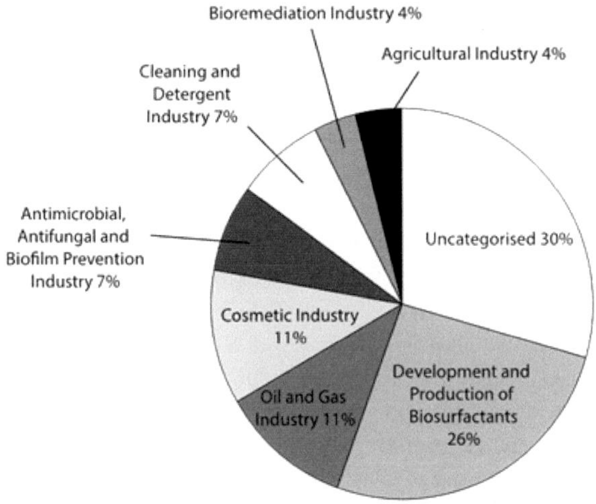

Figure 8. Shows the main categories in which patents have been granted during 2019.

4. Discussion

Companies that focus on innovative trends in the market are known to have a competitive advantage in their sector (Kim and Bae 2017). Therefore, many companies now undertake intellectual property analysis and forecasting as an insight into future technological growth areas, and importantly to determine which areas to increase or decrease investment (Jeong and Yoon 2015). Therefore, research of the type conducted in this project could help companies profit in this growing market.

Biosurfactants lie within the biotechnology sector which had a global market value of 417 billion USD in 2018 and is predicted to be worth 729 billion USD in 2025 (Ugalmugale and Swain 2019). The biosurfactant market itself was estimated by one market research company to be 1.6 billion USD in 2018 and is estimated to increase by 5.6% to 2.4 billion USD by 2025 (Kunal Ahuja and Sonal Singh 2019). Another market research prediction estimated the global market of biosurfactants to be worth 30.64 billion USD in 2016 and set to grow to approximately 40 billion USD by 2021 (Singh et al. 2019b). Although these predictions are conflicted on how much the market is worth, they do acknowledge that its value is increasing. Trends shown in Fig. 3 support this prediction. Despite a lull in the number of biosurfactant patents being granted between 2007 and 2011 (shown in Fig. 3a), there is still an increasing trend until 2019. This lull could be due to the lag between application and substantive search to grant, which can take four to five years. This shows that innovation has continuously occurred over the past three decades and would indeed suggest that there is increasing profitability in this sector from the manufacturing of these compounds.

According to Singh et al. (2019), the European sector is said to be expected to grow by 53% and have the largest market for biosurfactants. However, the research conducted in this investigation showed that Europe currently only has 9.2% of the total number of patents as shown in Fig. 4. Furthermore, BASF Cognis is one of the top manufacturers of surfactant compounds (Singh et al. 2019b) although neither of the German patents analyzed in this research belong to this company. This disparity could be due to firms choosing not to patent their works and instead use trade secrets or similar tactics to protect their IP. More research would be required to prove this. This could highlight a limitation in the data, or, it could be that BASF simply have not had any patents granted to them. In contrast Singh et al. (2019) found that the USA was the second leader in the biosurfactant industry. It is suggested that this is could be due to stricter guidelines in USA slowing technological development. Figure 4 shows that South Korea was found to have by far the highest number of granted relevant patents. This is surprising as Singh et al. suggest that Asia is to become an increased consumer of biosurfactants but does not yet hold high market value in this sector. This development may have already begun with the number of biosurfactant patents being granted in South Korea increasing from 30 in 2010 to 97 in 2019, this is an increase of 246%. In comparison, Europe, which is said to be the largest contributor to the market, did not appear to have any significant increase in the number of biosurfactant patents being granted in recent years in the raw data.

Recently there has been a great deal of research conducted regarding the applications of biosurfactants in different areas of industry. Biosurfactants can be used in many products as either a main active ingredient, or, mixed with other components (Akbari et al. 2018). In this investigation the patents analysed were divided into categories in order to explore whether patents were being granted in relation to certain sectors of the market more than others. As shown in Fig. 5 development and production of biosurfactants had the highest number of granted patents at 72. This category refers to the use of particular and, in many cases, novel strains of bacteria to produce a biosurfactant. Alternatively, it can refer to methods to produce a biosurfactant. Often these methods improve the yield of biosurfactant. This is supported by Sekhon Randhawa and Rahman (2014) as they indicate that one of the main reasons Rhamnolipids struggle to become leaders in this market is because the "economics of production is a major bottleneck". It can be suggested that this has been the case for many types of biosurfactant and therefore many methods that improve the possible commercialization of biosurfactants would be anticipated. Furthermore, in 2019 only 27 patents were granted that relate to biosurfactants, 26% of these patents were within this category as shown in Fig. 8.

Oil and gas recovery techniques had the second highest number of relevant patents. It is, therefore, unsurprising that four jurisdictions had the most patents relating to this category. These were China, Canada, the Russian Federation and India as seen in Fig. 6. Microbially enhanced oil recovery (MEOR) was first suggested in 1926 by Beckman. Zobell was the first to patent a method for MEOR in 1946 and since then many more have been filed (Geetha et al. 2018, United States Patent No. 2,413,278, 1946). Russia is the fourth largest oil exporting country and it is therefore not surprising that it would have many patents in this field as shown in Fig. 6. However, it is surprising that no patents are granted within this database in Kuwait, Saudi Arabia or the United Arab Emirates as these are the other leading nations in this respect (Basher et al. 2018).

It is predicted that market focus will shift towards more sustainable energy as technological advances make renewable energy more affordable and efficient. Although there is no direct link between public interest and research, many governments believe that there should, at least, be some societal benefit to research. As many governments around the world invest large amounts of money into research and development (R&D) there will always be an element of societal influence and political pressure on the nature of the R & D that is being conducted (Hill 2016). Climate change has been accepted as a serious problem by the majority of populations in developed countries, despite some minorities losing faith in climate scientists (Capstick et al. 2015). Therefore, oil companies will have to change their business model to incorporate more low carbon assets (Fattouh et al. 2019). This would suggest that where patents are published innovation is likely (Françoise and Glen 2017) although as a consequence there will be a decrease in the number of patents being filed in relation to oil recovery. The sustainability category may see an increase in the number of patents in the future. This category includes patents related to biofuels, bioplastics and the depolymerization of plastics within the sustainability industry. However, in 2019 oil and gas recovery techniques claimed 11% of the patents granted as shown in Fig. 8.

Climate change requires multiple disciplines to perform effective R&D. This includes sustainability research and applications for agriculture. In order to be able to produce enough energy, research into renewable energy sources and biofuels will be imperative. In addition, plant breeding in agriculture is imperative in mitigating the effects of climate change. In order to support this essential R&D governments will wish to implement taxation of carbon emissions and pollution in general (Tylecote 2019). Currently very few patents are granted in relation to agriculture, but South Korea and the USA have the most and therefore may be more likely to see an increase in this area. South Korea and Japan are currently the only jurisdictions with patents within the sustainability sector.

Due to the Covid-19 (SARS-CoV-2) pandemic there is likely to be an increase in the value of industries that otherwise may not have seen such a rise. In an effort to reduce the spread of this virus, the public have been asked to increase the washing of their hands together with disinfection of commonly touched surfaces. Meanwhile scientists have been researching which detergents are effective at killing the virus and biosurfactants will likely play an important role in the effective disinfection of commonly cleaned areas in conjunction with chemical detergents such as bleach products (Smith et al. 2020). The detergent industry was already expected to increase at a compound annual growth rate of 6% between 2018 and 2027 (Research and Markets 2020b). Now, the Detergent Industry is expecting a surge in demand due to it being essential for disinfecting surfaces, to slow the spread of the virus. This was shown by a 200% increase in sales in Italy in March 2020 alone (Research and Markets 2020a). In 2019 according to Fig. 8 the cleaning industry and the antimicrobial category claimed 7% each of the patents granted relating to biosurfactants that year. Biosurfactants are likely to be further researched as they are environmentally friendly alternatives to classical surfactants. This will be important to reduce the environmental impact of increased use of detergents, particularly in localized areas where frequent disinfection is imperative (Smith et al. 2020).

The pharmaceutical industry is experiencing high demand for the manufacturing of drugs to treat the symptoms of the illness and also to create vaccinations (Research and Markets 2020a). Therefore, patents within the antimicrobial, antifungal and biofilm prevention industry and the health and pharmaceutical are likely to increase. Biosurfactants play an important role in the effectiveness of these products. An example of this is that some biosurfactant vaccines have been shown to be able to induce aspects of the human immune system and similar effects have shown to be effective with viruses such as HIV-1 and foot-and-mouth (Smith et al. 2020).

The Covid-19 virus stores its proteins and RNA enclosed within a lipid membrane, as is the case with many coronaviruses that infect animals and humans. This leaves a breach in its defenses where biosurfactants are able to interact with the hydrophobic region in the membrane, in turn causing damage and disabling the virus. Biosurfactants could be an answer to the lack of treatments that attack the virus itself, rather than merely lessening the symptoms. They can be effective during later stages of the virus's life cycle when it is moving to infect a new host cell. This solution has more longevity than many others as the chance of the virus becoming resistant is less likely with this method than with available drugs (Smith et al. 2020).

Growth in the medical and pharmaceutical category, however likely, is not without its difficulties. It would have been expected that more patents would be granted in this category than have been found in this research. In the USA a large amount of government funding is awarded to National Institutes of Health however there are few patents filed in this industry. This could be due to the patent system acting as a hindrance to the commercialization of products. Often by the time the product is ready to be sold the patent protection will have expired. It has been suggested that to rectify these problems a patent reform is necessary (Tylecote 2019). In this research, the jurisdictions where the most patents were granted in the industry were Australia and Chinese Taipei so this could suggest that these jurisdictions could be leaders in this area in the future.

In this research 85 patents did not specify what type of biosurfactant was being referred to as shown in Fig. 7. This can be due to it being a novel biosurfactant and, therefore, it is not relevant to allocate it to a type of previously known biosurfactants. In addition, it could also be as the biosurfactant is only a component in a formula, for example, that a broad range of biosurfactants could be used in place of one another. Most patents related to rhamnolipids. However, surfactin was specified in only 23 patents. This data is supported as surfactin was the first biosurfactant to be purified and characterized and then rhamnolipids outperformed surfactin and gained popularity. This was due to rhamnolipids being found to have a wide range of applications (Sekhon Randhawa and Rahman 2014). Furthermore, the rhamnolipid market is expected to rise in value by 8% between 2016 and 2023 (Singh et al. 2019b). Sophorolipids were specified in 33 patents. It was expected that they would be mentioned in similar numbers of patents as rhamnolipids, as they hold the largest market share of all biosurfactants (Singh et al. 2019b). This could be due to them having a solubilization ratio similar to that of synthetic surfactants (Sen 2010). All but one of the top five most specified biosurfactants were from the glycolipid family with 124 patents referring to a glycolipid. Secondarily to glycolipids, lipopeptides such as surfactin and iturin were referenced in 42 patents. Lecithins are produced by plants and although they are brilliant emulsifiers, they tend to be expensive to produce, especially on an industrial scale (Sekhon Randhawa and Rahman 2014). It could be for this reason that there are less patents that refer to them (only nine). A similar situation is said for Saponins, where only one patent was found.

South Korea and China had the most diversity in the types of biosurfactants being referred to in their patents. South Korea refers to 21 different biosurfactants and China, 15. These all tended to be in the more popular types of surfactants within the biosurfactant industry, in Fig. 7. Although, according to Fig. 4 the USA has the second highest number of patents but these patents are not as diverse. The subject of the majority of USA patents are Rhamnolipids and Sophorolipids (19 out of a total of 31).

There are few prominent companies within certain jurisdictions. China, Japan and Russia have very few companies with more than two patents, however, their percentage claim of the overall number of patents in their jurisdiction is relatively high. Six granted patents belong to the University of Hunan in China, and this is 21.4% of all the patents granted in China related to biosurfactants. In Japan the National Institute of Advanced Industrial Science and Technology claim five different patents which is 19.2% of Japanese biosurfactant patents. In Russia Aktsionernaja Neftjanaja Kompa has had six patents granted and Ftjanaja Kompanija Bashneft has had five patents granted, together these two companies claim 42.3% of the Russian granted patents. Relative to the abundant number of patents granted in South Korea (97 patents), particularly notable companies are few and far between. Eight patents are granted by the Gyeongbuk Institute for Marine Bioindustry in South Korea, however, 20 other companies had only a few granted patents each. This shows that the South Korean patent landscape is varied with the maximum patents belonging to any one company being eight, only 8.25% of the patents granted in this Jurisdiction.

4.1 *Limitations of Data*

Espacenet was used in order to ensure consistency when collecting the data. However, Kim et al. (2017) highlighted that in their research no database or search engine returned all the relevant patents. Furthermore, in lots of jurisdictions, older patents are filed in an analogue system. These patents that are not machine readable will not have been included and so the number of patents analysed could be lower than the true figure. More evidence for this comes as Sekhon and Rahman (2014) reported that a possible 250 additional patents had been found that related to biosurfactants.

Another limitation of this data is that although steps were put in place, in the form of guidelines, there is still an element of subjectivity as to which patents should have been included or not. This means there may be errors.

This research only looked at granted patents. However, there were many patent application documents found under the key terms on Espacenet. These could provide further insight into the current innovation and could also provide a more complete report to future R&D.

5. Conclusions

In conclusion, research confirms that the market for biosurfactants is growing and this sector is likely to continue to see an increase in value in the immediate future. This is supported by the increasing trend in patents being granted in relation to biosurfactants over the past three decades. There is some disparity between this research and the current literature in terms of the leading regions in this field. This research suggests that South Korea has both the most patents and diversity in the industrial areas that these patents relate to. The European jurisdictions in this research were found to have fewest patents overall. Conversely, commercial literature suggests that the European market is the leader in this industry.

The most common category where the patents analyzed were found was the 'development and production of biosurfactants'. This was supported by the literature as this is an area of research that is considered to be essential to the commercialization of these compounds. The second most common category was the oil and gas recovery techniques. This was also expected as much research has been done on microbial enhanced oil recovery (MEOR) and over a long period of time. However, due to public concerns about climate change it is likely that this trend in granted patents will slow as more renewable energy will be considered by energy companies. This will also likely initiate an increase in patents being granted in the sustainability category and the agricultural sector.

Due to the Covid-19 pandemic it is likely that the detergent industry will increase dramatically in value as sales escalate to combat the spread of the virus. Therefore, it is likely that new patents will be granted in the cleaning and detergent industry category. In addition, it is likely there will be an increased interest in pharmaceutical manufacturing. However, according to this research this is an area currently lacking in patents and current literature suggests that the patent systems tend to slow the development and commercialization of products in this sector. The most common type of biosurfactants to be specified in the patents analyzed were rhamnolipids. This was followed by sophorolipids. These are both types of glycolipids which are incredibly popular as they have broad applications. This result was expected as rhamnolipids are the most popular biosurfactants and according to commercial literature, are expected to increase their market value in the coming years. Most patents referenced a novel biosurfactant, or, did not specify a particular type of biosurfactant if it was a component in a formulation. However, plant biosurfactants are lacking in patents, probably due to the lack of methods available to commercialize these compounds.

Future research could be conducted to investigate the data provided by patent application documents. These could provide a more complete picture of the future to these compounds and the diverse applications they provide.

References

Akbari, S., N. Abdurahman, R. Yunus, F. Fayaz, O. Alara et al. 2018. Biosurfactants—a new frontier for social and environmental safety: a mini review. Biotechnology Research and Innovation, pp. 81–90.

Alpin, T. and J. Davis. 2017. Intellectual Property Law. Text, Cases, and Materials. (3rd Ed.). Oxford: Oxford University Press.

Basher, S., A. Haug and P. Sadorsky. 2018. The impact of oil-market shocks on stock returns in major oil-exporting countries. Journal of International Money and Finance 86: 264–280.

Bently, L., B. Sherman, D. Gangiee and P. Johnson. 2018. Intellectual Property Law. Oxford: Oxford University Press.

Bogers, M., R. Bekkers and O. Granstrand. 2012. Intellectual property and licensing strategies in open collaborative innovation. pp. 37–58. *In*: Heredero, P. and D. López (eds.). Open Innovation in Firms and Public Administrations: Technologies for value creation. IGI global.

Capstick, S., L. Whitmarsh, W. Poortinga, N. Pidgeon et al. 2015. International trends in public perceptions of climate change over the past quarter century. Wiley Interdisciplinary Reviews: Climate Change 6(1): 35–61.

Chen, J. X. Song, H. Zhang, Y-B. Qu and J.-Y. Miao. 2006a. Sophorolipid produced from the new yeast strain Wickerhamiella domercqiae induces apoptosis in H7402 human liver cancer cells. Applied Microbiology and Biotechnology 72(1): 52–59.

Chen, J., X. Song, H. Zhang and Y. Qu. 2006b. Production, structure elucidation and anticancer properties of sophorolipid from Wickerhamiella domercqiae. Enzyme and Microbial Technology 39(3): 501–506.

Cortes-Sanchez, A., H. Hernandez-Sanches and M. Jaramillo-Flores. 2013. Biological activity of glycolipids produced by microorganisms: New trends and possible therapeutic alternatives. Microbial Research 168(1): 22–32.

European Patent Office. 2019, 8 26. Espacenet - country codes. Retrieved from European Patent Office: https://worldwide.espacenet.com/help?locale=en_EP&method=handleHelpTopic&topic=countrycodes.

European Patent Office. 2020, 1 17. Espacenet: Patent Database with over 110 million documents. Retrieved from European Patent Office: https://www.epo.org/searching-for-patents/technical/espacenet.html#tab-2.

Fattouh, B., R. Poudineh and R. West. 2019. The rise of renewables and energy transition: what adaptation strategy exists for oil companies and oil-exporting countries? Energy Transitions 3(1): 45–58.

Françoise, S. and G. Glen. 2017. Managing Biotechnology. Hoboken: Wiley.

Geetha, S.J., I.M. Banat and S.J. Joshi. 2018. Biosurfactants: Production and potential applications in microbial enhanced oil recovery (MEOR). Biocatalysis and Agricultural Biotechnology 14: 23–32.

Gudiña, E.J., C. Amorim, A. Braga, Â. Costa, J.L. Rodrigues, S. Silvério, L.R. Rodrigues. 2020. Biotech green approaches to unravel the potential of residues into valuable products. pp. 97–150. *In*: Inamuddin and A. Asiri (eds.). Sustainable Green Chemical Processes and their Allied Applications. Springer Nature Switzerland AG.

Hill, S. 2016. Assessing (for) impact: Future assessment of the societal impact of research. Palgrave Communications 2(1): 1–7.

Jeong, Y. and B. Yoon. 2015. Development of patent roadmap based on technology roadmap by analyzing patterns of patent development. Technovation, pp. 37–52.

Kim, G. and J. Bae. 2017. A novel approach to forecast promising technology through patent analysis. Technological Forecasting and Social Change 117: 228–237.

Kiran, G.S., A.N. Lipton, V. Pandian and J. Selvin. 2016. Rhamnolipid biosurfactants: evolutionary implications, applications and future prospects from untapped marine resource. Critical Reviews in Biotechnology 36(3): 399–415.

Kunal Ahuja, S.S. (2019, November). Biosurfactant Market Size, Value, Industry Share Report 2025. Retrieved from Global Market Insights: https://www.gminsights.com/industry-analysis/biosurfactants-market-report.

Mudhoo, A., S.K. Sharma and C.N. Mulligan. 2014. Biosurfactants: Research Trends and Applications. Boca Raton: CRC Press.

Pepper, I.L., C.P. Gerba and T.J. Gentry. 2015. Environmental Microbiology 3rd Edition. Academic Press.

Rahman, P.K. and P.K. Gakpe. 2008. Production, characterisation and applications of biosurfactants-Review. Biotechnology 7(2): 360–370.

Research and Markets. 2020a. COVID-19. Retrieved from Research and Markets: https://www.researchandmarkets.com/issues/covid-19.

Research and Markets. 2020b. Detergents - global market outlook (2018–2027). Retrieved from Research and Markets: https://www.researchandmarkets.com/reports/5017523/detergents-global-market-outlook-2018-2027.

Sechley, K.A. and H. Schroeder. 2002. Intellectual property protection of plant biotechnology inventions. Trends in Biotechnology, pp. 456–461.

Sekhon Randhawa, K. and P. Rahman. 2014. Rhamnolipid biosurfactants—past, present, and future scenario of global market. Frontiers in Microbiology 5: 454.

Sen, R. 2010. Biosurfactants (Vol. 672). New York: Springer Science & Business Media, LLC.

Sharma, S.K., A. Mudhoo and C. Mulligan. 2014. Biosurfactants: Research Trends and Applications. Boca Raton, Boca Raton: CRC Press, 2014: CRC Press.

Singh, H.B., C. Keswani and S.P. Singh. 2019a. Intellectual property issues in microbiology. Springer Nature Singapore Pte Ltd.

Singh, P., Y. Patil and V. Rale. 2019b. Biosurfactant production: emerging trends and promising strategies. Journal of Applied Microbiology 126(1): 2–13.

Smith, M.L., S. Gandolfi, P.M. Coshall, P.K. Rahman et al. 2020. Biosurfactants: A Covid-19 Perspective. Frontiers in Microbiology 11.

Soberón-Chávez, G. 2011. Biosurfactants: From Genes to Applications (Vol. 20).

Soberon-Chavez, G., A. Gonzalez-Valdez, M.P. Soto-Aceves and M. Cocotl-Yanez. 2020. Rhamnolipids produced by Pseudomonas: from molecular genetics to the market. Microbial Biotechnology.

Tseng, Y.-H., C.-J. Lin and Y.-I. Lin. 2007. Text mining techniques for patent analysis. Information Processing & Management, pp. 1216–1247.

Tylecote, A. 2019. Biotechnology as a new techno-economic paradigm that will help drive the world economy and mitigate climate change. Research Policy 48(4): 858–868.

Ugalmugale, S. and R. Swain. (2019, November). Biotechnology Market Share, Growth Forecasts Report 2025. Retrieved from Global Market Insights : https://www.gminsights.com/industry-analysis/biotechnology-market.

Yali, F. 2008. Building Biotechnology: business, regulations, patents, law, politics, science 3rd ed. Washington, D.C: thinkBiotech LLC.

4

Biosurfactant

A Biomolecule and its Potential Applications

Priyanka Patel,[1,*] *Shreyas Bhatt,*[1] *Hardik Patel,*[2] *R Z Sayyed*[3] and *Lilana Aguilar-Marcelino*[4]

1. Introduction

Biosurfactants are defined as biologically derived microbial surface-active materials. Biosurfactants are amphiphilic compounds produced by plants and microorganisms. Rising public awareness towards environmental pollution has led to the exploration and expansion of knowledge which could help in cleaning of organic and inorganic pollutants such as hydrocarbons and metals. A substitute and environment friendly technique of remediation of these pollutants is the application of biosurfactants and biosurfactant-producing microbes. Production of biosurfactants depends on physical and chemical factors, moreover it is directly proportional to the growth of microbes.

Biosurfactants are amphiphilic microbial molecules having hydrophilic and hydrophobic moieties that partition at the interfaces of fluid phases with different polarities like liquid/gas, liquid/liquid or liquid/solid, which reduce surface and interfacial tensions (Mouafo et al. 2018). Such characteristics help these biomolecules to play an important role in detergency, emulsification, foam formation and dispersal, which are qualities required by industries (Santos et al. 2016). A non-polar moiety often possesses a hydrocarbon chain, whereas a polar moiety may be possess ionic, non-ionic or amphoteric compounds.

[1] Department of Life Sciences, Hemchandracharya North Gujarat University - 384 265, India.
[2] Government Dental College and Hospital, Civil Hospital Campus, Ahmedabad - 380 016, India.
[3] Department of Microbiology, PSGVPM'S ASCCollege, Shahada - 425409 (MS), India.
[4] Centro Nacional de Investigacion Disciplinaria en Salud Animal Inocuidad, INIFAP, Km 11 Carreterra Federal Cuernavaca. No. 8534, Col. Progreso, Jiutepec, Morelos, C.P. 62550, Mexico.
* Corresponding author: pu2989@gmail.com

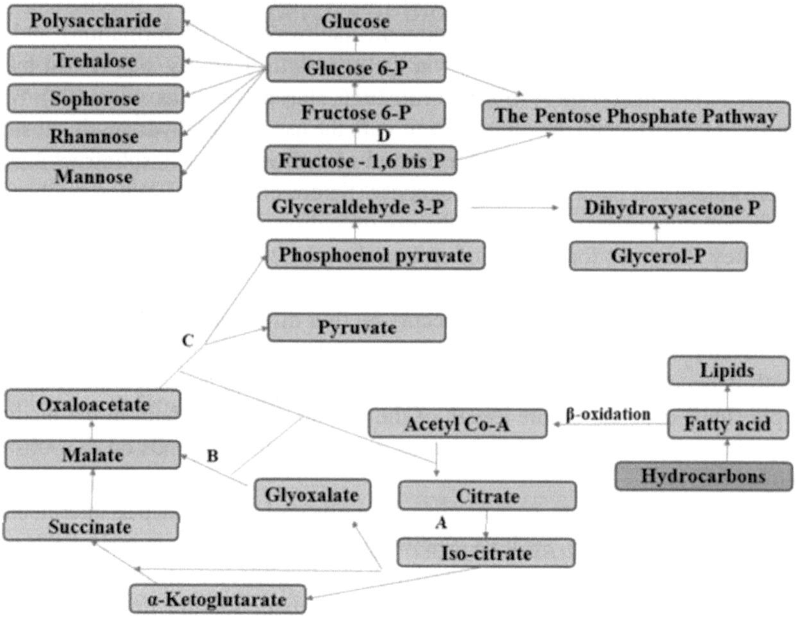

Figure 4. Synthesis of biosurfactant precursors with use of hydrocarbons as substrate (Courtesy: Santos et al. 2016).

When a hydrocarbon is used as the carbon source, the microbial mechanism is mainly directed to the lipolytic pathway and gluconeogenesis, thereby allowing it to be utilized for fatty acids or sugars production (Fig. 4). The gluconeogenesis pathway is activated for the sugar production. This pathway includes the oxidation of fatty acids through oxidation to acetyl-CoA. Initiating with the formation of acetyl-CoA, the reactions involved in the polysaccharide precursors synthesis, such as glucose 6-phosphate, are essentially the inverse of those involved in glycolysis. However, reactions catalysed by pyruvate kinase and phosphofructokinase-1 are irreversible. The formation of glucose 6-phosphate, is the main precursor of polysaccharides and disaccharides formed for the hydrophilic moiety of glycolipids production (Tokumoto et al. 2009).

8. Applications of Biosurfactants

There are various environmental and industrial applications of biosurfactants.

8.1 Environmental Applications of Biosurfactant

Environmental applications of biosurfactants are as follow:

8.1.1 Biosurfactants in Biodegradation Processes

The application of biosurfactants is a promising approach to develop bioremediation efficacy of hydrocarbon contaminated environments. There are two methods for enhancement of hydrocarbon bioremediation. The first comprises of the enhancement in bioavailability of substrates for microbes, while the second includes the interface with the cell surface which improves the hydrophobicity of the surface allowing the attachment of hydrocarbons with bacterial cells (Mulligan and Gibbs 2004). Moreover, by reducing surface and interfacial tensions, biosurfactants could be expected to increase the biodegradation of hydrocarbons by emulsification, mobilization and solubilization (Plociniczak et al. 2011).

8.1.2 Biosurfactants in MEOR (Microbial Enhanced Oil Recovery)

The studies on MEOR reveal the use of core substrates and columns containing the desired substrate, generally sand (Bordoloi and Konwar 2008). The substrate is utilized to reveal the effectiveness of biosurfactants in the recovery of oil from reservoirs. For experimental purposes, dry sand is packed into a glass column, which is then saturated with crude oil and an aqueous solution of the biosurfactant poured into it. The significance of biosurfactants in MEOR technique is estimated by quantifying the amount of oil released from the column after the process. This process can be carried out at different temperatures to estimate the oil recovery stimulated by the biosurfactant (Plociniczak et al. 2011).

8.1.3 Biosurfactants for Co-Contaminated Sites Remediation

The U.S. Environmental Protection Agency (USEPA) estimated about 40% of sites in the U.S. are co-contaminated with metals and organic pollutants (Sandrin and Maier 2003). The occurrence of toxic metals like arsenic, cadmium, lead, among others, causes inhibition of biodegradation of organic compounds (Maslin and Maier 2000). Biosurfactants produced by microbes is a promising approach for increasing biodegradation of organic compounds. The implementation of biosurfactants or microbes produced biosurfactants in *in situ* contaminated sites appears to be more eco-friendly than using modified clay complexes or metal chelators for bioremediation (*l.c.*).

8.2 Industrial Applications of Biosurfactants

There are certain environmental applications of biosurfactant which are depicted in Table 3. Biosurfactants have a varied range of applications in diverse industries such as agriculture, cosmetics, cleaning, environment, food, medicine, mining, nanotechnology, petroleum, textile, wastewater and several other industries. These compounds are known as multifunctional agents including anti-adhesive agents, antimicrobial, emulsifying, moisturizing, and stabilizing agents (Banat et al. 2000). In modern times, biosurfactants are getting potential attention for environmental applications in remediation of organic and inorganic contaminants, mainly in the cosmetics industry, enhanced oil recovery, heavy metal removal from soil and water, and pharmaceutical products (Akbari et al. 2018, Al-Wahaibi et al. 2014, Boruah and Gogoi 2013).

8.2.1 Biosurfactant in Agriculture

Microorganisms produce biosurfactants which are more beneficial compared to chemically synthesized surfactants in the agriculture field. The biosurfactants possess antimicrobial activities and can be mostly applied in the agricultural field for the

Table 3. Industrial applications of Biosurfactants.

Industry	Applications	Role of Biosurfactant	References
Agriculture	Bactericides, Biocontrol, Fertilizers, Fungicides, Heavy metal removal, Pesticide, Plant disease removing agent, Root colonization, Soil remediation	Wetting, dispersion, suspension of powdered pesticides and fertilisers, emulsification of pesticide solutions, facilitation of biocontrol mechanisms of microbes, plant pathogen elimination and increased bioavailability of nutrients for beneficial plant-associated microbes	Sachdev and Cameotra (2013)
Cosmetics	Acne treatment, Anti-adhesive agents, Anti-cancer agents, Anti-fungal agents, Anti-microbial agents, Anti-viral agents, Anti-wrinkle products, Body washes, Deodorants, Eye shadows, Foaming agents, Hair products, Health and beauty products Lotions, Lip colour, Skin smoothing	Emulsification, foaming agents, solubilisation, wetting agents, cleansers, antimicrobial agents, mediators of enzyme action	Vijayakumar and Saravanan (2015)
Cleaning	Washing detergents	Detergents and sanitisers for laundry, wetting, spreading, corrosion inhibition	Vijayakumar and Saravanan, (2015), Banat et al. (2010)
Environment	Bioremediation, Oil spill clean-up operations, Soil remediation and flushing	Emulsification of oils, lowering of interfacial tension, dispersion of oils, solubilisation of oils, wetting, spreading, detergency, foaming, corrosion inhibition in fuel oils and equipment, soil flushing	Silva et al. (2014), Plociniczak et al. (2011)
Food	Anti-adhesive agent, Anti-microbial agent, De-emulsification, Emulsification, Emulsifying agents, Food preservation	Solubilisation of flavoured oils, control of consistency, emulsification, wetting agent, spreading, detergency, foaming, thickener	Campos et al. (2013)

Table 3 contd. ...

...Table 3 contd.

Industry	Applications	Role of Biosurfactant	References
Medicine	Microbiological, Pharmaceuticals, Therapeutics	Anti-adhesive agents, antifungal agents, antibacterial agents, antiviral agents, vaccines, gene therapy, immune modulatory molecules	Mnif and Ghribi (2015), Banat et al. (2014), Lawniczak et al. 2013
Mining	Flotation, Heavy metal clean-up operations, Soil remediation	Wetting and foaming, removal of metal ions from aqueous solutions, soil and sediments, heavy metal sequestrants, spreading, corrosion inhibition in oils	Sarubbo et al. (2015)
Nanotechnology	Synthesis of nanoparticles	Emulsification, stabilisation	Vijayakumar and Saravanan (2015), Mulligan (2009)
Petroleum	Bio-adsorbent, *Bio composite agent*, De-emulsification, Enhanced oil recovery, Heavy metal removal agent	Emulsification of oils, *lowering of interfacial* tension, de-emulsification of oil emulsions, solubilisation of oils, viscosity reduction, dispersion of oils, wetting of solid surfaces, spreading, detergency, foaming, corrosion inhibition in fuel oil sand equipment	Souza et al. *(2014)*, Plociniczak et al. (2011)
Textiles	Bleaching assistant, Dyeing and printing, Finishing of textiles, Levelling agent, Lubricant, Preparation of fibres, Scouring agent	Wetting, penetration, solubilisation, emulsification, detergency and dispersion, wetting and emulsification in finishing formulations, softening	Helmy et al. (2011), Banat et al. (2010)
Waste water treatment	Bio-adsorbent, Bio composite agent, Heavy metal removal agent	Exploited as ion collectors	Sarubbo et al. (2015), Vecino et al. (2015), Akbari et al. (2018)

(Akbari et al. 2018, Santos et al. 2016, Usman et al. 2016, Plociniczak et al. 2011)

management of certain diseases of microbial origin (*l.c.*). Surfactants are supposed to help microbes to adsorb soil particles occupied by pollutants, thus reducing the diffusion route length between the site of absorption and bio-uptake by the microbes. Biosurfactants can be exploited for biodegradation of pollutants by which the condition of the soil can be improved. In agriculture, surfactants are used for hydrophilization of soils to obtain even spreading of fertilizer in the soil. They also prevent encrusting of certain fertilizers during storage and promote dispersion and penetration of the toxicants in pesticides. Moreover, biosurfactants are applied as a part of agri-business, where they play an important role in the biocontrol of microbes such as parasitism, antibiosis and prompted fundamental protection. Generally, rhamnolipid biosurfactant is produced by the genus *Pseudomonas* and is known to have potent antimicrobial activity (Roy 2017). Further, no adverse effects on humans or the environments are determined due to exposure to rhamnolipid biosurfactants (*l.c.*). Fengycins are also reported to possess antifungal action and they may be effective in biocontrol of plant diseases (Kulimushi et al. 2017). Some bacteria are able to produce lipopeptide biosurfactants with insecticidal activity against fruit fly *Drosophila melanogaster* and therefore can be used as a biopesticide (Mulligan 2005). Now a day's prices of new pesticides have risen because the application of biosurfactants is an enhanced environmentally friendly cost (*l.c.*).

8.2.2 Biosurfactants in Cosmetic Industries

In the cosmetic industry, due to their emulsifying property, foaming, water binding capacity, spreading and wetting properties effecting viscosity and product consistency, replacement of chemically synthesised surfactants by biosurfactants has been proposed (Bhadoriya et al. 2013). Biosurfactants are used as acne pads, antimicrobial agents, anti-dandruff products, baby products, bath products, cleansers, contact lens solutions, emulsifiers, foaming agents, lipsticks, mascara, mediators of enzyme action and toothpaste (Bhadoriya et al. 2013). Sophorolipids have large scale application in the beautifying agents' industry. They have exceptional attributes that incorporate hostile to radical properties, incitement of dermal fibroblast metabolism and hygroscopic properties to help sound skin physiology; future prospects of sophorolipid based items incorporate a small variety of facial make up, creams, excellence washes and hair items (*l.c.*). Biosurfactants have gained the attention of cosmetic and pharmaceutical industries due to their possible use as detergents, wetting, emulsifying, foaming and solubilizing agents, and many other useful properties (Marchant and Banat 2012). The usage of biosurfactants in these two industries is very extensive since they are one of the main components of importance in the production of products such as cleansers, creams, hair conditioners, moisturizers, shampoos, shower gels, soaps, toothpastes and many other healthcare and skin care products (Chakraborty et al. 2015, Satpute et al. 2010). Studies have reported that the tendency of chemical surfactants in cosmetic product formulations to create a possible skin allergy risk is a thought provoking problem among others (Akbari et al. 2018, Bujak et al. 2015). However, the outstanding characteristics of bio-based surfactants make them an admirable component as a green product.

8.2.3 Biosurfactants in Food Processing

Biosurfactants have been used for various food processing applications and play a role as a food formulation ingredient and anti-adhesive agents; as a food formulation ingredient they endorse the construction and stabilization of emulsions due to their ability to decrease surface and interfacial tension. Biosurfactants are also used to control the aggregation of fat globs, stabilization of aerated systems, improvement of constancy and texture of fat-based products, enhancement of shelf-life of products containing starch, adjustment of rheological properties of wheat dough, and others (*l.c.*). In bakery and ice-cream formulations, biosurfactants act by regulating the consistency and solubilizing the flavour oils. Addition of rhamnolipid surfactants could improve the stability of dough, volume, texture and conservation of bakery products (Van Haesendonck and Vanzeveren 2004). They could also improve the properties of butter, cream and frozen confectionery products. L-Rhamnose has substantial potential for flavouring food (Vijayakumar and Saravanan 2015).

8.2.4 Biosurfactants in Medicine

Biosurfactants are active therapeutic representatives. For example, amino acid-based surfactants, glycolipid biosurfactants and saccharide—fatty acid esters possess antimicrobial activity (Kitamoto et al. 2009). Glycolipid biosurfactants, Mono Acyl Glycerol (MAGs) and polyunsaturated fatty acids possess anticancer activity (Bhadoriya et al. 2013). Various applications of biosurfactants in medicine are depicted in Table 4.

Table 4. Applications of Biosurfactants in the medical field.

Microorganisms	**Biosurfactant type**	**Application**
Bacillus licheniformis	Lichenysin	Antibacterial activity and chelating properties that might explain the membrane-disrupting effect of lipopeptides
Bacillus pumilus	Pumilacidin (Surfactin analogue)	Antiviral activity against herpes simplex virus 1 (HSV-1) inhibitory activity against H+, K+-ATPase and protection against gastric ulcers in vivo
Bacillus subtilis	Iturin	Antimicrobial activity and antifungal activity against profound mycosis. Effect on the morphology and membrane structure of yeast cells
Bacillus subtilis	Surfactin	Antimicrobial and antifungal activities. Haemolysis and formation of ion channels in lipid membranes. Antiviral activity against human immunodeficiency virus 1 (HIV-1).
Candida antartica	Mannosylerythritol lipids	Antimicrobial, immunological and neurological properties.
Lactobacillus	Surlactin	Anti-adhesive activity against several pathogens including enteric bacteria.
Pseudomonas aeruginosa	Rhamnolipid	Antimicrobial activity against Mycobacterium tuberculosis.
Rhodococcuserythropolis	Treahalose lipid	Antiviral activity against HSV and influenza virus
Streptococcus thermophilus	Glycolipid	Anti-adhesive activity against several bacterial and yeast strains isolated from voice prostheses.

Rodrigues et al. (2006)

8.2.4.1 Antimicrobial activity

Different structures of biosurfactants give them the capability to exhibit useful performance. Due to their structure, biosurfactants exert their toxicity on cell membrane permeability similar to a detergent like impact (Gielen et al. 2008).

8.2.4.2 Anti-cancer activity

Microbial extracellular glycolipids cause cell differentiation instead of cell proliferation in the human promyelocytic leukaemia cell line and affords the basis for the utilization of microbial extracellular glycolipids as novel reagents for the cancer cell treatment (Garti 2003).

8.2.4.3 Anti-adhesive agents

Biosurfactants have been found to prevent the adhesion of pathogenic microbes to infection sites or solid surfaces. Rodrigues et al. 2006, demonstrated that pre-coating vinyl urethral catheters by running the surfactin solution before inoculation with media resulted in the decline in the amount of biofilm formed by *E. coli, Proteus mirabilis* and *Salmonella typhimurium*.

8.2.4.4 Immunological adjuvants

Bacterial lipopeptides comprise potent non-pyrogenic and non-toxic immunological adjuvants when combined with conventional antigens. An enhancement of the humoral human response was exhibited when low molecular mass antigens Iturin AL and herbicolin A were present (Gharaei-Fathabad 2011).

8.2.4.5 Antiviral activity

Researchers have reported Antibiotic properties and prevention of growth of Human Immunodeficiency Virus (HIV) in leucocytes by biosurfactants (Muthusamy et al. 2008). Moreover, the increased prevalence of HIV in women, began the need for a controlled, effectual and safe vaginal topical microbicide (Muthusamy et al. 2008). Sophorolipid surfactants produced by *C. Bombicola* and its structural analogues like the sophorolipid diacetate ethyl ester is the most effective spermicidal and virucidal agent.

8.2.4.6 Gene delivery

Studies reported that the formation of an economical and safe method for creating exogenous nucleotides into mammalian cells is very critical for basic sciences and clinical applications (*l.c.*).

8.2.5 Biosurfactants in Pharmaceuticals

The major role of biosurfactants in pharmaceutical processing is to increase the solubility of drugs, especially those which are very poorly soluble in water. It includes a large amount of new and developing bioactive agents (e.g., peptides, proteins, oligonucleotides, vaccines and vitamins), to allow there *in vivo* delivery. Biosurfactants also enhance the constancy of encapsulated drugs, their thermodynamic activity and rate of diffusion. The furthermost common use of surfactants is for self-assembly systems as drug delivery vehicles or carriers (Bhadoriya et al. 2013).

Biosurfactants are prominent apparatuses of certain cosmetics, dermatological and personal care products.

8.2.6 Biosurfactants in Petroleum Industries

Biosurfactants are a novel group of molecules and among the most powerful by-products that modern microbial technology can suggest in the fields such as bio-corrosion and bio-degradation of hydrocarbons within oil reservoirs, enzymes and biocatalysts for petroleum, and more (Vijayakumar and Saravanan 2015). Biosurfactants also perform the most important role in petroleum extraction, transportation, upgrading, refining and also manufacturing of petrochemicals. (Fakruddin 2012). Certain species of *P. aeruginosa* and *B. subtilis* can produce rhamnolipid and surfactin, a lipoprotein type biosurfactant, respectively; which possess the capability to increase solubility and bioavailability of a petrochemical mixture and also to stimulate indigenous microbes to promote biodegradation of diesel contaminated soil.

8.2.7 Biosurfactants in Textile Industries

Almost all surfactants, applied as commercial laundry detergents, are chemically synthesized and exert toxicity to fresh water living microbes. Biosurfactants like Cyclic Lipo-Peptide (CLP) are stable over a varied range of pH (7.0–12.0) and even heating them at higher temperature does not result in any kind of loss to their surface-active property (Helmy et al. 2011). They exhibited excellent emulsifying ability with vegetable oils and demonstrated great compatibility and stability with commercial laundry detergents facilitating their inclusion in the manufacturing of laundry detergents (Banat et al. 2010).

9. Advantages of Biosurfactants

- Biosurfactants are effective surface and interfacial tension reducers.
- They are easily biodegraded by microorganisms in the environment.
- They are specific in their action, hence play specific functions.
- They can be produced from a variety of relatively cheap raw materials.
- They can be produced from industrial waste and by-products and thus are key into acceptable production economics.
- Biosurfactants present lower toxicity than synthetic surfactants.
- Environmental factors like ionic strength, pH and temperature do not affect a majority of the biosurfactants.
- They are environment-friendly.
- They have higher specificity.
- Many biosurfactants are stable at extreme pH levels, salinities and temperatures.
- Surfactants of biological origin have the feature of compatibility, and thus have found effective application in pharmaceuticals, cosmetics, food industries and others.
- Use of low-cost substrates for their production.

10. Conclusion

Biosurfactants are extensively used as multi-functional compounds because of their advantageous properties and cost effectiveness as compared to synthetic surfactants. Biosurfactants could assist as green alternatives in numerous environmental application including biodegradation processes, MEOR, co-contaminated sites remediation as well as industrial applications including agriculture, cosmetic, food processing, medicine, pharmaceuticals, petroleum and textile sectors active to utilization of biosurfactants for various resolutions. This chapter is significantly compiled using a promising approach related to the classification, properties, production, various environmental applications, industrial applications and advantages of natural biosurfactants and their influence on the environment and human health. Biosurfactants have attracted extensive attention for present and future applications because of their safe and eco-friendly properties.

Acknowledgements

The authors are thankful to the department of Life Sciences, HNGU, Patan.

References

Akbari, S., N.H. Abdurahman, R.M. Yunus, F. Fayaz, O.R. Alara et al. 2018. Biosurfactants—a new frontier for social and environmental safety: a mini review. Biotechnol. Res. Innov. 2: 81–90.

Al-Wahaibi, Y., S. Joshi, S. Al-Bahry, A. Elshafie, A. Al-Bemani et al. 2014. Biosurfactant production by *Bacillus subtilis* B30 and its application in enhancing oil recovery. Colloids Surf. B: Biointerfaces. 114: 324–333.

Banat, I.M., R.S. Makkar and S.S. Cameotra. 2000. Potential commercial applications of microbial surfactants. Appl. Environ. Microbiol. 53: 495–508.

Banat, I.M., A. Franzetti, I. Gandolfi, G. Bestetti, M.G. Martinotti et al. 2010. Microbial biosurfactants production, applications. Appl. Microbiol. Biotechnol. 87: 427–444.

Banat, I.M., M.A.D. De Rienzo and G.A. Quinn. 2014. Microbial biofilms: Biosurfactants as antibiofilm agents. Appl. Microbiol. Biotechnol. 98: 9915–9929.

Barros, F.F.C., C.P. Quadros, M.R. Marostica, G.M. Pastore et al. 2007. Surfactin: chemical, technological and functional properties for food applications. Quim. Nova. 30(2): 1–14.

Batista, R.M., R.D. Rufino, J.M. Luna, J.E.G. Souza, L.A. Sarubbo et al. 2010. Effect of medium components on the production of a biosurfactant from *Candida tropicalis* applied to the removal of hydrophobic contaminants in soil. Water Environ. Res. 82: 418–425.

Bhadoriya, S.S., N. Madoriya, K. Shukla, M.S. Parihar et al. 2013. Biosurfactants: a new pharmaceutical additive for solubility enhancement and pharmaceutical development. Biochem. Pharmacol. 2(2): 1–5.

Bordoloi, N.K. and B.K. Konwar. 2008. Microbial surfactant-enhanced mineral oil recovery under laboratory conditions. Colloid Surf. B. 63: 73–82.

Boruah, B. and M. Gogoi. 2013. Plant based natural surfactants. Asian J. Home Sci. 8(2): 759–762.

Bujak, T., T. Wasilewski and Niziol-Lukaszewska, Z. 2015. Role of macromolecules in the safety of use of body wash cosmetics. Colloids Surf. B: Biointerfaces. 135: 497–503.

Campos, J.M., T.L.M. Stamford, L.A. Sarubbo, J.M. Luna, R.D. Rufino et al. 2013. Microbial biosurfactants as additives for food industries. Biotechnol. Prog. 29: 1097–1108.

Chakraborty, J. and S. Das. 2014. Biosurfactant-based bioremediation of toxic metals. pp. 167–201. *In*: Microbial Biodegradation and Bioremediation.

Chakraborty, S., M. Ghosh, S. Chakraborti, S. Jana, K.K. Sen et al. 2015. Biosurfactant produced from *Actinomycetes nocardiopsis*A17: characterization and its biological evaluation. Inter. J. Biol. Macromol. 79: 405–412.

Deleu, M. and M. Paquot. 2014. From renewable vegetables resources to microorganisms: new trends in surfactants. C. R. Chimie. 7: 641–646.

Fakruddin, M.D. 2012. Biosurfactant: production and application. J. Pet. Environ. Biotechnol. 3(4): 1–5.

Fenibo, E.O., S.I. Douglas and H.O. Stanley. 2019. A review on microbial surfactants: production, classifications, properties and characterization. J. Adv. Microbiol. 18(3): 1–22.

Fox, S.I. and G.A. Bala. 2000. Production of surfactant from *Bacillus subtilis* ATTCC 21332 using potato substrates. Bioresour. Technol. 75(3): 235–240.

Garti, N. 2003. Microemulsions as microreactors for food applications. Curr. Opin. Colloid Interface Sci. 8: 197–211.

Gharaei-Fathabad, E. 2011. Biosurfactants in pharmaceutical industry: a mini—review. Am. J. Drug Discov. Dev., pp. 1–11.

Gielen, D., J. Newman and M.K. Patel. 2008. Reducing industrial energy use and CO_2 emissions: the role of materials science. MRS Bulletin 33: 471–477.

Gusmao, C.A.B., R.D. Rufino and L.A. Sarubbo. 2010. Laboratory production and characterization of a new biosurfactant from *Candida glabrata* UCP 1002 cultivated in vegetable fat waste applied to the removal of hydrophobic contaminant. World J. Microbiol. Biotechnol. 26: 1683–1692.

Haritash, A.K. and C.P. Kaushik. 2009. Biodegradation aspects of polycyclic aromatic hydrocarbons (PAHs): A review. J. Hazard. Mater. 169: 1–15.

Helmy, Q., E. Kardena, N. Funamizu and W. Wisjnuprapto. 2011. Strategies toward commercial scale of biosurfactant production as potential substitute for it's chemically counterparts. Int. J. Biotechnol. 12: 66–86.

Kalogiannis, S., G. Iakovidou, M. Liakopoulou-Kyriakides, D.A. Kyriakidis, G.N. Skaracis et al. 2003. Optimization of xanthan gum production by *Xanthomonas campestris* grown in molasses. Process Biochem. 39: 249–256.

Kitamoto, D., T. Morita, T. Fukuoka, T. Imura, M. Konishi et al. 2009 Self-assembling properties of glycolipid biosurfactants and their potential applications Curr. Opin. Colloid Interface Sci. 14: 315–328.

Kulimushi, P.Z., A.A. Arias, L. Franzil, S. Steels, M. Ongena et al. 2017. Stimulation of fengycin—type antifungal lipopeptides in *Bacillus amyloliquefaciens* in the presence of the maize fungal pathogen *Rhizomucor variabilis*. Front. Microbiol. 8(850): 1–12.

Lan, G., Q. Fan, Y. Liu, C. Chen, G. Li et al. 2015. Rhamnolipid production from waste cooking oil using *Pseudomonas* SWP-4. Biochem. Eng. J. 101: 44–54.

Lawniczak, L., R. Marecik and L. Chrzanowski. 2013. Contributions of biosurfactants to natural or induced bioremediation. Appl. Microbiol. Biotechnol. 97: 2327–2339.

Li, J., M. Deng, Y. Wang and W. Chen. 2016. Production and characteristics of biosurfactant produced by *Bacillus pseudomycoides* BS6 utilizing soybean oil waste. Int. BiodetBiodegrad 112: 72–79.

Luna, J.M., R.D. Rufino, C.D.C. Albuquerque, L.A. Sarubbo, G.M. Campos-Takaki et al. 2011a. Economic optimized medium for tenso-active agent production by *Candida sphaerica* UCP 0995 and application in the removal of hydrophobic contaminant from sand. Int. J. Mol. Sci. 12: 2463–2476.

Luna, J.M., R.D. Rufino, L.A. Sarubbo, L.R.M. Rodrigues, J.A.C. Teixeira et al. 2011b. Evaluation antimicrobial and antiadhesive properties of the biosurfactant Lunasan produced by *Candida sphaerica* UCP 0995. Curr. Microbiol. 62: 1527–1534.

Luna, J.M., R.D. Rufino, L.A. Sarubbo and G.M. Campos-Takaki. 2013. Characterisation, surface properties and biological activity of a biosurfactant produced from industrial waste by *Candida sphaerica* UCP 0995 for application in the petroleum industry. Coll. Surf. B Biointerfaces 102: 202–209.

Magalhaes, E.R.B., F.L. Silva, M.A.D.S.B. Sousa and E.S. Dos Santos. 2018. Use of different agro-industrial waste and produced water for biosurfactant production. Biosci. Biotechnol. Res. Asia 15: 17–26.

Makkar, R.S. and S.S. Cameotra. 2002. An update on the use of unconventional substrates for biosurfactant production and their new applications. Appl. Micobiol. Biotechnol. 58: 428–434.

Maneerat, S. 2005. Production of biosurfactants using substrates from renewable-resources. Songklanakarin J. Sci. Technol. 27(3): 675–683.

Marchant, R. and I.M. Banat. 2012. Microbial biosurfactants: Challenges and opportunities for future exploitation. Trends Biotechnol. 30(11): 558–565.

Marchant, R., S. Funston, C. Uzoigwe, P.K.S.M. Rahman et al. 2014. Production of biosurfactants from non-pathogenic bacteria. pp. 73–81. *In*: Biosurfactants.

Maslin, P.M. and R.M. Maier. 2000. Rhamnolipid enhanced mineralization of phenanthrene in organic metal co-contaminated soils. Bioremed. J. 4: 295–308.

Mnif, I. and D. Ghribi. 2015. Microbial derived surface-active compounds: Properties and screening concept. World J. Microbiol. Biotechnol. 31: 1001–1020.

Mouafo, T.H., A. Mbawala and R. Ndjouenkeu. 2018. Effect of different carbon sources on biosurfactants' production by three strains of *Lactobacillus* spp. BioMed. Res. Inter, pp. 1–15.

Mulligan, C.N. and B.F. Gibbs. 2004. Types, production and applications of biosurfactants. Proc. Indian Nat. Sci. Acad. 1: 31–55.

Mulligan, C.N. 2005. Recent advances in the environmental applications of biosurfactants. Curr. Opin. Coll. Interface Sci. 14: 372–378.

Muthusamy, K., S. Gopalakrishnan, T.K. Ravi, P. Sivachidambaram et al. 2008. Biosurfactants: properties, commercial production and application. Curr. Sci. 94: 736–747.

Nitschke, M., C. Ferraz and G.M. Pastore. 2004. Selection of microorganisms for biosurfactant production using agro-industrial wastes. Braz. J. Microbiol. 435: 81–85.

Patil, S.B., Y.P. Sawant, L.H. Kamble, C.J. Raorane et al. 2016. Primary screening of actinomycetes in prospects with biosurfactant production from animal fat waste. Int. J. Curr. Microbiol. Appl. Sci. 5: 92–97.

Plociniczak, M.P., G.A. Plaza, Z.P. Seget and S.S. Cameotra. 2011. Environmental applications of biosurfactants: recent advances. Int. J. Mol. Sci. 12: 633–654.

Radzuan, M.N., I.M. Banat and J. Winterburn. 2017. Production and characterization of rhamnolipid using palm oil agricultural refinery waste. Biores. Technol. 225: 99–105.

Rahman, P.K.S.M. 2008. Production, characterisation and applications of biosurfactants-review. Biotechnol. 7(2): 360–370.

Rautela, R. and S.S. Cameotra. 2014. Role of biopolymers in industries: their prospective future applications. pp. 133–142. *In*: Fulekar, M., B. Pathak and R. Kale (eds.). Environment and Sustainable Development.

Rodrigues, L.R., I.M. Banat, J.A. Teixeira and R. Oliveira. 2006. Biosurfactants: potential applications in medicine. J. Antimicrob. Chemother. 57: 609–618.

Roy, A. 2017. A review on the biosurfactants: properties, types and its applications. J. Fundam. Renew. Energy. Appl. 8(1): 1–5.

Rubio-Ribeaux, D., R.F. Da Silva Andrade, G.S. Da Silva, R.A. De Holanda, M.A. Pele et al. 2017. Promising biosurfactant produced by a new *Candida tropicalis* UCP 1613 strain using substrates from renewable-resources. Afr. J. Microbiol. Res. 11: 981–991.

Rufino, R.D., J.M. Luna, G.I.B. Rodrigues, G.M. Campos-Takaki, L.A. Sarubbo et al. 2011. Application of a yeast biosurfactant in the removal of heavy metals and hydrophobic contaminant in a soil used as slurry barrier. Appl. Environ. Soil Sci., pp. 1–7.

Sachdev, D.P. and S.S. Cameotra. 2013. Biosurfactants in agriculture. Appl. Microbiol. Biotechnol. 97: 1005–1016.

Sandrin, T.R. and R.M. Maier. 2003. Impact of metals on the biodegradation of organic pollutants. Environ. Health Perspect. 111: 1093–1100.

Santos, D.K.F., R.D. Rufino, J.M. Luna, V.A. Santos, A.A. Salgueiro et al. 2013. Synthesis and evaluation of biosurfactant produced by *Candida lipolytica* using animal fat and corn steep liquor. J. Pet. Sci. Eng. 105: 43–50.

Santos, D.K.F., Y.B. Brandao, R.D. Rufino, J.M. Luna, A.A. Salgueiro et al. 2014. Optimization of cultural conditions for biosurfactant production from *Candida lipolytica*. Biocatal. Agric. Biotechnol. 3: 48–57.

Santos, D.K.F., R.D. Rufino, J.M. Luna, V.A. Santos, L.A. Sarubbo et al. 2016. Biosurfactants: multifunctional biomolecules of the 21st century. Int. J. Mol. Sci. 17(401): 1–31.

Sarubbo, L.A., R.B. Rocha, Jr., J.M. Luna, R.D. Rufino, V.A. Santos et al. 2015. Some aspects of heavy metals contamination remediation and role of biosurfactants. Chem. Ecol. 31(8): 707–723.

Satpute, S.K., S.S. Bhuyan, K.R. Pardesi, S.S. Mujumdar, P.K. Dhakephalkar et al. 2010. Molecular genetics of biosurfactant synthesis in microorganisms. Adv. Exp. Med. Biol. 672: 14–41.

Sellami, M., A. Khlifi, F. Frikha, N. Miled, L. Belbahri et al. 2016. Agro-industrial waste-based growth media optimization for biosurfactant production by *Aneurinibacillusmigulanus*. J. Microbiol. Biotechnol. Food Sci. 5: 578–583.

Silva, R.C.F.S., D.G. Almeida, J.M. Luna, R.D. Rufino, V.A. Santos and L.A. Sarubbo. 2014. Applications of biosurfactants in the petroleum industry and the remediation of oil spills. Int. J. Mol. Sci. 15: 12523–12542.

Silva, S.N.R.L., C.B.B. Farias, R.D. Rufino, J.M. Luna, L.A. Sarubbo et al. 2010. Glycerol as substrate for the production of biosurfactant by *Pseudomonas aeruginosa* UCP0992. Coll. Surf. B Biointerfaces 79: 174–183.

Singh, P., Y. Patil and V. Rale. 2018. Biosurfactant production: emerging trends and promising strategies. J. Appl. Microbiol. 126: 2–13.

Souza, A.F., D.M. Rodriguez, D.R. Ribeaux, M.A. Luna, T.A. Lima e Silva et al. 2016. Waste soybean oil and corn steep liquor as economic substrates for bio emulsifier and biodiesel production by *Candida lipolytica* UCP 0998. Int. J. Mol. Sci. 17: 1608.

Souza, E.C., T.C. Vessoni-Penna and R.P. Souza Oliveira. 2014. Biosurfactant-enhanced hydrocarbon bioremediation: An overview. Int. Biodeterior. Biodegrad. 89: 88–94.

Sudhakar-Babu, P., A.N. Vaidya, A.S. Bal, R. Kapur, A. Juwarka et al. 1996. Kinetics of biosurfactants production by *Pseudomonas aeruginosa* strain from industrial wastes. Biotechnol. Lett. 18: 263–268.

Thompson, D.N., S.L. Fox and G.A. Bala. 2000. Biosurfactants from potato process effluents. Appl. Biochem. Biotechnol. 84: 917–930.

Tokumoto, Y., N. Nomura, H. Uchiyama, T. Imura, T. Morita et al. 2009. Structural characterization and surface-active properties of a succinyl trehalose lipid produce by *Rhodococcus* sp. SD-74. J. Oleo Sci. 58: 97–102.

Usman, M.M., A. Dadrasnia, K.T. Lim, A.F. Mahmud, S. Ismail et al. 2016. Application of biosurfactants in environmental biotechnology; remediation of oil and heavy metal. AIMS. Bioeng. 3(3): 289–304.

Van Haesendonck, I.P.H. and E.C.A. Vanzeveren. 2004. Rhamnolipids in bakery products. International Application Patent (PCT), Washington, DC., USA.

Vandana, P. and D. Singh. 2018. Review on biosurfactant production and its application. Int. J. Curr. Microbial. App. Sci. 7(8): 4228–4241.

Varjani, S.J. and V.N. Upasani. 2017. Critical review on biosurfactant analysis, purification and characterization using rhamnolipid as a model biosurfactant. Bioresour. Technol. 232: 389–397.

Vecino, X., J.M. Cruz and A.B. Moldes. 2015. Wastewater treatment enhancement by applying a lipopeptide biosurfactant to a lignocellulosic biocomposite. Carbohydr. Polym. 131: 186–196.

Vijayakumar, K. and V. Saravanan. 2015. Biosurfactants—types, sources and applications. Res. J. Microbiol. 10(5): 181–192.

5

Biosurfactants from *Pseudomonads*

Applications in Food Industry

Dibyajit Lahiri,[1] *Moupriya Nag*,[1] *Sougata Ghosh*,[2,3] *Ritwik Banerjee*[1] and *Rina Rani Ray*[4,*]

1. Introduction

Many microbial species possess the ability of synthesizing surface-active molecules termed as Biosurfactants (BS) (Maneerat 2004). These (BS) are the groups of amphiphilic compounds being created upon the surfaces of the microbial cells comprising of polar and non-polar moieties that help them to develop micelles at the Junction of fluids possessing various types of polarities (Adamczak and Bednarski 2000). These molecules contain side groups or various allied structures made up of polysaccharide-lipid composites, mycolic corrosive phospholipids, lipopeptide/lipoprotein and glycolipid (Arima et al. 1968). BS is one of the most important by-products being synthesized by the microbial cells which possess a wide application in various industries comprising of food, agriculture, environmental, petroleum, cosmetics, petroleum and pharmaceutical industries. These biologically synthesized compounds, specifically those of microbial origin, have several benefits which cannot be exhibited by synthetic compounds. BS possess properties that include low levels of toxicity, greater biodegradability, environmental compatibility, various types of surface activities at high temperature, salinity and pH (Hajfarajollah et al. 2018, Anvari et al. 2015). These compounds have a wide variety of useful applications

[1] Department of Biotechnology, University of Engineering & Management, Kolkata.
[2] Department of Chemical Engineering, Northeastern University, Boston, MA-02115, USA.
[3] Department of Microbiology, School of Science, RK University, Rajkot – 360020, Gujarat, India.
[4] Department of Biotechnology, Maulana Abul Kalam Azad University of Technology, West Bengal.
* Corresponding author: raypumicro@gmail.com

like restitution of the organic- and metal-contaminated sites, improved bacterial transport, and oil recovery (Desai et al. 1997, Makkar and Cameotra 1998) are used for bioremediation. They are also used as cosmetic additives in pharmaceutical industries, food processing and also in regulating biology (Ghasemi et al. 2019). Recent studies have shown interesting data that revealed the role of biosurfactants in the promotion of microbial growth, survival and defence in the environment (Bodour et al. 2003).

Pseudomonas is a genus of Gram-negative bacteria known to cause a number of biofilm associated acute and chronic infections, produce a broad range of biosurfactant molecules like glycolipids (mainly rhamnolipids), Lipopeptide and viscosins (Singh and Bhubendra 2016). Amongst these molecules rhamnolipids are the major type of biosurfactant.

In the present chapter the biochemical and functional attributes of rhamnolipids from pseudumonads will be discussed with a special reference to their roles in biofilm development and how these compounds can be utilised for human welfare.

2. Biochemical nature of Rhamnolipids (RLs)

RLs are chemical glycolipids that are being synthesized by *P. aeruginosa* and act effectively as an important biosurfactant (Bazsefidpar et al. 2019, Partovi et al. 2013, Rahman et al. 2009). It has shown a higher rate of acceptance due to its requirement of short incubation time and easy cultivating mechanism of the organisms. The structure of rhamnolipids varies due to the presence of rhamnose moiety and the length of the carbon chain. It has been further studied that RLs comprise of a rhamnose moiety known to be a glycon part and lipid moiety known as aglycon part (Abdel-Mawgoud et al. 2010). The rhamnose moiety comprise of one or two rhamnose being linked with one another by mono-RLs and di-RLs respectively. The lipid counterpart comprises of mono and poly unsaturated fatty acids comprising of hydroxyl fatty acid chains which are linked with one another. Various studies resulted in the development of approximately 60 homologous structures (Abdel-Mawgoud et al. 2010). The differences among the homologues exist due to the variation in interaction between the glycon and aglycon parts. Extensive research has been conducted on microorganisms which has shown that biosurfactants are mostly produced under the mesophilic conditions (Rosenberg et al. 1979, Cooper and Goldenberg 1987). Few studies have also shown that biosurfactants can be produced at high temperatures. *Bacillus* sp. produces a heat-resistant biosurfactant (at almost 323K) on a hydrocarbon rich medium and can be used in Microbial Enhanced Oil Recovery (MOER) techniques and sludge clean-up (Davey et al. 2003). Other thermotolerant biosurfactants have also been isolated having a wide variety of applications such as in the dairy industry, and as bio-emulsifiers too.

Although RLs has numerous advantages yet its mass production could not be conducted till 2016 due to its low yields and difficult processes. The company that premiered the production of RLs was Evonik Industry. They implemented *Pseudomonas putida* within butane for the synthesis of rhamnolipids.

3. Role of Biosurfactants in Pseudomonad Biofilm

Microorganisms like bacteria have a tendency to move towards a solid surface during adverse environmental conditions. They deposit polymeric substances to form biofilms as a survival strategy which protects them from the external environment (Banat et al. 2014). Varied research has shown that biosurfactants play an important role in the formation of biofilms by various types of microbial species (Mireles et al. 2001, Walencka et al. 2008, Rivardo et al. 2009). Studies have shown that lipopeptide biosurfactant known as putisolvins is responsible for the formation of biofilm by *P. putida* (Kuiper et al. 2004). Putisolvins possess the ability of inhibiting the formation of biofilms by other species of *Pseudomonas.* It has been also observed that the biosurfactants produced by *Streptococcus thermophiles* is responsible for decreasing the biofilm being formed by yeast on the surface of prosthesis (Rodrigues et al. 2006a). It has been observed that BSs have the ability to alter the surface properties and hence interfere with the mechanism of cellular adhesion on solid surfaces (Flemming and Wingender 2010, Rendueles and Ghigo 2012). *Pseudomonas aeruginosa* is a group of opportunistic pathogens possessing a biosurfactant property that helps in biofilm formation which are mostly studied modelled organisms producing rhamnolipids. RLs play an important role as a virulent factor responsible for the metastasis of cells being present within the biofilm and help in colonization at newer sites resulting in the development of biofilms (Schooling et al. 2004). They also act as anti-phagocytic agents and protect *P. aeruginosa* (Van Gennip et al. 2009).

3.1 Role of RLs at various stages of Biofilm Formation in Pseudomonads

RLs play an important role in regulating the structure of the biofilm formed by *P. aeruginosa.* Studies also showed that RLs deposition exogenously results in the initial attachment of sessile microcolonies upon the biotic and abiotic surfaces but has no role in the biofilm that has already been formed (Davey et al. 2003). RLs being produced endogenously interfere with the final stage of the formation of biofilm (Davey et al. 2003). It indicates that biofilm stability is dependent on the amount of RL synthesis. The synthesis of RLs is intricately regulated by timely expression of genes. Studies showed that the rhlAB operon in *P. aeruginosa* is under stage specific regulation for the production of RLs (Lequette and Greenberg 2005). The enzymes that are required for the production of RLs are expressed by the genes being present within the organism (Ochsner et al. 1994a). RLs production in *P. aeruginosa* is a density dependent process and is regulated by the process of quorum sensing among the sessile microcolonies that are thriving within the biofilm (Ochsner et al. 1994b). It is also partially regulated by RpoS that help in the gene expression taking place within the biofilm (Medina et al. 2003).

3.1.1 Attachment of Microcolonies to a Solid Surface

The surface adherence of the planktonic cells marks the initial step of colonization (Palmer et al. 2007). Cellular adhesion is dependent on the biotic and abiotic surface properties that include hydrophobicity comprising of molecules like lectins and adhesins that enhance the mechanism of attachment with the surface (Dunne 2002).

RLs are the amphipathic molecules that play an essential role in the mechanism of cell to cell or cell to surface interactions decreasing the ability of being adhered upon the surfaces. It has also been observed that excess production of rhamnolipids inhibits the formation of the biofilm (Davey et al. 2003). Research has shown that low amounts of RLs enhances cellular hydrophobicity by the production of lipopolysaccharide from the surface of the cell hence increasing the affinity of the cells towards a solid surface (Zhang and Miller 1994, Al-Tahhan et al. 2000). The adherence of the bacterial cells with the surface enunciates surface movement and clonal propagations thus developing microcolonies (O'Toole and Kolter 1998). It has been observed that RLs are necessary for the development of microcolonies with the enhancement of hydrophobicity at low concentrations (Pamp and Tolker-Nielsen 2007).

3.1.2 Development of Differentiated Biofilms by Cellular Proliferation

The maturation of biofilm is marked by the development of mushroom like structures possessing the microcolonies as their base (Klausen et al. 2003a, Klausen et al. 2003b). The bacterial cells belonging to the planktonic population then develop the ability of propagate out of the mushroom stalks and form a cap like structure with the help of flagellar motility and type IV pilli (Klausen et al. 2003a, Klausen et al. 2003b). Studies have shown that reduction in the cap formation of the biofilm occurs in the presence of excess RL production and also involves the rhlA mutant and presence of pilA/rhlA (Pamp and Tolker-Nielsen 2007).

3.1.3 Detachment of the Sessile Communities and Metastasis

The phenomenon of the development of biofilm involves the process of detachment and dispersal of the sessile cells (Kaplan et al. 2010). This comprises of active or passive or shear dependent dispersion that is a highly regulated mechanism (McDougald et al. 2012). RLs play a vital role in initiating the mechanism of dispersion at later stages of the formation of biofilm (Boles et al. 2005). Studies have also shown that RLs bring about detachment by the formation of cavities at the centre of biofilm structures (Boles et al. 2005). The mechanism of swarming motility is greatly influenced by the surface-active molecules (Caiazza et al. 2005).

4. Biosynthesis of RLs

In liquid cultures *P. aeruginosa* is was found to produce two types of rhamnolipids that comprise of rhamnosyl-β-hydroxydecanoyl-β-hydroxydecanoate (Rha-C-10-C10), monorhamnolipid and dirhamnolipid, rhamnosyl-rhamnosyl-β-hydroxydecanoyl-β-hydroxydecanoate (Rha-RhaC10-C10) (Deziel et al. 2000). Various types of RLs have been described on the basis of their chain length, the extent of saturation and on the basis of the existing chain of hydrocarbon in dTDP-L-rhamnose (nonspecific for carbon chains) (Deziel et al. 1999).

Three steps are involved in the production of RLs that comprise of synthesis of fatty acid moieties of the RLs and free 3-(3-hydroxyalkanoyloxy) alkonic acid catalysed by RhlA, rhamnosyltransferases RhlB and RhlC that help in the catalysis of the transfer of dTDP-L-rhamnose to HAA or to mono-rhamnolipids respectively

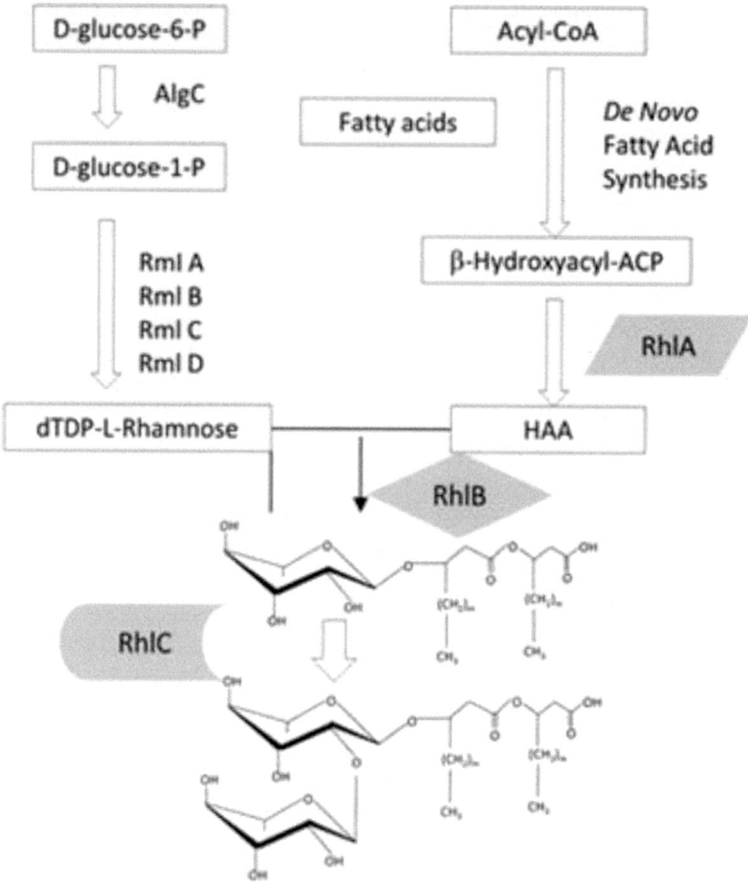

Figure 1. Biosynthese of rhamnolipid in *Pseudomonas* sp. (Chong and Li 2017).

(Deziel et al. 2003). Studies have also shown that RhlA is responsible for the diversion of β-hydroxydecanoyl-ACP from FASII cycle (López-Lara et al. 2010). The conversion of β-hydroxyacyl ACP into HAA is mediated by RhlA (Fig. 1).

4.1 *Synthesis of L-rhamnose*

Rhamnose is a type of deoxy-hexose sugar which is profoundly observed to be present in plants, animals and bacteria. The glucose scaffold is responsible for the development of L-rhamnose and yields deoxythymidine-di-phospho (dTDP)-L-rhamnose. The enzyme which converts thymidylmonophosphate nucleotide to glucose-1-phosphate is glucose-1-phosphate thymidyltransferase via dTDP-L-rhamnose pathway. It has been observed that the catalytic activity of RmlA is allosterically coordinated by the formation of the end product via the dTDP-L-rhamnose pathway (Blankenfeldt et al. 2000b). Chemically RmlA is a type of homo-tetramer comprising of three functional domains and one core domain possessing sequence similarity with that of nuceotidyltransferases and other domains possessing the recognition and binding site for sugarphosphate (Blankenfeldt et al. 2000b). Another potent enzyme which is involved in the process of biosynthesis is dTDP-D-glucose-4,6-dehydratase (RmlB) that helps in the catalysis of hydroxyl group present at the 4th carbon of D-glucose residue with subsequent dehydration resulting in the development of dTDP-4-keto-6-deoxy-D-glucose (Allard et al. 2001). Another type of enzyme dTDP-4-keto-6-deoxy-D-glucose-3,5-epimerase (RmlC) is responsible for performing double epimerization reactions (Graninger 1999). The fourth type of enzyme is dTDP-4-keto-6-deoxy-L-mannose reductase (RmlD) which helps in the formation of dTDP-L-rhamnose (Graninger et al. 1999). The operon is present within *P. aeruginosa* (rmlBDAC) codes for all the four types of enzymes that are responsible for the synthesis of RLs.

4.2 *Regulation of Biosynthesis of RLs*

It is known that dTDP-L-rhamnose pathway (Table 1) plays an important role in the synthesis of RLs. The availability of RmlA is reduced through allosteric inhibition

Table 1. Regulation of rhamnolipid production.

Regulatory factors	Role in production of RLs	Stimulus of Environment	Reference
GidA	Helps in the modulation of rhlR	Stimulus is not yet known	Gupta et al. 2009
DksA	Helps in the regulation of rhlAB and rhlR which is dependent on growth	Stimulus is not yet known	Jude et al. 2003
GacS-GacA (RsmA)	Activates the production of RLs by supressing QS	Stimulus is not yet known	Heurlier et al. 2004
QscR	Helps in the inactivation of heterodimer in the presence of LasR and Rlh R	It is a cell density dependent process	Ledgham et al. 2003
VqsM	QS regulator	It is a cell density dependent process	Dong et al. 2005
VqsR	QS regulator	It is a cell density dependent process	Juhas et al. 2004
PhoB	QS modulation at low phosphorous concentration	Phosphorous concentration regulated process	Jensen et al. 2006
BqsS-BqsR	Helps in the regulation of QS	Stimulus is not yet known	Dong et al. 2008
RpoS	Regulation of QS	Synthesis occurs in stationary phase which is a nutrient dependent process	Medina et al. 2003a, Medina et al. 2003b
Las R	Regulation of QS	It is a cell density dependent process	Dekimpe and Deziel 2009
PQS	It helps in global regulation of QS	It is a cell density dependent process	McKnight et al. 2000

by dTDP-L-rhamnose (Blankenfeldt et al. 2000b). It has been observed that dTDP-L-rhamnose is further transported to extracellular structures like exopolysaccharide (EPS) and lipopolysaccharide (LPS). The heterologous mono-rhamnolipids being synthesized by *Escherichia coli* co-expresses the operon rhlAB of *P. aeruginosa* (Cabrera-Valladares et al. 2006).

4.3 *Role of Quorum Sensing (QS) in the Production of RLs*

The QS systems are an important mechanism of bacterial communication that is marked by the synthesis of small signal molecules known as auto inducers. It is dependent on the density of the bacterial species existing within the system. Several bacteriological functions like formation of biofilm, bioluminescence, virulence and conjugation are controlled by the QS system (Williams et al. 2009). The regulatory protein and the autoinducer complex help in regulating the QS genes. *P. aeruginosa* comprise of two important QS systems named Las and Rhl. These QS systems are responsible for the production of homoserine lactones (Dekimpe and Deziel 2009). Another QS component is composed of phnAB and pqsABCDE, being regulated by transcriptional factor PqsR responsible for the synthesis of 4-hydroxy-2alkylquinolones and Pseudomonas quinolone signal (PQS). PQS in turn is responsible for the production of pyocyanin, elastase, RLs and lectins (Deziel et al. 2005). Thus, the mechanism of QS is dependent on rhl regulons containing rhlAB, responsible for RLs biosynthesis, LaecA for the synthesis of lectin, phzABCDEFG for the production of phenazine biosynthesis and hcnABC for the production of HCN (Schuster et al. 2007). The mechanism of QS is responsible for the development of pathogenesis with *P. aeruginosa* where LasR is mainly responsible for the virulence. The production of RLs is directly regulated by the mechanism of QS in presence of transcriptional factor RhlR which actually help in the activation of rhlAB transcription complex in the absence of the autoinducers (Medina et al. 2003c). RhlR is directly dependent on LasR under the limiting condition of phosphate availability. It is further regulated by transcriptional activator Vfr,RhlR and sigma r54 in the presence of low phosphate concentration. This results in the production of RLs in the presence of acylhomoserine lactones (AHL) (Jensen et al. 2006).

5. Modes of Detection of the Production of RLs

For a better understanding of the structural and functional features of RLs qualitative and quantitative detection is needed. Several methods (Table 2) have been used for the detection of RLs.

6. Production of RLs

6.1 *Fermentation Conditions for the Production of RLs*

Various studies have been performed for optimization of fermentation conditions to get the maximum yield of RLs. The type of feed, solubilisation of carbon sources, temperature, pH, dissolution of oxygen, rate of aeration, density of the cells, capability

Table 2. Various methods for rhamnolipid detection.

Name of the Method	Mechanism of Detection	Reference
Measurement of Surface Tension	The lowering of surface tension determines the effectiveness of the biosurfactant. This is an ideal method for identifying the presence of RLs. But sometimes it has been also found that some bacterial species produce fatty acids that also reduce the surface tension.	Busscher et al. 1996
Methylene Blue Active Substance Assay (MBAS)	It can be also used for the determination of RLs in water. It is an economical method, simple but the procedures that are involved are time consuming. As chloroform is used for the purpose of extraction, RLs produced are often toxic for the purpose of human use.	Hayashi 1975
Drop Collapse Assay Method	It is also an effective methodology in detecting the presence of RLs. The assay is performed in a microwell plate. The presence of an oil coating within the well determines the synthesis of RLs. It is a rapid method and does not involve any specialized mechanism foe detection of RLs.	George and Jayachandran 2009
Cetyltrimethyl ammonium bromide (CTAB) agar test method	The productions of anionic RLs is determined by the formation of a dark halo around the methylene blue agar and the organism can be identified as a RL producing strain. It is an easy process of isolating the biosurfactant.	Sarab al-Shamaa et al. 2019
Oil Spread Test	The biosurfactant can be analysed by the development of clear zones and the surface activity. The amount of clear zone is directly related to the amount of BS being produced. This is one of the most common assays being used for the determination of RLs.	Morikawa et al. 2000
Spectrophotometric Analysis	Rls can be determined by the use of the Orcinol method. Freshly prepared orcinol is used for the analysis. The mixture of the sample was warmed and stirred followed by keeping it at room temperature. The absorbance is measured and the concentration of RLs is determined by the use of the standard curve.	Rikalovic et al. 2012
Chromatographic Method of analysing RLs	Thin Layer Chromatography (TLC) The presence of carbohydrates, proteins and lipids can be analysed by the use of thin layer chromatography (TLC). Chloroform is used as the solvent for resuspending RLs and then appled on TLC plates. After the sample is dry, the TLC plate is placed within the mobile phase containing (methanol:chloroform). Then iodine vapours are used for the detection of lipids. Another plane is sprayed with anthrone and warmed over an oven to determine the presence of rhamnose by blue-green spots.	Das et al. 2009
	High Performance Liquid Chromatography It is the most common analytical method for the determination of RLs by the use of C-18 reverse-phase columns. Water acetonitrile is used as a mobile phase for the gradient run. This helps in the analysis of mono and di-rhamnolipids.	D′eziel et al. 2000

Table 2 contd. ...

...Table 2 contd.

Name of the Method	Mechanism of Detection	Reference
Fourier-transform infrared spectroscopy (FTIR)	This technique is specifically implemented for the determination of the functional groups present within RLs. This analytical technique is based on the principle of IR absorption bands that includes carboxyl, ester and hydroxyl groups.	Abbasi 2012
Nuclear Magnetic Resonance (NMR)	This is one of the most commonly used analytical method which works on the principle of absorption of the radio-frequencies for various types of atoms when exposed to magnetic fields and provide detailed information of the chemical groups and atoms being present within the compound.	Abdel-Mawgoud et al. 2010a

to remove the products influence the rate of production of RLs by *P. aeruginosa* (LaBauve and Wargo 2012). LB broth also proved to be an ideal media for the growth of *P. aeruginosa*. *In vitro* experiments showed that three RLs producing strains include *P. stutzeri, P. aeruginosa* SG and *P. aeruginosa* PrhlAB and can give the best yield if cultivated at 37°C for a period of 5 days at a shaking condition of 200 rpm (Zhao et al. 2019). It has been also suggested that improvisation of growth conditions could be achieved by the use of additive salts like KH_2PO_4, K_2HPO_4, glycerol, $MgSO_4.7H_2O$, KCl, $NaNO_3$,$CaCl_2$ and NaCl.

6.2 *Extraction and Purification of RLs*

Downstream processing is an ideal technique for maintaining the quality and minimising the cost of the product. Thus, proper purification techniques need to be assessed (Table 3) to get the maximum yield (Jadhav et al. 2018). The analytical methods that are being used for the purification of RLs involve adsorption chromatography, foam fraction, ultrafiltration and ion-exchange chromatography. Table 3 shows a detailed concept of various downstream processes that are involved in the purification of RLs.

7. Application of RLs

The properties of RLs like anti-adhesiveness, antibiofilm property, biodegradability, formation of pores, antibacterial activity, emulsification and de-emulsifications have made them important compounds in wide fields of application. They can be used in industries dealing with cosmetics, oils, agriculture, medicine, and special chemical foods (Table 4).

Food industry has marked application of RLs as compounds with antibacterial activity and an ability to perform emulsification (Madhu and Prapulla, 2014). They can also be used as a suitable agent for preventing contamination, transmission of diseases and food spoilage. Food industries have a serious problem of biofilm formation thus pre-treatment of surfaces with biosurfactants can act as an effective

Table 3. Processes of purification of rhamnolipids.

Types of Downstream Processing	Conditions for the separation of BS	Advantages	Reference
Adsorption Chromatography	It is based on the mechanism of adsorption of RLs on the surfaces of resins	It involves low solvent preparation, results in higher purification and is an economical method	Jadhav et al. 2018
Ultrafiltration	Aggregation of foam above CMC	Results in higher purity of BS and is an inexpensive method	Witek-Krowiak et al. 2011
Ion-Exchange Chromatography		It results in faster recovery of the product with higher rate of purification	Satpute et al. 2010
Centrifugation	Helps in the accumulation of RLs in the presence of the centrifugal forces	Better rate of purification	Mukherjee et al. 2006
Precipitation using Ammonium Sulphate Technique	Protein reach BS can be precipitated using salting out technique	It is an effective technique in recovering BS	Satpute et al. 2010
Acid Precipitation	Low pH results in precipitation	It provides a low cost and effective recovery of RLs	Mukherjee et al. 2006
Extraction using organic solvents	The RLs can be dissolved within organic solvents due to the presence of hydrophobic ends	Higher amount of purification and reusability	Satpute et al. 2010

Table 4. Functional attributes of rhamnolipids.

Applications of RLs	Functions	Reference
Mining process	Helps in the extraction of metals by the mining process	Campos et al. 2013
Bioremediation	Removal of contaminants from the soil by the process of desorption, influences the adhesion of microorganisms with the surface, it helps in the bioremediation of petroleum, pesticides and contaminated water	Singh et al. 2007, Juwarkar et al. 2008, Bragg et al. 1994
Pest control	Enhancement of agrochemical solubility and pesticides	Sachdev and Cameotra 2013
Oil Recovery	RLs help in the purpose of oil recovery	Li 2002
Food Processing	Helps in improvisation of structure and texture of dough.	Long 2014

mechanism in preventing the attachment of sessile microcolonies with them (Vatsa et al. 2010). They actively require emulsion for the preparation of butter, cream, margarine, chocolates and hotdogs (Campos et al. 2013). The RLs help in the maintenance of the texture of bakery products and the stability of the dough (Haesendonck and Vanzeveren 2006).

It has been observed that biosurfactants possess haemolytic activity to erythrocytes lesser in comparison to that of the synthetic surfactants. They also don't show any negative impact on lungs, liver, heart and kidney (Ivshina 1998). The efficacy of a surfactant is determined based on parameters that include emulsification index and concentration of micelle. It has been further observed that BSs possessing low concentration values of micelles are more efficient in lowering the interfacial and surface tension. These properties have made RLs applicable specifically in food industries (Fig. 2).

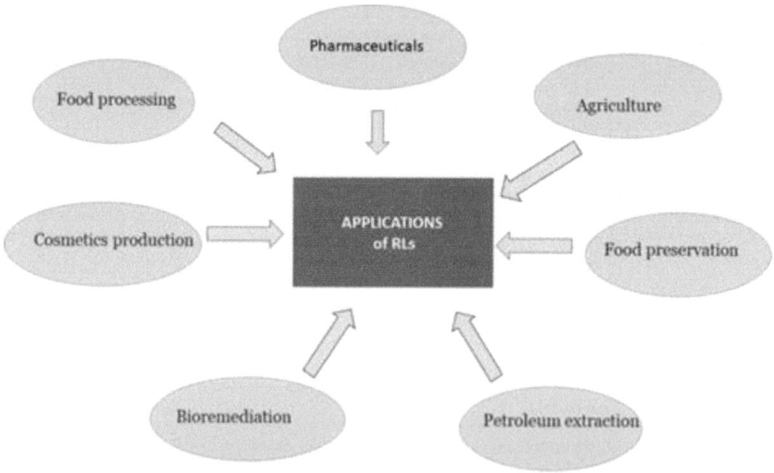

Figure 2. Applications of RLs in various sectors.

8. Conclusion

The chemical substances that can be used for the purpose of reducing the surface tension between two surfaces in industrial processes are known as surfactants. Chemical surfactants have several negative impacts on the environment and human health. Thus, an alternatives to such chemical substances are referred to as biosurfactants. Rhamnolipids are the most popularly produced biosurfactants from pseudomonads. RLs are biodegradable and eco-friendly biosurfactants that can be used in various industries. Extensive research is going on to develop recombinant RLs that possess a lower haemolytic effect and are not as detrimental for human consumption.

Acknowledgement

Dr. Sougata Ghosh acknowledges the Department of Science and Technology (DST), Ministry of Science and Technology, Government of India and Jawaharlal Nehru Centre for Advanced Scientific Research, India for funding under the Post-doctoral Overseas Fellowship in Nano Science and Technology (Ref. JNC/AO/A.0610.1(4) 2019–2260 dated August 19, 2019).

References

Abbasi, H., M.M. Hamedi, T.B. Lotfabad, H.S. Zahiri, H. Sharafi et al. 2012. Biosurfactant-producing bacterium, *Pseudomonas aeruginosa* MA01 isolated from spoiled apples: physicochemical and structural characteristics of isolated biosurfactant. J. Biosci. Bioeng. Feb. 113(2): 211–9.

Abdel-Mawgoud, A.M., R. Hausmann, F. Lépine, M.M. Müller, E. Déziel et al. 2010a. Rhamnolipids: detection, analysis, biosynthesis, genetic regulation and bioengineering of production. *In*: Sobéron-Chavez, G. (ed.). Biosurfactants, 1st edn., VII. Springer, Berlin, 216 pp. (hardcover edn.).

Abdel-Mawgoud, A.M., F. Lépine and E. Déziel. 2010. Rhamnolipids: diversity of structures, microbial origins and roles. Appl. Microbiol. Biotechnol. 86(5): 1323–36.

Adamczak, M. and W. Bednarski. 2000. Influence of medium composition and aeration on the synthesis of biosurfactants produced by *Candida antartica*. Biotechnol. Lett. 22: 313–316.

Allard, S.T., M.F. Giraud, C. Whitfield, M. Graninger, P. Messner et al. 2001. The crystal structure of dTDP-D-Glucose 4, 6-dehydratase (RmlB) from *Salmonella enterica* Serovar Typhimurium, the second enzyme in the dTDP-Lrhamnose pathway. J. Mol. Biol. 307(1): 283–295.

Al-Tahhan, R.A., T.R. Sandrin, A.A. Bodour and R.M. Maier. 2000. Rhamnolipid-induced removal of lipopolysaccharide from *Pseudomonas aeruginosa*: effect on cell surface properties and interaction with hydrophobic substrates. Appl. Microbiol. Biotechnol. 66: 3262–3268.

Anvari, S., H. Hajfarajollah, B. Mokhtarani and K.A. Noghabi. 2015. Physiochemical and thermodynamic characterization of lipopeptide biosurfactant secreted by *Bacillus tequilensis* HK01† RSC Adv. 5: 91836–91845.

Arima, K., A. Kakinuma and G. Tamura. 1968. Surfactin, a crystalline peptidelipid surfactant produced by *Bacillus subtilis*: Isolation, characterization and its inhibition of fibrin clot formation. Biochem. Biophys. Res. Commun. 3: 488–494.

Banat, I.M., M.A.D. De Rienzo and G.A. Quinn. 2014. Microbial biofilms: biosurfactants as antibiofilm agents. Appl. Microbiol. Biotechnol. 98(24): 9915–9929. doi:10.1007/s00253-014-6169-6.

Bazsefidpar, S., B. Mokhtarani, R. Panahi and H. Hajfarajollah. 2019. Recent advancements in the production of rhamnolipid biosurfactants by *Pseudomonas aeruginosa* Biodegradation, pp. 1–11.

Blankenfeldt, W., M.F. Giraud, G. Leonard, R. Rahim, C. Creuzenet et al. 2000b. The purification, crystallization and preliminary structural characterization of glucose-1-phosphate thymidylyltransferase (RmlA), the first enzyme of the dTDP-L-rhamnose synthesis pathway from *Pseudomonas aeruginosa*. Acta Crystallogr. D. Biol. Crystallogr. 56(Pt 11): 1501–1504.

Bodour, A.A., K.P. Drees and R.M. Maier. 2003. Distribution of biosurfactant-producing bacteria in undisturbed and contaminated arid southwestern soils. Appl. Environ. Microbiol. 69(6): 3280–3287.

Boles, B.R., M. Thoendel and P.K. Singh. 2005. Rhamnolipids mediate detachment of *Pseudomonas aeruginosa* from biofilms. Mol. Microbiol. 57: 1210–1223.

Bragg, J., R. Prince and E. Harner. 1994. Effectiveness of bioremediation for the *Exxon Valdez* oil spill. Nature 368: 413–418.

Busscher, H.J., M. van der Kuijl-Booij and H. van der Mei. 1996. Biosurfactants from thermophilic dairy streptococci and their potential role in the fouling control of heat exchanger plates. J. Ind. Microbiol. 16: 15–21.

Cabrera-Valladares, N., A.P. Richardson, C. Olvera, L.G. Trevino, E. Deziel et al. 2006. Monorhamnolipids and 3-(3-hydroxyalkanoyloxy)alkanoic acids (HAAs) production using *Escherichia coli* as a heterologous host. Appl. Microbiol. Biotechnol. 73(1): 187–194.

Caiazza, N.C., R.M. Shanks and G.A. O'Toole. 2005. Rhamnolipids modulate swarming motility patterns of Pseudomonas aeruginosa. J. Bacteriol. 187: 7351–7361.

Campos, J.M., T.L. Stamford, L.A. Sarubbo, J.M. de Luna, R.D. Rufino et al. 2013. Microbial biosurfactants as additives for food industries. Biotechnol. Prog. 29(5): 1097–108.

Cooper, D.G. and B.G. Goldenberg. 1987. Surface-active agents from two *Bacillus* species. Appl. Environ. Microbiol. 53: 224–229.

Chong, H. and Q. Li. 2017. Microbial production of rhamnolipids: opportunities, challenges and strategies. Microb. Cell Fact 16: 137.

Das, P., S. Mukherjee and R. Sen. 2009. Substrate dependent production of extracellular biosurfactant by a marine bacterium. Bioresour. Technol. 100(2): 1015–9.

Davey, M.E., N.C. Caiazza and G.A. O'Toole. 2003. Rhamnolipid surfactant production affects biofilm architecture in *Pseudomonas aeruginosa* PAO1. J. Bacteriol. 185: 1027–1036.

Dekimpe, V. and E. Deziel. 2009. Revisiting the quorum-sensing hierarchy in *Pseudomonas aeruginosa*: the transcriptional regulator RhlR regulates LasRspecific factors. Microbiology 155(Pt 3): 712–723.

Desai, J.D. and Banat, Ibrahim. 1997. Microbial production of surfactants and their commercial potential. Microbiology and Molecular Biology Reviews: MMBR. 61: 47–64. 10.1128/.61.1.47-64.1997.

Deziel, E., F. Lepine, D. Dennie, D. Boismenu, O.A. Mamer et al. 1999. Liquid chromatography/mass spectrometry analysis of mixtures of rhamnolipids produced by *Pseudomonas aeruginosa* strain 57RP grown on mannitol or naphthalene. Biochim. Biophys. Acta 1440(2-3): 244–252.

Deziel, E., F. Lepine, S. Milot and R. Villemur. 2000. Mass spectrometry monitoring of rhamnolipids from a growing culture of *Pseudomonas aeruginosa* strain 57RP. Biochim. Biophys. Acta 1485(2-3): 145–152.

Deziel, E., F. Lepine, S. Milot and R. Villemur. 2003. RhlA is required for the production of a novel biosurfactant promoting swarming motility in *Pseudomonas aeruginosa*: 3-(3-hydroxyalkanoyloxy) alkanoic acids (HAAs), the precursors of rhamnolipids. Microbiology 149(Pt 8): 2005–2013.

Deziel, E., S. Gopalan, A.P. Tampakaki, F. Lepine, K.E. Padfield et al. 2005. The contribution of MvfR to *Pseudomonas aeruginosa* 6382 R.S. Reis et al./Bioresource Technology 102(2011): 6377–6384. pathogenesis and quorum sensing circuitry regulation: multiple quorum sensing-regulated genes are modulated without affecting lasRI, rhlRI or the production of N-acyl-L-homoserine lactones. Mol. Microbiol. 55(4): 998–1014.

Dong, Y.H., X.F. Zhang, J.L. Xu, A.T. Tan, L.H. Zhang et al. 2005. VqsM, a novel AraC-type global regulator of quorum-sensing signalling and virulence in *Pseudomonas aeruginosa*. Mol. Microbiol. 58(2): 552–564.

Dong, Y.H., X.F. Zhang, S.W. An, J.L. Xu, L.H. Zhang et al. 2008. A novel two-component system BqsS-BqsR modulates quorum sensing-dependent biofilm decay in *Pseudomonas aeruginosa*. Commun. Integr. Biol. 1(1): 88–96.

Dunne, W.M. Jr. 2002. Bacterial adhesion: seen any good biofilms lately? Clin. Microbiol. Rev. 15: 155–166.

Flemming, H.C. and J. Wingender. 2010. The biofilm matrix. Nat. Rev. Microbiol. 8: 623–633.

George, S. and K. Jayachandran. 2009. Analysis of rhamnolipid biosurfactants produced through submerged fermentation using orange fruit peelings as sole carbon source. Appl. Biochem. Biotechnol. 158(3): 694–705.

Ghasemi, A., M. Moosavi-Nasab, P. Setoodeh, G. Mesbahi, G. Yousefi et al. 2019. Biosurfactant production by Lactic Acid Bacterium *Pediococcus dextrinicus* SHU1593 grown on different carbon sources: strain screening followed by product characterization. Sci. Rep. 9(1).

Graninger, M., B. Nidetzky, D.E. Heinrichs, C. Whitfield, P. Messner et al. 1999. Characterization of dTDP-4-dehydrorhamnose 3, 5-epimerase and dTDP-4- dehydrorhamnose reductase, required for dTDP-L-rhamnose biosynthesis in *Salmonella enterica* serovar Typhimurium LT2. J. Biol. Chem. 274(35): 25069–25077.

Gupta, R., T.R. Gobble and M. Schuster. 2009. GidA posttranscriptionally regulates rhl quorum sensing in *Pseudomonas aeruginosa*. J. Bacteriol. 191(18): 5785–5792.

Haesendonck, I.V. and E. Vanzeveren. 2006. US Pat., US20060233935A1.

Hajfarajollah, H., P. Eslami, B. Mokhtarani and K. Akbari Noghabi. 2018. Biosurfactants from probiotic bacteria: A review. Biotechnol. Appl. Biochem. 65: 768–783.

Hayashi. K.A. 1975. Rapid determination of sodium dodecyl sulfate with methylene blue. Anal. Biochem. 67(2): 503–6.

Heurlier, K., F. Williams, S. Heeb, C. Dormond, G. Pessi, D. Singer, M. Camara, P. Williams and D. Haas. 2004. Positive control of swarming, rhamnolipid synthesis, and lipase production by the posttranscriptional RsmA/RsmZ system in *Pseudomonas aeruginosa* PAO1. J. Bacteriol. 186(10): 2936–2945.

Jadhav, J., S. Dutta, S. Kale and A. Pratap. 2018. Fermentative production of rhamnolipid and purification by adsorption chromatography. Prep. Biochem. Biotechnol. 16; 48(3): 234–241.

Jensen, V., D. Lons, C. Zaoui, F. Bredenbruch, A. Meissner et al. 2006. RhlR expression in *Pseudomonas aeruginosa* is modulated by the *Pseudomonas quinolone* signal via PhoB-dependent and -independent pathways. J. Bacteriol. 188(24): 8601–8606.

Jude, F., T. Kohler, P. Branny, K. Perron, M.P. Mayer et al. 2003. Posttranscriptional control of quorum-sensing-dependent virulence genes by DksA in *Pseudomonas aeruginosa*. J. Bacteriol. 185(12): 3558–3566.

Juhas, M., L. Wiehlmann, B. Huber, D. Jordan, J. Lauber et al. 2004. Global regulation of quorum sensing and virulence by VqsR in *Pseudomonas aeruginosa*. Microbiology 150(Pt 4): 831–841.

Juwarkar, A.A., K.V. Dubey, A. Nair and S.K. Singh. 2008. Bioremediation of multi-metal contaminated soil using biosurfactant—a novel approach. Indian J. Microbiol. 48(1): 142–146.

Kaplan, J.B. 2010. Biofilm dispersal: mechanisms, clinical implications, and potential therapeutic uses. J. Dent. Res. 89: 205–218.

Klausen, M., A. Aaes-Jorgensen, S. Molin and T. Tolker-Nielsen. 2003a. Involvement of bacterial migration in the development of complex multicellular structures in *Pseudomonas aeruginosa* biofilms. Mol. Microbiol. 50: 61–68.

Klausen, M., A. Heydorn, P. Ragas, L. Lambertsen, A. Aaes Jorgensen et al. 2003b. Biofilm formation by *Pseudomonas aeruginosa* wild type, flagella and type IV pili mutants. Mol. Microbiol. 48: 1511–1524.

Kuiper, I., E.L. Lagendijk, R. Pickford, J.P. Derrick, G.E.M. Lamers et al. 2004. Characterization of two *Pseudomonas putida* lipopeptide biosurfactants, putisolvin I and II, which inhibit biofilm formation and break down existing biofilms. Mol. Microbiol. 51: 97–113.

LaBauve, A.E. and M.J. Wargo. 2012. Growth and laboratory maintenance of *Pseudomonas aeruginosa*. Curr. Protoc. Microbiol. May; Chapter 6: Unit 6E.1.

Ledgham, F., I. Ventre, C. Soscia, M. Foglino, J.N. Sturgis et al. 2003. Interactions of the quorum sensing regulator QscR: interaction with itself and the other regulators of *Pseudomonas aeruginosa* LasR and RhlR. Mol. Microbiol. 48(1): 199–210.

Lequette, Y. and E.P. Greenberg. 2005. Timing and localization of rhamnolipid synthesis gene expression in *Pseudomonas aeruginosa* biofilms. J. Bacteriol. 187: 37–44.

Li, Q. 2017. Rhamnolipid synthesis and production with diverse resources. Front. Chem. Sci. Eng. 11: 27–36.

Long, X., Q. Meng and G. Zhang. 2014. Application of biosurfactant rhamnolipid for cleaning of UF membranes. J. Membr. Sci. 457: 113–119.

López-Lara, I.M. and O. Geiger. 2010. Formation of fatty acids. pp. 385–393. *In*: Timmis, K.N. (ed.). Handbook of Hydrocarbon and Lipid Microbiology. Springer, Berlin, Heidelberg.

Madhu, A.N. and S.G. Prapulla. 2014. Evaluation and functional characterization of a biosurfactant produced by *Lactobacillus plantarum* CFR 2194. Applied Biochemistry and Biotechnology. Feb. 172(4): 1777–1789.

Makkar, R.S. and S.S. Cameotra. 1998. Production of biosurfactant at mesophilic and thermophilic conditions by a strain of *Bacillus subtilis*. J. Ind. Microbiol. and cxz Biotechnology 20(1): 48–52.

Maneerat, S., T. Nitoda, H. Kanzaki and F. Kawai. 2004. Bile acids is new products of a marine bacterium, *Myroides* sp. strain SM1. Appl. Microbiol. Biotechnol. 67: 683–699.

McDougald, D., S.A. Rice, N. Barraud, P.D. Steinberg, S. Kjelleberg et al. 2012. Should we stay or should we go: mechanisms and ecological consequences for biofilm dispersal. Nat. Rev. Microbiol. 10: 39–50.

McKnight, S.L., B.H. Iglewski and E.C. Pesci. 2000. The Pseudomonas quinolone signal regulates rhl quorum sensing in *Pseudomonas aeruginosa*. J. Bacteriol. 182(10): 2702–2708.

Medina, G., K. Juarez and G. Soberon-Chavez. 2003. The *Pseudomonas aeruginosa* rhlAB operon is not expressed during the logarithmic phase of growth even in the presence of its activator RhlR and the autoinducer Nbutyryl-homoserine lactone. J. Bacteriol. 185: 377–380.

Medina, G., K. Juarez, R. Diaz and G. Soberon-Chavez. 2003a. Transcriptional regulation of *Pseudomonas aeruginosa* rhlR, encoding a quorum-sensing regulatory protein. Microbiology 149(Pt 11): 3073–3081.

Medina, G., K. Juarez, B. Valderrama and G. Soberon-Chavez. 2003c. Mechanism of *Pseudomonas aeruginosa* RhlR transcriptional regulation of the rhlAB promoter. J. Bacteriol. 185(20): 5976–5983.

Mireles, J.R. 2nd, A. Toguchi and R.M. Harshey. 2001. *Salmonella enterica* serovar typhimurium swarming mutants with altered biofilm-forming abilities: surfactin inhibits biofilm formation. J. Bacteriol. 183: 5848–5854.

Morikawa, M., Y. Hirata and T. Imanaka. 2000. A study on the structure-function relationship of lipopeptide biosurfactants. Biochim. Biophys. Acta. 15; 1488(3): 211–8.

Mukherjee, S., P. Das and R. Sen. 2006. Towards commercial production of microbial surfactants. Trends Biotechnol. 24(11): 509–15.

Ochsner, U.A., A. Fiechter and J. Reiser. 1994a. Isolation, characterization, and expression in *Escherichia coli* of the *Pseudomonas aeruginosa* rhlAB genes encoding a rhamnosyltransferase involved in rhamnolipid biosurfactant synthesis. J. Biol. Chem. 269: 19787–19795.

Ochsner, U.A., A.K. Koch, A. Fiechter and J. Reiser. 1994b. Isolation and characterization of a regulatory gene affecting rhamnolipid biosurfactant synthesis in *Pseudomonas aeruginosa*. J. Bacteriol. 176: 2044–2054.

O'Toole, G.A. and R. Kolter. 1998. Flagellar and twitching motility are necessary for *Pseudomonas aeruginosa* biofilm development. Mol. Microbiol. 30: 295–304.

Palmer, J., S. Flint and J. Brooks. 2007. Bacterial cell attachment, the beginning of a biofilm. J. Ind. Microbiol. Biotechnol. 34: 577–588.

Pamp, S.J. and T. Tolker-Nielsen. 2007. Multiple roles of biosurfactants in structural biofilm development by *Pseudomonas aeruginosa*. J. Bacteriol. 189: 2531–2539.

Partovi, M., T.B. Lotfabad, R. Roostaazad, M. Bahmaei, S. Tayyebi et al. 2013. Management of soybean oil refinery wastes through recycling them for producing biosurfactant using *Pseudomonas aeruginosa* MR01. World J. Microbiol. Biotechnol. 29(6): 1039–1047.

Rahman, P.K., G. Pasirayi, V. Auger and Z. Ali. 2009. Development of a simple and low cost microbioreactor for high-throughput bioprocessing. Biotechnol. Lett. 31(2): 209–214.

Rendueles, O. and J.M. Ghigo. 2012. Multi-species biofilms: how to avoid unfriendly neighbors. FEMS Microbiol. Rev. 36: 972–989.

Rikalovic, M., G. Gojgic-Cvijovic, M. Vrvic and I. Karadzic. 2012. Production and characterization of rhamnolipids from *Pseudomonas aeruginosa* san ai. J. Serb. Chem. Soc. 77(1): 27–42.

Rivardo, F., R.J. Turner, G. Allegrone, H. Ceri and M.G. Martinotti et al. 2009. Anti-adhesion activity of two biosurfactants produced by *Bacillus* spp. prevents biofilm formation of human bacterial pathogens. Appl. Microbiol. Biotechnol. 83: 541–553.

Rodrigues, L., H. van der Mei, I.M. Banat, J. Teixeira and R. Oliveira et al. 2006a. Inhibition of microbial adhesion to silicone rubber treated with biosurfactant from *Streptococcus thermophilus* A. FEMS Immunol. Med. Microbiol. 46: 107–112.

Rosenberg, E., A. Ziclerberg, C. Rubinowitz and D.L. Gutnick. 1979. Emulsifier of Arthrobacter RAG-1: isolation and emulsifying properties. Appl. Environ. Microbiol. 37: 402–408.

Sachdev, D.P. and S.S. Cameotra. 2013. Biosurfactants in agriculture. Appl. Microbiol. Biotechnol. Feb. 97(3): 1005–16.

Satpute, S.K., A.G. Banpurkar, P.K. Dhakephalkar, I.M. Banat and B.A. Chopade. 2010. Methods for investigating biosurfactants and bioemulsifiers: a review. Crit. Rev. Biotechnol. 30(2): 127–44.

Schooling, S.R., U.K. Charaf, D.G. Allison and P. Gilbert. 2004. A role for rhamnolipid in biofilm dispersion. Biofilms 1: 91–99.

Schuster, M. and E.P. Greenberg. 2007. Early activation of quorum sensing in *Pseudomonas aeruginosa* reveals the architecture of a complex regulon. BMC Genomics 8: 287.

Shamaa, S. and S. Bahjat. 2019. Detection of Rhamnolipid Production in *Pseudomonas aeruginosa*. J. Phys.: Conf. Ser. 1294 062083.

Singh, A., J.D. Van Hamme and O.P. Ward. 2007. Surfactants in microbiology and biotechnology: Part 2. Application aspects. Biotechnol. Adv. Jan.–Feb. 25(1): 99–121.

Singh, P. and B.N. Tiwary. 2016. Isolation and characterization of glycolipid biosurfactant produced by a *Pseudomonas otitidis* strain isolated from Chirimiri coal mines, India. Bioresour. Bioprocess. 3: 42.

Van Gennip, M., L.D. Christensen, M. Alhede, R. Phipps, P.O. Jensen et al. 2009. Inactivation of the rhlA gene in *Pseudomonas aeruginosa* prevents rhamnolipid production, disabling the protection against polymorphonuclear leukocytes. APMIS 117: 537–546.

Vardor-Suhan, F. and N. Kosaric. 2000. Biosurfactants, 2nd ed. Encyclopedia of Microbiology. Academic Press 1: 618–635.

Vatsa, P., L. Sanchez, C. Clement, F. Baillieul et al. 2010. Rhamnolipid biosurfactants as new players in animal and plant defense against microbes. Int. J. Mol. Sci. 11(12): 5095–5108.

Walencka, E., S. Rozalska, B. Sadowska and B. Rozalska. 2008. The influence of *Lactobacillus acidophilus*-derived surfactants on staphylococcal adhesion and biofilm formation. Folia Microbiol. (Praha) 53: 61–66.

Williams, P. and M. Camara. 2009. Quorum sensing and environmental adaptation in *Pseudomonas aeruginosa*: a tale of regulatory networks and multifunctional signal molecules. Curr. Opin. Microbiol. 12(2): 182–191.

Witek-Krowiak, A., J. Witek, A. Gruszczynska, R.G. Szafran, T. ´ Ko´zlecki et al. 2011. World J. Microbiol. Biotechnol. 27: 1961–1964.

Zhang, Y. and R.M. Miller. 1994. Effect of a *Pseudomonas rhamnolipid* biosurfactant on cell hydrophobicity and biodegradation of octadecane. Appl. Environ. Microbiol. 60: 2101–2106.

Zhao, F., H. Jiang, H. Sun, C. Liu, S. Han et al. 2019. Production of rhamnolipids with different proportions of mono-rhamnolipids using crude glycerol and a comparison of their application potential for oil recovery from oily sludge. RSC Adv. 9(6): 2885–2891.

6

Functional Properties and Potential Application of Biosurfactants as a Natural Ingredient in the Food Industry

Dayang Norulfairuz Abang Zaidel,[1,4,*] *Hassan Ahmadi Gavlighi,*[2] *Nozieana Khairuddin,*[3] *Noorazwani Zainol,*[1] *Zanariah Hashim,*[4] *Nor Azizah Mohammad*[4] and *Nurul Asmak Md Lazim*[4]

1. Introduction

Surfactants are chemical compounds that contain hydrophilic (polar) and hydrophobic (nonpolar) moieties called amphipathic molecules. Typically, surfactants are used to decrease the interfacial and surface tensions between liquid-solid or liquid-liquid molecules. There are two basic types of surfactants, namely chemical surfactants and biosurfactants. Chemical surfactants are derived from petroleum, whereas surface-active compounds produced from microorganisms are called biosurfactants. Biosurfactants have greater efficiency, stability, biocompatibility, biodegradability, and cost-effectiveness, along with lower toxicity, compared to chemical surfactants. Plus, similar to their chemical counterparts, biosurfactants can also be used for various applications. Hence, biosurfactants have drawn great attention among researchers, including the chemical industry. In this chapter, the different sources of biosurfactant production and processing are discussed in detail in Section 2. Properties of biosurfactants are highlighted in Section 3 and the mechanisms of

[1] Institute of Bioproduct Development, UTM Johor Bahru, Malaysia.
[2] Department of Food Science and Technology,Tarbiat Modares University, Iran.
[3] Faculty of Agriculture and Food Science, UPM Bintulu, Malaysia.
[4] School of Chemical and Energy Engineering, UTM Johor Bahru, Malaysia.
* Corresponding author: dnorulfairuz@utm.my

biosurfactant activities in food systems are described in Section 4. In Section 5, each type of biosurfactant is highlighted for potential food applications in various food industries. The challenges and strategies to improve the production of biosurfactant are highlighted in Section 6.

2. Production of Biosurfactants from Various Sources

Many living organisms synthesize amphipathic molecules. These molecules typically comprise hydrophilic and hydrophobic moieties, otherwise known as surface-active components or surfactants (Fig. 1). Surfactants can display surface-active properties, which enable them to reduce micelle or pseudo-micelle formation, the surface tension of the air-to-water boundary, and the interfacial tension of the oil-to-water boundary. Hence, surfactants exhibit high dispersion, emulsification, detergency, and foaming characteristics. Currently, there are two major types of surfactants available in the market, namely chemical surfactants and biosurfactants. On the other hand, surface-active chemicals derived from microorganisms are termed biosurfactants, as discussed in detail in Section 2 (Sekhon et al. 2012).

Biosurfactants are surfactants or amphipathic molecules that contain both hydrophobic and hydrophilic moieties within a single compound. There are two types of fatty acids (hydroxy (α-alkyl or β-hydroxy) or long-chain saturated/unsaturated or linear/branched) in the hydrophobic component of a biosurfactant. Meanwhile, the hydrophilic component could be a mono-/di-/poly-saccharide based carbohydrate, cyclic peptide (cation or onion), phosphate, alcohol, amino or carboxylic acid. Generally, artificial surfactants are categorized based on the type of polar group. However, microbial biosurfactants are categorized based on biochemical structure and microbial source (Sekhon et al. 2012, Singh and Tiwary 2016, Rane et al. 2017, Fontes et al. 2012, Araujo et al. 2019). The microorganisms used to produce biosurfactants include bacteria, yeast, and filamentous fungi. The production of biosurfactants completely uses both water-immiscible hydrocarbons and/or water-soluble compounds. These compounds are used to provide carbon (nutrients) and energy sources to establish microbial growth (Singh and Tiwary 2016, Plaza and Acha 2020, Singh et al. 2018) (Fig. 2).

Biosurfactants are typically selected based on their capacity to ensure high biodegradability, stability, and specificity along with low activity, toxicity, and impact on the environment under severe conditions, based on extensive industrial applications and a variety of structures (Olasanmi and Thring 2018). In 2011, the

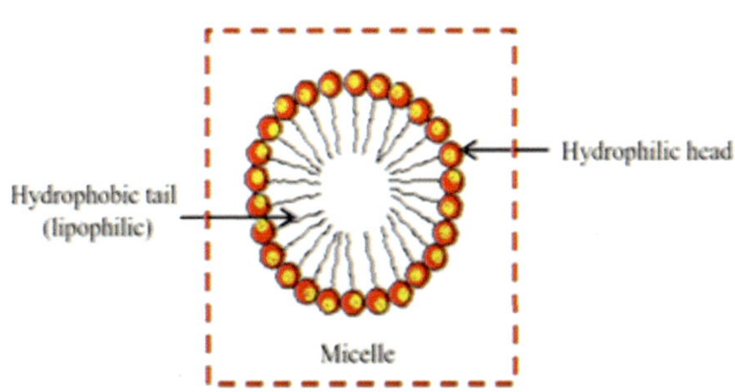

Figure 1. Basic structure of surfactants.

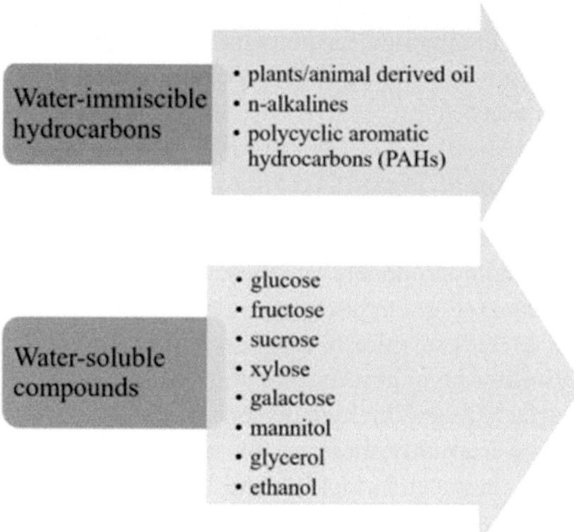

Figure 2. Microorganisms consumption compounds.

global market for biosurfactants was estimated at US $ 1.736 billion, whereas Transparency Market Research reports a market value of US $ 2.211 billion in 2018 corresponding to a 3.5% mean annual growth rate between 2011 and 2018 (Sekhon et al. 2012).

Biosurfactants can be divided into several categories:

i. Ionic-based charges (cationic, non-ionic, anionic, or neutral),
ii. Molecular weight (low or high),
iii. Type of secretion (extracellular, intracellular, or followed by microbial cells).

2.1 *Method of Biosurfactant Production*

Biosurfactants are generally produced from various microorganisms through extracellular, intracellular secretions or via attachment to specific cell components mainly during growth in water-insoluble substrates (Ocampo et al. 2017, Lobato et al. 2013, Singh et al. 2018). The entire process is depicted in Fig. 3.

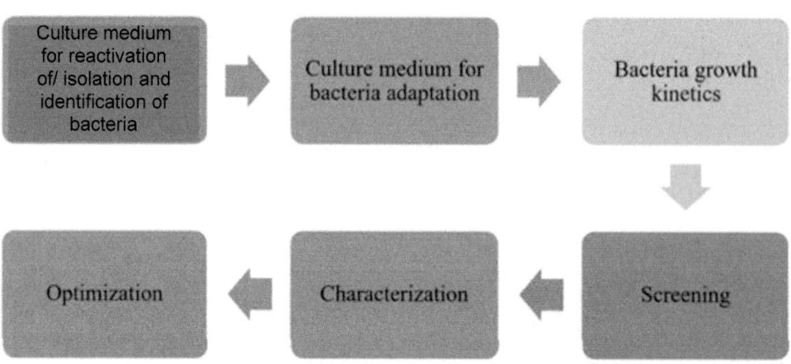

Figure 3. Procedure process for biosurfactants production.

2.1.1 *Culture Medium for Reactivation of Bacteria/Isolation and Identification of Bacteria*

The culture medium for reactivation is generally applicable to researchers who possess the functional bacteria (i.e., selective bacteria) to produce biosurfactants. Deactivated bacteria are reactivated by culturing in a fresh medium to ensure growth. Conversely, the isolation and identification processes are applicable to new, unknown bacteria species discovered from wastes or other sources before the strain is identified using 16s RNA sequencing. The bacteria are identified and selected based on the capability to produce biosurfactants after confirmation of the sequence in GenBank (Nayarisseri et al. 2018).

2.1.2 *Culture Medium for Bacteria Adaptation*

For adaptation, once the bacteria are confirmed (from the literature review) to possess biosurfactant properties, a fresh new-formulation medium is prepared. The new-formulation medium is subsequently prepared based on the purpose of the study. If a medium is combined with any agro-industrial waste, the ratio of the formulation will be different from a normal fresh medium. Numerous low-cost substrates such as edible oils, waste oils, plant-based oils, distillery wastes, and lactic whey are used to enhance the production of biosurfactants. Other substrates used include starch, molasses, animal fats, soap stock, and olive oil mill effluents (Singh and Tiwary 2016). Then, colonies with clear halos are selected and inoculated to multiply replication in a fresh broth.

2.1.3 *Bacteria Growth Kinetics*

The production of biosurfactants typically occurs during the exponential or static stages of bacteria, which is when the nutrient-limiting conditions happen in the growth medium (Fig. 4). At this stage, the development of bacteria is measured from dry cell weight/biomass.

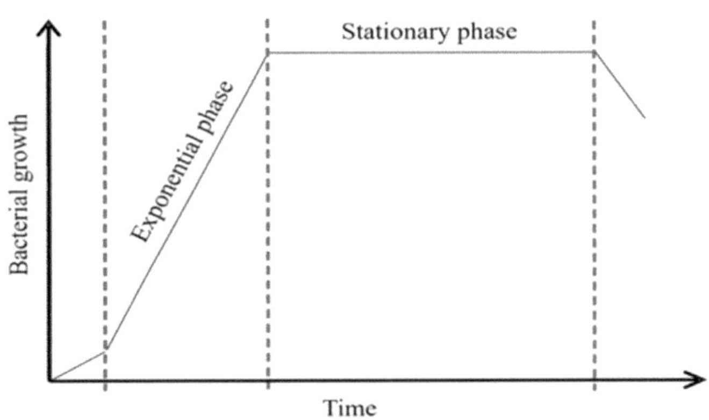

Figure 4. Graph of bacteria growth.

2.1.4 Screening Process

During the screening process, the bacteria are tested for the following parameters; emulsification index (EI_{24}), surface tension (ST), cell-free supernatant (CFS), and reducing sugars based on the DNS technique. The milky appearance of the medium indicates that the bacteria has produced the biosurfactant. The properties of the surfactant can be examined using ST and EI tests.

2.1.5 Characterization Process

The fundamental constituents of the biosurfactant is characteristically detected using thin-layer chromatography (TLC). The major constituents such as lipids, carbohydrates, and amino acids are measured and recorded accordingly.

2.1.6 Cultivation Optimization

The microorganisms that possess a high potential as a biosurfactant are subsequently optimized for commercial cultivation (Sekhon et al. 2012, Muthusamy et al. 2008). The list of microorganisms and their production of biosurfactants is given in Table 1.

Depending on chemical structure, biosurfactants can be divided into six classes: (i) glycolipids, (ii) phospholipids, (iii) lipopeptides and lipoproteins, (iv) fatty acids, (v) neutral acids, and (vi) polymeric- and particulate-based surfactants (Olasanmi and Thring 2018). Some studies categorize the compounds based on low or high molecular weight and chemical structure. Examples of biosurfactants with low molecular weight are lip peptides (gramicidin, polymyxin) and glycolipids (trehalolipids, sophorolipids, rhamnolipids). In contrast, biosurfactants with high molecular weight comprise lipo polysaccharides, polysaccharides, amphiphilic mixtures, proteins, lipoproteins, or other complex polymeric mixtures (Martinez-Toledo et al. 2018).

Numerous works have purposefully formulated novel media comprising low-cost renewable substrates to produce effective microorganisms that can be used as a biosurfactant. For example, olive oil mill effluent (OOME), frying oil, soap stock, molasses, animal fat, milk serum and other starch-rich residues, such as manioc liquid residue, soy seed, and beet, have been found suited for this purpose. Other notable examples include sweet potato, potato sugars, wheat bran, rice and wheat stalks, soy, corn and rice husks, and sugar cane. Notable waste residues have also been examined such as pineapple and carrot processing waste, oil mill wastes (e.g., coconut, soy, peanut and canola), banana waste, coffee processing residues (e.g., coffee pulp), fruit processing residues (e.g., apple and grape pulp) and manioc bagasse (Lobato et al. 2013).

2.2 Extraction of Biosurfactants from Agro-industrial Residue

Agro-industrial organic waste has recently attracted much attention from the biosurfactant community because it is an inexpensive, essential, and highly available substrate that can be used as the raw material for biosurfactant production. Moreover, various studies have reported a higher production of biosurfactants from agro-industrial residues compared to other media that contain glucose as a carbon source. Various waste residues and by-products derived from agro-industrial processes have

Table 1. Different types of microorganisms and biosurfactants production.

Microorganisms	Biosurfactants type
Glycolipids	
Rhodococcus erythropolis, Arthrobacter paraffineu, Mycobacterium phlei, Nocardia erythropolis	Trehalose mycolates
Mycobacterium fortium, Micromonospora sp., *M. smegmatis, M. paraffinicum, Rhodococcus erythropolis*	Trehalose esters
Corynebacterium diptheriae, Mycobacterium smegmati, Arthrobacter sp.	Trehalose mycolates of mono, di, trisaccharides
Pseudomonas sp.	Rhamnolipids
Torulopsis bombicola/apicola, Torulopsis petrophilum, Candida sp.	Sophorolipids
Serratia rubidaea	Rubiwettins R1 and RG1
Lactobacillus fermenti	Diglycosyl digyycerides
Schizonella melanogramma	Schizonellins A and B
Ustilago maydis and Geotrichum candidum	Ustilipids
Bacillus sp.	Amino acid lipids
Pseudomonas flocculosa	Flocculosin
Ustilago zeae, Ustilago maydis	Cellobiolipids
Phospholipid and Fatty acids	
Candida sp., *Corynebacterium* sp., *Micrococcus* sp., *Acinetobacter* sp., *Thiobacillus thiooxidans, Asperigillus* sp., *Pseudomonas* sp., *Mycococcus* sp., *Penicillium* sp.	Phospholipids, Fatty acids
Lipopeptides and lipoproteins	
Bacillus brevis	Gramicidins
Bacillus licheniformis	Peptide lipids
Bacillus polymyxa	Polymyxin E1
Pseudomonas rubescens, Thiobacillus thiooxidans	Ornithine-lipid
Pseudomonas fluorescens	Viscosin
Serratia marcescens	Serrawettin
Glucunobacter cerius	Cerilipin
Agrobacterium tumefaciens	Lysine-lipid
Bacillus subtilis	Surfactin, subtilysin, subsporin
Bacillus licheniformisIM 1307	Lichenysin G
Pseudomonas sp., *Thiobacillus* sp., *Agrobacterium* sp., *Gluconobacter* sp.	Ornithine lipid
Streptomyces canus	Amphomycin
Diheterospora chlamydosporia	Chlamydocin
Tolypocladium inflatum	Cyclosporin A
Streptomyces fungicidicus	Enduracidin A
Streptomyces globocacience	Globomycin

Table 1 contd. ...

...Table 1 contd.

Microorganisms	Biosurfactants type
Bacillus subtilis	Bacillomycin L
Bacillus subtilis	Iturin A
Pseudomonas putida	Putisolvin I and II
Arthrobacter	Arthrofactin
Bacillus thuringiensis CMB26	Fengycin
Bacillus subtilis	Mycobacillin
Polymeric surfactants	
Acinetobacter calcoaceticus RAG-1, Arethrobacter calcoaceticus	Lipoheteropolysaccharide (Emulsan)
Acinetobacter calcoaceticus A_2	Heteropolysaccharide (Biodispersan)
Acinetobacter calcoaceticus strains	Polysaccharide protein
Saccharomyces cerevisiae	Manno-protein
Candida petrophillum, Endomycopsis lipolytica	Carbohydrate-protein
Candida tropicalis	Mannan-lipid complex
Shizonella melanogramma, Ustiloga maydis	Mannose/erythrose lipid
Pseudomonas fluorescences, Debaryomyces polymorphus	Carbohydrate-protein-lipid complex
Candida lipolytica	Liposan
Acinetobacter calcoaceticus	Alasan
Pseudomonas aeruginosa	Protein PA
Particulate biosurfactants	
Acinetobacter sp. *H01-N*	Membrane vesicles
Acinetobacter calcoaceticus	Fimbriae, whole cells

been used for biosurfactant production, as described in the literature. For example, the literature reports utilizing industrial waste such as animal fat, oil processing, fruits and vegetables, starch, sugar, and distilleries (Table 2).

Typically, the physicochemical characterization of agro-industrial waste is performed before further analysis is done. In this step, the nitrogen content is determined using the Kjeldahl method, whereas the protein content is obtained from the Nitrogen factor. The carbohydrate and fat content are deduced via the Gravimetric method according to the APHA standard procedure. Next, the culture is cultivated in a selected medium to evaluate the formation of biosurfactant. A translucent halo is identified as the potential biosurfactant and methylene blue (MB) is used as the indicator. The reduction of MB to white indicates that the cultivator can produce a translucent halo of biosurfactant (Villarreal et al. 2016). However, the extraction of the biosurfactant must be done by centrifuging the culture media for 20 minutes at 4000 rpm to recover the bacteria cell. The acidification technique (pH = 2.0, HCI) is then applied to precipitate the resulting supernatant. The subsequent mixture is then stored overnight at 4ºC. The above step is then repeated for the precipitate (centrifuge speed = 4000 rpm at time = 20 min). A chloroform-to-methanol mixture in the ratio of

Table 2. Various types of waste used for biosurfactant production and cultivation modes (adapted from Angeles Dominguez Rivera et al. 2019).

Substrate	Microorganism	Cultivation mode	References
By-products and vegetable oil processing waste	*T. versicolor*	Solid state fermentation	Lourenco et al. (2018)
Frying oil waste	*P. cepacia* CCT6659	Batch bioreactor	Soares da Silva RCF et al. (2019)
Cashew apple juice	*Y. lipolytica*	Batch shake flask	Fontes et al. (2012)
Potato peel powder	*Klebsiella* sp. RJ-03	Batch shake flask	Jain et al. (2013)
Lignocellulose hydrolysates	*L. pentosus*	Batch shake flask	Portilla-Rivera et al. (2007)
	B. tequilensis ZSB10	Batch shake flask	Cortes-Camargo et al. (2016)

2:1 is then added and a vortex is initiated to mix the solution. The resultant content is centrifuged yet again, based on the previous conditions, after evaporating the solvent by drying in open air at 40°C (Suresh et al. 2012, Iroha et al. 2015).

3. Properties of Biosurfactants

3.1 *Chemical and Surface Properties of Biosurfactants*

The classification of microbial surfactants or biosurfactants, either as anionic or neutral, relies on the chemical composition and species of microbes of the biosurfactant (Kralova and Sjöblom 2017). The hydrophilic part of the biosurfactant is usually made up of carbohydrates, amino acids, phosphates, or cyclic peptides, while long-chain fatty acids or fatty acid derivatives characterize the hydrophobic part of the biosurfactant. Examples of biosurfactants with low-molecular-mass are glycolipids, lipopeptides, and phospholipids (Rosenberg and Ron 1999). On the other hand, polymeric and particulate surfactants are examples of high-molecular-mass biosurfactants. Surfactants with a high molecular mass are efficient at stabilizing emulsions while those with a low-molecular mass are powerful at lowering surface and interfacial tension.

Biosurfactants isolated from *C. utilis* demonstrated a high fatty acid content, 68.63% oleic acid content, and exhibited excellent stability at temperatures below 200°C via thermogravimetric analysis (Ribeiro et al. 2020b). Glycolipid biosurfactants produced from *S. cerevisiae* has have been found to be resistant to high temperatures. They also exhibited no toxic effects (Ribeiro et al. 2020a). Carla et al. (2020) explored the production of biosurfactants from non-cultivable microorganisms through metagenomic methodologies. After functional screening, the researchers found that clone 3C6 exposed the biosurfactant protein properties and the open reading frame (ORF). Since it highly resembled the sequences that encode a hypothetical protein from the *Halobacteriaceae* family, the clone was named metagenomic biosurfactant protein 1 (MBSP1). *Candida utilis* generated a biosurfactant with a carbohydrate-protein-lipid complex. This biosurfactant had 6.39% of sugar content, 7.10% of protein content, and 65.88% of lipid content(Campos et al. 2019). The surface tension was decreased by 26.63 mN/m when the biosurfactant was added. Tripathi et al. (2019) synthesized a rhamnolipid biosurfactant from a new genus, *Marinobacter*, of which *Pseudomonas aeruginosa* is a common source related to pathogenicity. *Marinobacter* sp. MCTG107b is a glycolipid biosurfactant that showed good emulsification properties with no pathogenicity after evaluation via a *Galleria mellonella* infection model. This glycolipid biosurfactant was composed of 95.39% di-rhamnolipid content from a mixture of rhamnolipid, as confirmed by Orcinol assays and a HPLC-MS analysis. A potential cell associated-biosurfactant (CABS) was investigated by Satpute et al. (2019). The CABS was derived from *Lactobacillus acidophilus* NCIM 2903 with prominent bands of 45 kDa molecular weight and characterized as part of the glyco lipoprotein class.

Thin-layer chromatography (TLC) and Fourier Transform Infrared Spectroscopy (FT-IR) analysis revealed that the biosurfactant DCS1 derived from *Bacillus methylotrophicus* was a lipopeptide. This compound showed excellent antioxidant and antimicrobial potential (Jemil et al. 2017). Rufino et al. (2014) produced a lipopeptide biosurfactant from *Candida lipolytica*, with a 50% protein, 20% lipid, and 8% carbohydrate composition. The predominant fatty acid in the biosurfactant was lauric acid (75.34%) while the minor components were palmitoleic acid (4.23%), oleic acid (6.36%), caprylic acid (7.96%), myristic acid (3.85%), and palmitic acid (2.25%).

The biosurfactants reported to have low or no toxicity must be explored in more depth to determine their potential use in food applications. For example, *Candida lipolytica* UCP0988 was assessed to have no toxicity against the fish, *Poecilia vivipara*, hence reinforcing its suitability for the food industry (Santos et al. 2017). Although the biosurfactant contained various chemical compositions and properties, it still offered comparable potential and advantages (such as biocompatibility and digestibility; specificity; surface and interfacial activity; tolerance to temperature, pH, and ionic strength; availability; biodegradability; low toxicity; emulsion stabilizing and destabilizing capability) compared to conventional and synthetic surfactants (Campos et al., 2013). Table 3 lists some biosurfactants from various classes, together with the chemical and functional properties of each, indicating the high potential of these biosurfactants in food applications.

3.2 *Functional Properties of Biosurfactants for Food Application*

3.2.1 *Biosurfactant as Potential Antimicrobial and Antifungal*

Giri et al. (2019) used *Bacillus subtilis* VSG4 and *Bacillus licheniformis* VS16 to produce biosurfactants. They proved that the biosurfactants had effective anti-adhesive, anti-biofilm, and antimicrobial potential. BS-VSG4 proficiently inhibited *E. coli* MTCC 65 growth while BS-VS16 was highly effective at inhibiting *S. aureus* ATCC 29523. Moreover, both biosurfactants had more than 1.5 mg/mL anti-adhesive potential against *Bacillus cereus*, *Salmonella typhimurium*, and *Staphylococcus aureus*. As for anti-biofilm activity, BS-VSG4 exhibited a 63.9%–80.03% eradication percentage while BS-VS16 showed a 61.1%–68.4% eradication

Table 3. Chemical and functional properties of biosurfactants for potential uses in food application (year 2012–2020).

Surfactant class	**Microorganism species**	**Chemical properties**	**Functional properties**
Glycolipids			
Glycolipid	*S. cerevisiae* URM 6670 (Ribeiro et al. 2020a)	High linoleic acid (50.58%), surface (26.64 ± 0.06 mN/m) and interfacial tension (9.12 ± 0.04 mN/m)	Antioxidant and Emulsifier
	C. Utilis (Ribeiro et al. 2020b)	High oleic acid (68.63%)	Antioxidant and Emulsifier
Rhamnolipids	*Serratia rubidaea* SNAU02 (Nalini and Parthasarathi 2018)	Not determined	Antifungal Biocontrol agents against *Fusarium* wilt disease of eggplant
Rhamnolipids	*Marinobacter* sp. MCTG107b (Tripathi et al. 2019)	95.39% di-rhamnolipid	Emulsifier High surface activity, reduce surface tension and stabilize emulsion
Surfactin	Surfactin-based nano emulsion, AUSN-1 (Joe et al. 2012)	72.5 nm droplet size, high thermodynamic stability, 0.03 N/m surface tension, 19.03 N s/m^2 viscosity, 691.10 kg/m^3 density and 12.21 mL/100 mL yield	Antimicrobial and antifungal
Lipopeptides			
Lipopeptides	*Candida lipolytica* UCP 0988 (Rufino et al. 2014)	50% protein, 20% lipid, 8% carbohydrate	Emulsifier: Highly efficient in reducing surface tension, low CMC and no toxicity
Lipopeptides	Biosurfactant DCS1 from *Bacillus methylotrophicus* (Jemil et al. 2017)	Protein units and lipid moieties (TLC analysis)	Antioxidant, antimicrobial, antifungal, anti-adhesive, anti-biofilm
Lipopeptides and phospholipopeptide	VSG4 (lipopeptide) VS16 (phospho lipopeptide) (Giri et al. 2019)	Low surface tension	Antioxidant, antimicrobial, and anti-adhesive or anti-biofilm potential
Carbohydrate-protein-lipid complex	*Candida utilis* (Campos et al. 2019)	65.88% lipids, 7.10 protein, 6.39% sugars	Emulsifier: High emulsification activity and emulsion stabilizing capability

Table 3 contd. ...

...*Table 3 contd.*

Surfactant class	Microorganism species	Chemical properties	Functional properties
Polymeric			
Polymeric protein	MBSP1 (Araújo et al. 2020)	A single gene encoding a protein	Emulsifier: Highly efficient in reducing surface tension and stabilizing emulsions for long periods
Not determined	*Lactobacillus casei*, BS-LB1 and BS-LZ9 (Merghni et al. 2017)	Polymeric	Antioxidant, antiproliferative, anti-adhesive, anti-biofilm
Polymeric	*A. piechaudii* CC-ESB2 (Chen et al. 2015)	Not determined	Antioxidant and Emulsifier
Polymeric	*L. acidophilus* NCIM 2903 (Satpute et al. 2019)	Low surface and interfacial tension, 23.6 mg/mL of CMC	Antimicrobial, anti-adhesive, anti-biofilm, emulsifier
Polymeric	*Pediococcus acidilactici and Lactobacillus plantarum* (Yan et al. 2019)	higher surface activity	Antimicrobial, anti-adhesive and anti-biofilm

percentage. Additionally, *Pediococcus acidilactici* and *Lactobacillus plantarum* were used to isolate a biosurfactant, which proved effective at preventing *Staphylococcus aureus* biofilm-related infections. This is because the biosurfactant had a stronger anti-adhesive activity, so it efficiently reduced the *S. aureus* biofilm growth. This result was further supported by an SEM analysis (Yan et al. 2019). Additionally, the biosurfactant also impeded the synthesis of AI-2, which relies on the dose and the influence of biofilm-related gene expression levels, such as cid A, ica A, dlt B, agr A, Sortase A and sar A. Moreover, CABS exhibited antibacterial and anti-adhesive activity against *S. aureus*, *B. subtilis*, *E. coli*, and *P. vulgaris*. CABS also exhibited anti-biofilm and anti-adhesive potential against *P. vulgaris* and *B. subtilis* (Satpute et al. 2019).

Nalini and Parthasarathi (2018) found a biosurfactant that could substitute a chemical surfactant. It also improved agricultural practices, specifically inhibiting a disease caused by fungus. The authors fermented mahua cake oil to produce a rhamnolipid biosurfactant produced from *Serratia rubidaea* SNAU02. The biosurfactant successfully inhibited the growth of *Fusarium oxysporum* f. sp. *melongenae*. The evaluation proved that 250 mg/mL of *Serratia rubidaea* SNAU02 biosurfactant was environmentally friendly, with high biodegradability and low toxicity. Moreover, it could also fight a disease that attacks eggplant known as *Fusarium* wilt. Another study investigated two strains of *Lactobacillus casei* and isolated biosurfactants BS-LB1 and BS-LZ9 from both strains. The anti-adhesive and anti-biofilm activity of these biosurfactants were then examined (Merghni et al. 2017). Biosurfactant BS-LB1 effectively inhibited the growth of *S. aureus* with an IC_{50} value of 1.92 ± 0.26 mg/mL and BS-LZ9 effectively inhibited the growth of

S. aureus at an IC_{50} value of 2.16 ± 0.12 mg/mL. As for anti-biofilm activity, biosurfactant BS-LB1 exhibited 80.22 ± 1.33% to 86.21 ± 2.94% eradication percentage while BS-LZ9 exhibited 53.38 ± 1.77% to 64.42 ± 2.09% eradication percentage.

Jemil et al. (2017) discovered the highly effective antimicrobial activity of DCS1 lipopeptides. These biosurfactants also had powerful anti-adhesive properties and could potentially impede biofilm formation. DCS1 lipopeptides possessed the highest antifungal activity against *Aspergillus niger* and *Aspergillus flavus* and the highest antibacterial activity against *K. pneumoniae*. The following step in their study involved the incubation of the lipopeptides with proteolytic enzymes for 1 h at various temperatures (40 – 100°C) and pH (3.0 – 10.0). The findings showed that the lipopeptides retained 100% antibacterial activity against *K. pneumoniae* under two conditions: (1) pH 8 and pH 10 and (2) temperatures of up to 100°C for 20 min. Furthermore, the adhesion of *C. albican* was highly inhibited at 89.3%. The DCS1 lipopeptides, at effective doses of 0.36 mg/mL, 2 mg/mL, 0.4 mg/mL, and 0.015 mg/mL, caused 50% adhesion inhibition of *S. typhimurium*, *K. pneumoniae*, *B. cereus*, and *S. aureus*, respectively. Similarly, the lipopeptides showed powerful anti-biofilm activity due to the low concentration of effective dose required (ranging from 0.096 to 2 mg/mL) against all pathogens tested.

Sunflower oil produced the smallest droplet size (72.52 nm) in the formulation of surfactin-based nano emulsion when compared to castor oil, coconut oil, groundnut oil, and sesame oil (Joe et al. 2012). AUSN-1 created from sunflower oil and surfactin-based nano emulsion exhibited high thermodynamic stability and low surface tension, as well as antimicrobial and antifungal effects. AUSN-1expressed antimicrobial effect against food-borne bacterial species, including *Staphylococcus aureus*, *Listeria monocytogenes*, and *Salmonella typhi*, with the most prominent effect shown against *S. typhi*. AUSN-1 also showed high sporicidal activity against *Bacillus cereus* and *Bacillus circulans* and great fungicidal activity, mostly against *Rhizopus nigricans* followed by *Aspergillus niger* and *Penicillium* sp. Following that, AUSN-1, when applied on vegetables, raw chicken, milk, and apple juice, resulted in the degradation of microbials and fungi. Therefore, these findings reinforced the application of AUSN-1 as a food preservative.

3.2.2 *Biosurfactant as a Potential Antioxidant*

A biosurfactant derived from yeast presented acceptable antioxidant potential with no toxicity identified on cell strains, L929 and RAW 264.7, via an MTT assay (Ribeiro et al. 2020, Giri et al. 2019) found that *Bacillus subtilis* VSG4 and *Bacillus licheniformis* VS16 could be used to isolate biosurfactants with considerable antioxidant potential. The study found that, at 5 mg/mL concentration, using DPPH and a hydroxyl radical scavenging assay, biosurfactant VSG4 and biosurfactant VS16 exhibited 69.1%–73.5% and 63.3%–69.8% antioxidant potential, respectively. Biosurfactants BS-LB1 and BS-LZ9 isolated from two strains of *Lactobacillus casei* displayed appreciable antioxidant and antiproliferative activity (Merghni et al. 2017). Biosurfactants BS-LB1 and BS-LZ9 expressed 74.6% and 77.3% inhibition of DPPH radical scavenging activity, respectively, when evaluated at a concentration of 5.0 mg/mL. Meanwhile, biosurfactants BS-LB1 and BS-LZ9 demonstrated IC_{50}

value ranging from 109.1 ± 0.84 mg/mL and 129.7 ± 0.52 mg/mL, respectively, using Methylthiazole tetrazolium (MTT) reduction assay on an epithelial cell line (Hep-2).

DCS1 lipopeptides demonstrated potent antioxidant properties based on five different assays; DPPH radical scavenging, ferric reducing antioxidant power, ferrous ion-chelating activity, β-carotene bleaching, and inhibition of linoleic acid peroxidation (Jemil et al. 2017). The lipopeptides showed 80.6% inhibition at a concentration of 1 mg/mL, reached a maximum of 3.0 ($OD_{700\ nm}$) at a concentration of 2 mg/mL, strongly chelated at 79.8% at a concentration of 4 mg/mL, obtained an IC_{50} value of 42 μg/mL, and reached 76.8% inhibition at a concentration of 0.1 mg/mL, after 9 days. Biosurfactants with potent antioxidant and emulsifying properties were successfully produced from *A. piechaudii* CC-ESB2 via Chinese medicinal herb fermentation (CMHF) (Chen et al. 2015).

3.2.3 Biosurfactant as Potential Emulsifier

Nitsche and Silva (2016) reviewed several biosurfactants as potential food emulsifiers. They found that biosurfactants isolated from *Bacillus subtilis* MTCC441 could emulsify edible oils and help produce better cookies (Suresh et al. 2012). Meanwhile, when compared with glycerol monostearate—a commercial emulsifier—0.1% biosurfactant from *Bacillus subtilis* SPB1 promoted dough-like texture properties, such as increased springiness, adhesiveness, and cohesiveness and reduced hardness (Zouari et al. 2016). Campos et al. (2013) summarized that yeast such as *Candida valida*, *Candida utilis*, *Hansenula anomala*, *Rrhodotorula graminis*, *Rhodospiridium diobovatum*, the red alga *Porphiridium cruentum*, and *Klebsiella* sp. and *Acinetobacter calcoaceticus* bacteria could also produce biosurfactants with efficient emulsifying properties based on Barros et al.'s (2007) study. The resulting biosurfactant was found to have a higher stabilizing activity than carboxymethyl cellulose and gum Arabic. Usually, fungi-derived biosurfactants are efficient at reducing surface and interfacial tension within fluids or colloids and can therefore form a stable emulsion (Silva et al. 2018). However, not all biosurfactants have good emulsifying properties, even if they can reduce surface and interfacial tension, for example, sophorolipids, a low-molecular mass biosurfactant (Cavalero and Cooper 2003). Similarly, Campos et al. (2013) found that polymeric surfactants such as liposan efficiently acted as an emulsifier but did not exhibit the potential to reduce surface tension.

Ribeiro and Sarubbo (2020) also agreed that a surfactant's emulsification efficacy is not commonly associated with the ability to lower the surface tension of water or hydrocarbon because the coalescence in an emulsion could disrupt and weaken the surface energy between the phases, which then creates stereo static and electrostatic barriers.

The biosurfactant from yeast could be used to produce cookies with favorable physical, physicochemical, and textural properties. In effect, this biosurfactant could replace egg yolk (Ribeiro et al. 2020a). A single gene-encoding protein named MBSP1 produced independently from *Escherichia coli* possessed biosurfactant activity that led to hydrocarbon degradation (Araújo et al. 2020). Biosurfactant MBSP1 expressed an emulsification index of more than 50% and proved stable in kerosene emulsion for longer than one year in storage. MBSP1 obtained 6 N/m average interfacial tension, whereas synthetic surfactant (SDS 1%) and water showed about 2.5 N/m and 35 N/m, respectively. Thus, MBSP1 could promote stable emulsions and reduce interfacial tension. *Marinobacter* sp. MCTG107b proved to be the biosurfactant with the most potential to stabilize emulsions compared to MCTG167, MCTG106, MCTG161(2c4), and MCTG4b. MCTG107b reduced the surface tension of the culture medium to 31 N/m, the greatest surface activity of all the other biosurfactants (Tripathi et al. 2019). CABS demonstrated proficient emulsification capacity and stability associated in line with its capability to reduce surface tension from 71 mN/m to 26 mN/m, attaining 23.6 mg/mL of CMC, simultaneously decreasing the interfacial tension of several hydrocarbons, reducing the contact angle and leading to efficient spreading, whilst exhibiting a stable anionic nature at various pH levels and temperatures (Satpute et al. 2019). Giri et al. (2019) reported that biosurfactants from VSG4 and VS16 exhibited emulsification and surface tension stability under various pH levels and temperatures. For various pH levels, BS-VSG4 and VS16 showed 64.1% and 61.1% maximum emulsification indices, respectively, and exhibited 27.2 mN/m and 32.2 mN/m minimum surface tension levels, respectively, at pH 7. BS-VSG4 and VS16 achieved 53.4% to 61.2% emulsification index and 57.8% to 52.9% emulsification indices, respectively, and expressed (29.6–34.2) mN/m and (27.1–31.4) mN/m surface tension ranges, respectively, when evaluated at a temperature of up to 100°C.

Rufino et al. (2014) discovered an efficient biosurfactant derived from *Candida lipolytica* that exhibited lower critical micelle concentration (CMC) values than the other biosurfactants produced from the same genus (Luna et al. 2009, Adamczak and Bednarski 2000). Besides, the biosurfactants also had a low toxicity level and an anionic character and, at a concentration of 0.03%, could greatly decrease the surface tension of water from 70 mN/m to 25 mN/m. *Candida utilis* created a biosurfactant that proved beneficial for food applications. Most of the yeast strains such as *Candida*, *Pseudozyma*, and *Yarrowia* have been identified as GRAS (generally recognized as safe) (Campos et al. 2014). Moreover, Campos et al. (2015) also found that biosurfactants from *Candida utilis* and added Guar gum successfully emulsified sunflower oil to formulate mayonnaise. The product was observed to remain stable for about 30 days at 4°C. The biosurfactants showed no toxicity from an *in vivo* toxicity test involving rats. In a different study, biosurfactant was created from *Candida utilis* through a mineral medium. The biosurfactant was then subjected to waste frying oil and 6% glucose (Campos et al. 2019). In alkaline pH, it showed high emulsification activity and formed a stable emulsion with vegetable oil. However, the biosurfactant could not survive extreme conditions, although it did exhibit a stable surface tension. The result of the study showed that 0.7% of the biosurfactant combined with Guar gum is highly suitable for the formulation of food emulsions, especially for salad dressings.

4. Mechanisms of Biosurfactant Activity in a Food System

Biosurfactants, also known as microbial surfactants, commonly consist of lipids, carbohydrates, or proteins. In general, based on their synthesis with a specific microorganism, a biosurfactant can be categorized as components with low molecular weight or high molecular weight (Rosenberg and Ron 1999). As such, they act differently when introduced to a food system, as both categories have different structures and chemical components. Additionally, both are produced by different microorganisms, which also lead to differences in functional properties. The mechanism of these biosurfactants is very complicated, especially when it comes to food application. So, they are still under development, but there is a high demand for their applications in the food industry. Therefore, only established data obtained from commercially available biosurfactants applied in food studies are discussed. In general, the food mechanism that could benefit from biosurfactants is not limited to beverages, bakery and dairy products, as described further in this section.

4.1 Emulsificatiosn and Instability of Emulsion

Emulsification is a process in which a fat droplet is emulsified in the presence of a biosurfactant to obtain a stable emulsion. During the emulsification process, generally, the hydrophilic part and the hydrophobic part of the biosurfactant molecules will interact with a water medium and an oil medium, respectively, and as a result prevent the medium from separating. It is reported that biosurfactants, namely glycolipids, fatty acids, and phospholipids, which are categorized as low-molecular-weight components, are effective at lowering the tension between two immiscible phases. Meanwhile, biosurfactants categorized as high-molecular-weight components show promising emulsifying properties but are not effective at lowering surface tension (Cooper and Cavalero 2003). Examples of biosurfactants with high molecular weight are polysaccharides, proteins, and a mixture of these polymers. These biosurfactants are important because they bind and/or coat the interface of two immiscible media and allow emulsification to occur (Rosenberg and Ron 1999). Biosurfactants, in particular, play an important role, either in lowering the interfacial tension or contributing to the stability of the emulsion, depending on its structural properties (Campos et al. 2013).

Instability may occur when emulsification is not achieved. Common instability mechanisms include creaming, coalescence, flocculation, and phase separation in emulsions. One of the reasons for instability is the failure of the biosurfactant to interact or emulsify the fat droplets, hence resulting in the aggregation of the droplets, the rupture of the droplets, and/or the merging of bigger droplets before separating into two media.

4.2 Adsorption at the Interface

The presence of hydrophobic components in the biosurfactant in the prepared medium will allow adsorption activity to occur during the emulsification process (Satpute et al. 2010). The adsorption activity, which occurs at the interface, will provide a better understanding of the interaction between the surfactant and the molecules. The adsorption activity could be determined based on the interfacial value, which will also indicate the physical character of the biosurfactant. The interfacial value determines the forces generated between the hydrophobic molecules and a certain structure (Muthusamy et al. 2008). A group of researchers modeled the interfacial tension to envisage the properties of additional surfactant mixtures (Vincent et al. 2011). The mechanism of adsorption and desorption were also used for the purification process to increase the recovery process of the biosurfactant (Dubey et al. 2005).

4.3 *In vitro* Lipid Interaction

The interaction of biosurfactants with lipids produces a multiplex polymer with different forms and curvatures due to the chemical components in the hydrocarbon chain and the hydrophilic chain. As such, the newly formed layers would be able to penetrate the lipid bilayer more and modulate the protein structure (Daniel 2017). Due to the different hydrophobicity and hydrophilicity of the glycolipid, a biosurfactant could be classified into three types, based on the structural properties: (a) Rhamnolipid and sophorolipid, in which its sugar groups, rhamnose and sophorose, respectively, are attached to an alkyl carbon tail; (b) Surfactin, which possesses cyclic lipopeptide components at its alkyl carbon tail; (c) Saponins, which possess a triterpenoid or steroid component with sugar components attached to it via a glycoside chain (Ron and Rosenberg 2001). In particular, surfactin, when interacting with phospholipid groups such as phosphatidylcholin (POPC) and dipalmitoylphosphatidylcholin (DPPC), tends to form a bi-layer at the headgroup that allows it to flow more (Carillo et al. 2003) and then tends to form a cylindrical shape in an inverted cone without leaving an open pore in the membrane. This then allows it to have a permeability effect (Epand and Vogel. 1999). This-similar shaped biosurfactant was also reported in a study done on microbial peptides using interfacial modeling (William 2011). On the other hand, saponin showed a different mechanism when interacting with the DPPC bi-layer compared to the synthetic surfactant; in this case, the saponin increased the flexibility of the formed layers without affecting the DPPC multiple monolayers while the synthetic surfactant decreased the flexibility and substituted the DPPC in the bi-layer (Prades et al. 2014).

4.4 Modifying Starch and Interaction with Protein

Starch modification is a complex process that involves changing gelatinization, visco-elasticity properties, and stability; therefore, modified starches can be applied in many food products. Some have been used in pasta and confectionery products, with the addition of the biosurfactant to enhance consistency, to decrease the staling of pastries (Nitschke and Costa 2007, Campos et al. 2014), and to improve the rheological and textural properties of dough (Marchant and Banat 2012, Kralova and Sjoblom 2009). The interaction between the ionic molecules in a biosurfactant with protein components will modify the rheological properties of the entire structure, thereby forming modifiable components. This feature has been widely applied in confectionery products, as it increases the elasticity of the protein and hence increases the mass production of confectionery products (Shepherd et al. 1995, Zajic et al. 1976).

Enzymes in the presence of biosurfactants are reported to exhibit better binding activity compared to synthetic surfactants. The enzymes help increase the hydrophobicity of the biosurfactant, in turn, increasing the enzymatic properties (Ishii et al. 2012), and further increasing its ability to create a binding site for specific purposes (Otzen 2017) and enhancing its potential to be used as a stain remover (Liu et al. 2012).

4.5 *Dispersion and Solubility*

The addition of biosurfactant may increase the dispersibility and solubility of non-soluble material in a medium although both have different applications. A dispersant is a material used to prevent insoluble particles from accumulating. The dispersion mechanism can be applied in the adsorption process while other applications will alter the dispersion properties of one medium and then mixing it with another medium to obtain a clear solution (Nitschke and Costa 2007). One of the main issues related to the application of cocoa powder in beverages is that cocoa powder is less soluble in water. Hence, it is suggested that the hydrophilic part of a biosurfactant be reacted with cocoa powder, so that the former coats the lecithin inside the latter to enhance its overall wettability and dispersibility (Campos et al. 2013). Solubility is a process whereby an insoluble material dissolves in a water medium. In some cases, the insoluble material will be entrapped in encapsulated micellar forms if the mass of the surfactant is too high (Satpute et al. 2010).

5. Application of Biosurfactants in the Food Industry

5.1 *Biosurfactants as Food Ingredients*

Being less-toxic, biodegradable, and having a selective specific interface in comparison to a synthetic surfactant, biosurfactants show promising applications as food ingredients. In general, a food ingredient is defined as a substance added to food without any nutritional value, particularly to increase and/or maintain the shelf life of a product. Additionally, a biosurfactant helps stabilize the food matrix and thus enhances the quality of the food product. Food additives fall under food ingredients and are important for food product preparation. The functional properties of food ingredients are mainly dependent on chemical structure—whether cationic, anionic, neutral, or amphoteric (Satpute 2010). These structures, in particular, are surface-active components that have the ability to minimize the tension between two immiscible layers and increases the stability of an emulsion. In emulsion-based foods, the surface-active properties of a product are highly dependent on rheological properties, which can range from low solidity to high solidity. Biosurfactants with high emulsifying properties and are categorized as 'generally regarded as safe' (GRAS), in particular, those produced by yeast and bacteria, with the most common being *Bacillus* and *Psuedomonas*, are the most common mass-produced biosurfactants (Silva et al. 2014, Santos et al. 2016). *Saccharomyces cerevisiae* and *Candida Utilis* are normally used to produce biosurfactants with good emulsifying properties and broad temperature and pH stability, so they can be widely used as an emulsification medium to develop emulsion-based foods (Barros et al. 2007, Banat et al. 2000). Besides being used as emulsifiers, biosurfactants are also used as thickening and gelling agents, and stabilizers with anti-microbial, anti-fungal properties, and anti-oxidant properties (Campos et al. 2014, Yalcin and Cavusoglu 2010).

The current synthetic stabilizing agents used in food products are gum Arabic and carboxymethyl cellulose. However, it has been reported that biosurfactants produced by the *Candida* species can perform better as stabilizers than both the synthetic stabilizers above (Baros et al. 2007). Rhamnolipids produced by *Pseudomonas species* have been reported to impact the texture of dough and enhance the shelf life of bakery products (Muthusamy et al. 2008). Surfactin, one of the lipopeptides produced by *Bacilus subtilis*, has reportedly has shown antimicrobial properties against microorganisms. Other biosurfactants, which also possess antimicrobial properties, are fengycin, iturin, bacillomycins, and mycosubtilins (Das et al. 2008). The most commonly reported antimicrobial activity is related to the surface coating of utensils used in the preparation of food products such as the use of a biosurfactant as a coating agent on a heat exchanger used for milk processing (Bouwman et al. 1982). Recently, a biosurfactant isolated from the *Bacillus* strain showed promising results as an antioxidant agent and antibacterial agent, showing a scavenging activity higher than 50% on average. The biosurfactant also exhibited positive anti-bacterial activity against both Gram-positive and Gram-negative bacteria (Giri et al. 2019). However, it is challenging to produce commercial biosurfactants, but the demand to apply biosurfactants as food ingredients in the food industry is growing.

5.2 *Biosurfactants as Food Preservatives*

Biosurfactants can act as efficient food preservative agents, owing to their antibacterial, antifungal, and antiviral activities. These activities are dependent on the chemical structure of the biosurfactant, which is composed of a hydrophilic part and a hydrophobic part. Similar to chemical-based surfactants such as SDS (sodium dodecyl sulfate), a biosurfactant can bind to both the hydrophobic lipid layers of a cell membrane and hydrophilic proteins, enabling it to lyse and thus break cells. Biosurfactants from the glycolipid group, particularly the rhamnolipids and the lipopeptides, have notable antimicrobial and antifungal activities. For example, Fernandes et al. (2007) demonstrated that lipopeptides produced from *Bacillus subtilis* R14 had a broad spectrum of actions against 29 bacteria, including those with multidrug-resistant profiles. As for antifungal activity, rhamnolipids have been reported to be effective against *Enicillium crysogenum*, *Rhizoctonia solani*, the phytopathogenic *Botrytis cinereal*, *Aureobasidium pullulans*, *Aspergillus niger*, and *Chaetonium globosum* and (Abalos et al. 2001). Rodrigues et al. (2006a) also reviewed various biosurfactants including trehalose lipids from *Rhodococcus erythropolis*, glycolipids from *Streptococcus thermophilus*, rhamnolipids from *Pseudomonas aeruginosa*, and surfactin and iturin from *B. subtilis*, and noted promising antibacterial, antifungal, and antiviral activities.

Additionally, due to their low or non-toxic properties, biosurfactants could be favorably applied as preservatives in food. For example, Bezerra de Souza Sobrinho et al. (2013) found no toxicity exhibited by a glycolipid biosurfactant produced by *C. sphaerica* against the seeds of *Brassica oleracea*, *Chicoria intybus*, *Lactuca sativa* and *Solanum gilo*. The same glycolipid biosurfactant employed as a bioindicator showed no toxicity against the microcrustacean *Artemia salina*. The biosurfactant is also biodegradable, making it superior to chemical-based surfactants. For instance, a previous study showed that rhamnolipids are superior to Triton X-100 (Mohan et al. 2006).

Biosurfactants are effective biofilm control agents, so they can be used as food preservatives. The food industry faces challenges of food spoilage when bacteria or other pathogenic microbes attach onto food surfaces, hence leading to food deterioration and the transfer of diseases. In this case, biosurfactants can be used to prevent the microbes from adhering to or forming biofilms on food surfaces by creating a barrier between the bacteria and the food surface. Once applied to a food surface, the biosurfactant could also prevent food from being contaminated. One study discovered that polystyrene conditioned with surfactin and rhamnolipids managed to lower the hydrophobicity of the surface (do Valle Gomes and Nitschke 2012). Although the anti-adhesive property of biosurfactants varies across different types of pathogens, in general, studies have proven pathogenic food bacteria could be prevented from adhering to surfaces that have been pre-conditioned with biosurfactants, even in a nutrient-rich environment. Similarly, Dusane et al. (2011) reported that glycolipids derived from *Serratia marcescen* managed to disrupt the preformation of biofilm by *C. albicans* BH, *P. aeruginosa* PAO1 and *B. pumilus* TiO1 cultures. Meanwhile, Haesendonck and Vanzeveren (2004) successfully improved bread preservation during storage by reducing its susceptibility to microbial contamination when added with a preservation agent, specifically a rhamnolipid derived from a *Pseudomonas* sp., which the researchers subsequently patented. Moreover, De Araujo et al. (2011) found that the *Listeria monocytogenes* food pathogen was prevented from adhering to polystyrene surfaces with the application of rhamnolipids, which showed a higher adhesion reduction rate compared to SDS.

5.3 *Biosurfactant in Meat Industry*

5.3.1 *Biosurfactant as an Antioxidant in the Meat Industry*

The lipid oxidation of meat products, such as beef patties during storage, is a concern affecting the meat industry. Oxidation can create radicals that affect human safety. It also causes an off-flavor and negative consumer perceptions. On the other hand, applying a synthetic compound in food formulation is also a concern to the consumer. So now meat producers are looking for a natural replacement. Recently, a natural, novel biosurfactant was produced from *Enterobacter cloacae* C3 strain with a lipopetidic structure. First, DPPH radical-scavenging assay, β-carotene bleaching assay, metal chalation, ferric-reducing activity, and linoleic acid peroxidation were used to evaluate the antioxidant activity of the biosurfactant *in vitro*. Also, to validate the real antioxidant of the biosurfactant activity in food, a beef patty was used and a TBARS test was run. The highest radical scavenging activity was obtained at 1 mg/ml of lipopeptide. The higher content of hydrophobic amino acids (glycine, alanine, valine, leucine, and isoleucine) and charged amino acids (Aspartic and glutamic acid) in the structure of the lipopeptide were the main reasons leading to its good antioxidant activity. Off the antioxidant tests, the reducing ferric power of the lipopeptide was similar (2 mg/ml) to that of BHA. This higher reducing power could be due to the hydroxyl group in the structure of the lipopeptide. In general, the lipopeptide's *in vitro* antioxidant activity is comparable to that of chemical antioxidants. Next, the efficiency of the lipopeptide at 0.5% concentration was evaluated against BHA to evaluate the secondary products of oxidation. The oxidative stability of the lipopeptide and BHA was found similar and effective for controlling oxidation during 14 days of storage. The TBARS value of the natural lipopeptide antioxidant and the control (without any antioxidant) were 0.41 MDA/kg of meat and 3.5 mg MDA/kg of meat, respectively. The antioxidant activity of the lipopeptides can be explained by the metal chelation, electron, and hydrogen donation of the lipopeptide (Hmidet et al. 2020).

The higher content of polyunsaturated fatty acids of poultry meat makes it susceptible to lipid oxidation. There are many strategies to control existing oxidation, with one being the use of natural ingredients extracted from plants such as essential oils. This compound is used because hydrophobicity or solubility and stability is a limiting factor that can be overcome using encapsulation techniques such as nano emulsions. Doost et al. (2019) evaluated thymol nano emulsions prepared by Quillaja Saponin (QS) as a biosurfactant and its oxidative stability in fresh chicken breast (FCB). The FCB was dipped in 4% thymol oil-in-water nano emulsion (QS-to-oil ratio of 0.25) and stored at 4°C for 14 days. The MDA value of the FCB treated with nano emulsion increased gradually compared to the control, albeit recording a 70% lower MDA value than the control. The solubility and the bio accessibility of thymol as a natural antioxidant together with the biosurfactant as an emulsifier improved the FCB lipid oxidation.

5.3.2 *Biosurfactant as an Antibacterial in the Meat Industry*

The antimicrobial effect of rhamnolipid against Gram-positive (*L. monocytogenes, B. cereus and S. aureus*) bacteria was investigated under different pH conditions. Rhamnolipids exhibited higher efficiency in acidic conditions for all Gram-negative bacteria. The MBC of *Bacillus cereus was* 39.1 µg/mL whereas *L. monocytogenes* exhibited 312.5 µg/mL at pH 7. The decreasing cell surface hydrophobicity and broken cytoplasmic membrane were the primary mechanisms that led to the bacterial sensitivity to RL in acidic conditions (de Freitas Ferreira et al. 2019). The main bacteria microflora of fresh chicken meat stored in a refrigerator were *Pseudomonas* strains that caused spoilage and off-flavor via lipase or protease activity. The skin of poultry meat has a high fat content and may produce a biofilm that can feed on the accessible and cause the growth of microorganisms. Mellor et al. (2011) isolated the biosurfactant from *Pseudomonas fluorescens* grown on chicken thighs and investigated the effect of biosurfactant addition on the spoilage rate of refrigerated chicken meat. The total aerobic counts of the sample treated with biosurfactant was higher than 10^6 cfu/g after 24 h, which is unacceptable for consumption. This higher content could be due to the solubilizing of fat and accessibility to the bacterial strains. So, a lower content of psychrotrophs with a lower production of biosurfactant could reduce the spoilage of chicken meat.

5.3.3 *Biosurfactant as an Antiadhesive in the Meat Industry*

Fresh meat composition consists of all nutrients ideal for the growth of microbes that can attach to surfaces and subsequently lead to spoilage and consumer illness. The antiadhesive properties of biosurfactant can inhibit the growth of pathogen microorganisms. One study evaluated the antiadhesive property of biosurfactant derived from *Lactobacillus paracasei* subsp. tolerans N_2. The study tested 0.01 mg/mL to 10 mg/mL of biosurfactant concentration against different pathogen bacteria isolated from fresh beef. The antiadhesive activity against all bacteria strains decreased as the concentration of biosurfactant increased. The researchers observed that a 10 mg/mL concentration of biosurfactant led to the complete inhibition of the adhesiveness of *Bacillus* sp. BC1, *S. aureus* STP1, and *S. xylosus* STP2. This reduced adhesiveness could be caused by a change in the hydrophobicity between the surface and the strains due to the addition of the biosurfactant (Mouafo et al. 2020).

5.4 *Biosurfactant in the Dairy Industry*

5.4.1 *Biosurfactant as a Stabilizer in the Dairy Industry*

Rhamnolipids are groups of glycolipids produced from different Pseudomonas strains with one or two rhamnose linked to one or two beta-hydroxy 3-hydroxy fatty acids. In the food industry, rhamnolipids can be used as an emulsion stabilizer, to prevent bread from becoming stale, to enhance the creamy properties of butter, to control the texture and consistency of ice cream, to solubilize flavor, and as a precursor to the flavor of Furaneol (a product of the hydrolysis of rhamnolipids) (Irfan-Maqsood and Seddiq-Shams 2014).

A self-generated fermentation product of a by-product of the corn milling industry, known as corn steep liquor (CSL), was extracted by organic solvents such as chloroform and ethyl acetate, and exhibited a surface reduction activity of more than 30 mN/m. The structural composition of CSL consists of fatty acids (C_{16} and C_{18}) with flavonoids (vanillic acid, p-coumaric acid, ferulic acid, sinapic acid, and quercetin). This composition makes it a biosurfactant with good antioxidant activity. The effect of CSL as a natural biosurfactant extracted using ethyl acetate to stimulate the growth of probiotic strain (*Lactobacillus casei*) in drinkable yoghurt was evaluated. The incorporation of different concentrations of biosurfactant (0–0.5 g/L) into drinkable yoghurt increased the growth rate of the *L. casei* probiotics (LópezPrieto et al. 2019).

5.4.2 *Biosurfactant as an Antiadhesive in the Dairy Industry*

The biofilm is a colony of microorganisms that grows around a cell and attaches to the cell surface to protect itself from external environmental conditions. *Staphylococcus aureus* is a biofilm-forming pathogenic microorganism that uses nutrients in the environment to grow on surfaces or equipment in the food industry such as in the dairy and meat industry. Therefore, it is a great concern because it leads to industrial problems and the contamination of food during processing (Galie et al. 2018). Silva et al. (2017) investigated the effect of a rhamnolipid biosurfactant in different concentrations (0.5%–10%) and temperatures on the biofilm production of *S. aureus*. Even at a low concentration (0.5%), at 4°C,the RL was efficient at removing more than 86% of biofilm formed. Carbohydrates, with higher hydrophilic properties, form the main structure of the biofilm in the dairy industry. The RL can interact with the carbohydrates and solubilize them, therefore reducing the surface tension between the biofilm and easily removing the biofilm from the surface of the equipment.

5.5 *Biosurfactant in the Cereal Industry*

5.5.1 *Biosurfactant as an Antioxidant in the Cereal Industry*

A marine sponge associated with actinomycetes *Nesterenkonia* sp. MSA31 with a lipopeptide structure was produced and fortified into a muffin formulation as a fat replacer (0.5%, 0.75%, and 1%). The textural properties and antioxidant activity were evaluated using 2,2-diphenyl-1-picryl hydrazyl (DPPH) radical scavenging assay at (0.5–6) mg/mL. Hardness, chewiness, and gumminess decreased but springiness (freshness) and cohesiveness increased when 0.75% of lipopeptide MSA31 was added. The antioxidant activity of lipopeptide was dose-dependent and reached approximately 90% radical scavenging activity, higher than BHT as the control. This antioxidant activity may be related to the unsaturated fatty acids in the structure of this lipopeptide (Kiran et al. 2017).

5.5.2 *Biosurfactant as an Emulsifier in the Cereal Industry*

Biosurfactants can also be applied in the cereal industry to replace traditional emulsifiers. Emulsifiers are used in food for various reasons; from controlling texture, extending shelf-life, improving consistency and agglomeration, to modifying rheological properties. Lecithin and gum Arabic are examples of food emulsifiers derived from plants. These emulsifiers, although widely used in the industry, face functional limitations. In particular, they cannot withstand microwave processing and irradiation (Campos et al. 2013). Meanwhile, GM-soy-derived lecithin has also caused much debate in recent decades. According to Campos et al. (2013), extracellular bio emulsifiers derived from yeast (e.g., the red alga *Porphiridium cruentum*, *Rhodospiridium diobovatum*, bacteria belonging to the *Klebsiella* sp. and *Acinetobacter calcoaceticus*, *Hansenula anomala*, *Rrhodotorula graminis*, *Candida valida*, and *Candida utilis*) have shown better stabilizing activity than gum Arabic and carboxymethyl cellulose. Therefore, existing additives could potentially be replaced with microbial emulsifiers and biosurfactants while offering superior performance than plant-based additives.

In a recent report, Punrat et al. (2019) demonstrated the successful modification of rice flour properties using a sophorolipid biosurfactant derived from *Wickerhamomyces anomalus* MUE24. The bio emulsifier improved retrogradation and increased the water-holding capacity and swelling power of rice flour. Flour modification is important to make it more effective for the food industry. For example, retrogradation is a useful property in flour noodle-making and can reduce production time. In another study, the addition of 0.1% *B. subtilis* SPB1-derived biosurfactant improved the profile of cookie dough and resulted in a better-quality cookie, compared to when a commercial emulsifier known as glycerol monostearate was used (Zouari et al. 2016).

Despite the lack of reports related to the actual use of biosurfactant in the cereal industry or related products, the successful application of bio emulsifiers produced from microbes including *Candida utilis* and *Saccharomyces cerevisiae* have been reported in food such as salad dressings and bakery products. Further use of biosurfactants in cereal products requires more specialized and directed investigations involving the various stages of cereal processing—from the primary and secondary processes to the end-product.

5.6 *Biosurfactant in Food Emulsion*

5.6.1 *Biosurfactant as an Antioxidant in Food Emulsion*

There are seven amino acids linked to 13–14 fatty acids in biosurfactant surfactin, which is mainly produced from *Bacillus subtilis*, and is a lipopeptide. This chemical structure lends surface activity and natural antibiotic properties to the biosurfactant. The hydrophobic part of surfactin is made up of amino acids and fatty acids while the hydrophilic part is made up of two amino acids (Glutamic and aspartic acid) (Kadaikunnan et al. 2015). Surfactin has many applications in the food industry, amongst others, as an antibacterial, an antifungal, an emulsifier, or as a natural antioxidant. One study evaluated the effect of peptidic surfactin on the oxidative stability of 0.2 mmol/L concentration microemulsion formulation of docosahexaenoic acid single cell oil (DHASCO). Peroxide and TBARS values were used to indicate oxidative stability. The results showed that surfactin efficiently controlled the O/W microemulsion of DHASCO, based on the lowest peroxide value obtained. The thicker interfacial layer of surfactin around DHASCO created a protective area for oxygen and light that improved oxidative stability. In addition, the TBARS values showed that the microemulsion extended the oxidative stability of DHASCO for 90 days compared to the control. So, natural surfactin in a small amount could be used as an oxidative stabilizer in an emulsion system (He et al. 2017).

5.6.2 *Biosurfactant as a Stabilizer in Food Emulsions*

Sugar esters are nonionic surfactants (biosurfactants) composed of sugars in their hydrophilic part and fatty acids in their hydrophobic part. This surfactant has good surface tension and is therefore a good emulsion stabilizer. One study tested the emulsion stability of coconut milk with the addition of fructose, sucrose, and lactose esters produced by lipase from *Candida antarctica* type B (Neta et al. 2012). All sugar esters exhibited surface tension activity but lactose esters had the highest surface tension (38.0 N/m) and emulsifying power with a lower ratio of sugar esters to coconut milk (1:1) after 48 hours. The addition of biosurfactant could reduce the droplet diameters of coconut milk compared to the control (without biosurfactant) and consequently the biosurfactant replace the protein on the surface of the emulsion to protein-induced flocculation (Tangsuphoom and Coupland 2009).

6. Challenges of Biosurfactant Application in the Food Industry

6.1 *Safety Issues of Biosurfactants in the Food Industry*

Whether or not introducing microbe-derived products in food is safe is the main concern regarding food application. Therefore, more research is being directed towards biosurfactants with low toxicity and good surface activity. To overcome the issue of safety, the structure of the biosurfactant is modified or non-pathogenic producing strains such as yeasts and probiotics are used to produce biosurfactants with reduced toxicity (Kralova and Sjöblom 2009). To this end, important toxicity tests are run to ensure that the biosurfactants are safe to consume. These tests include those that test for acute toxicity, allergy, reproductive toxicity, and mutagenicity (Kralova and Sjöblom 2009). Hwang et al. (2008) studied the mutagenic and toxicity of surfactin and showed that it could be used at 500 mg/kg body weight per day without any concerns for humans.

The minimum concentration at which a biosurfactant will provide the maximum reduction of surface water tension to initiate micelle formation is called the critical micelle concentration (CMC). The lesser amount of surfactant required to reduce surface tension, the more efficient the surfactant, and thus the lower its CMC value. Even at low concentrations, biosurfactants could provide higher emulsification activity than the equivalent synthetic surfactants. In fact, the CMC of biosurfactants is generally 10–40 times lower than the latter. The biosurfactants' low CMC makes them ideal for use in different industrial and biotechnological applications (Desai and Banat 1997).

6.2 Sustainability and Environmental Impact of Biosurfactants

At the end of life, the disposal of surfactants is normally via wastewater. This wastewater eventually ends up in water bodies; therefore, threatening the ecosystem (Drakontis and Amin 2020). The concentration and hydrophobicity of the surfactants are the main indicators of their toxicity. For instance, a more hydrophobic surfactant is more toxic. As for concentration, if a surfactant is used above a certain level, its toxicity levels could be absorbed by animals, especially fish via its gills and bodies. This toxicity then circulates in the fish blood. When humans consume the fish, the toxicity could transfer and then damage the human body because of enzymatic activity. Given the toxicity of current surfactants, increasing research is being done to look for alternatives, namely biosurfactants. Biosurfactants are considered more biocompatible and biodegradable than chemical surfactants. Biosurfactants have been tested in the past and found to reach 80% biodegradability under specific circumstances. In this case, biodegradability was measured as a percentage and calculated based on the increase in surface tension in the samples of crude extracts (Drakontis and Amin 2020).

6.3 Challenges in Scaling-up the Production of Biosurfactants

Many research works and companies have looked into the large-scale production of biosurfactants. The cost of raw materials and the purification process of the biosurfactants are the main inhibitors to cheaply produced biosurfactants at a large scale. This high cost has been a hindrance to the biosurfactants' replacement of chemical surfactants, preventing them from being commercially used in many industries. The cost of producing biosurfactants is usually 10 to 12 times higher than that of chemical surfactants, so they usually are more expensive (Helmy et al. 2011). According to Helmy (2011) and Singh and Saini (2013), the high cost of production of biosurfactant stems from inefficient downstream processing, the low yield of biosurfactants from microorganisms, and the high cost of raw materials are the main causes for applications of biosurfactants in high-end products, high-end cosmetics, and medicines.

A bioreactor is needed to produce a high yield of biosurfactants at a large scale over a sustained period, but designing one is a significant challenge. There are certain operating problems inherent in the production of biosurfactants at the bioreactor level. For example, a basic problem with the bioreactor is the excessive formation of foam when the levels of aeration and agitation are raised. However, if the aeration rate is reduced, microbial growth will be significantly impacted and in turn impact the production of the biosurfactant. Moreover, it is not cost-effective to maintain the pH and temperature of a large-scale bioreactor. Therefore, an efficient bioreactor that can produce biosurfactant at a low cost while still being efficient is needed. However, not many studies have explored the scaling up of biosurfactant production, for example, from a shake flask to a bio reactor (Singh and Saini 2014).

6.4 Strategies for the Improved Production of Biosurfactants

Recent years have seen a considerable interest in natural surfactants. Besides the demand for global synthetic surfactants and growing industrial demand, the market opportunities for biosurfactants is also huge. Nevertheless, despite the interest in biosurfactants, their high production cost, low yield, and less-sophisticated product recovery are significant hindrances, preventing them from competing with chemically-synthesized compounds in the market (Helmy 2011). Not much attention has been placed on the economics of biosurfactant production despite its many advantages. Hence, to ensure that biosurfactants can economically compete with chemical surfactants, the economic strategies to produce biosurfactants must be prioritized (Rufino et al. 2014). The next section highlights the three strategies that can be implemented to improve the production of biosurfactants.

6.4.1 Utilization of Low-cost Raw Materials

Past works have found that almost 50% of the total cost of producing biosurfactant stems form the cost of raw materials (Singh and Saini 2013). To reduce the raw material cost, cheap and agro-based raw materials could be used as substrates for biosurfactant production. Previous research has also reported the possibility of using a variety of cheap raw materials for biosurfactant production, including lactic whey and distillery wastes, oil wastes, plant-derived oils, and starchy substances (Table 4).

Some works have found that vegetable oils and plant-derived oils, for example, rapeseed oil (Trummler et al. 2003), Turkish corn oil (Pekin et al. 2005), and soybean oil (Kim et al. 2006) are potential cheap and effective raw materials for producing biosurfactants such as rhamnolipids, sophorolipids and mannosylerythritol. On the other hand, castor oil, jatropha oil, and jojoba oil are some examples of plant-derived oils that are available at much cheaper rates, but all have unfavorable color, composition and odor, making them unsuitable for human consumption (Mukherjee et al. 2006). Dairy wastewater and oil waste from vegetable oil refineries (Table 4) are also good, cheap alternatives substrates that can be used to produce biosurfactants. *Bacillus subtilis* has been used with a substrate derived from cassava wastewater, which is generated in large amounts during the preparation of cassava flour, to produce surfactin (Nitschke and Pastore 2003, Nitschke and Pastore 2004, Nitschke and Pastore 2006). Waste from sugar industry such as sugar beet molasses has been

Table 4. List of inexpensive raw materials for the production of biosurfactants by various microbial strains (adapted from Mukherjee et al. 2006, Singh and Saini 2014).

Low cost or waste raw material	Biosurfactant type	Producer microbial strain	Max yield (g/l)	References
Rapeseed oil	Rhamnolipids	*Pseudomonas* species DSM 2874	45	Trummler et al. (2003)
Turkish corn oil	Sophorolipids	*Candida bombicola* ATCC 22214	400	Pekin et al. (2005)
Soybean oil	Mannosylerythritol lipid	*Candida* sp. SY16	95	Kim et al. (2006)
Waste frying oils	Rhamnolipids	*Pseudomonas aeruginosa* 47T2 NCIB 40044	2.7	Haba et al. (2000)
Oil refinery waste	Glycolipids	*Candida Antarctica* and/or *Candida apicola*	10.5/13.4	Bednarski et al. (2004)
Soybean oil refinery wastes	Rhamnolipids	*Pseudomonas aeruginosa* AT10	9.5	Abalos et al. (2001)
Cassava flour wastewater	Lipopeptides	*Bacillus subtilis* ATCC 21332 and *Bacillus subtilis* LB5a	2.2–3.0	Nitschke and Pastore (2003, 2004, 2006)
Sugar beet molasses	Rhamnolipids	*Pseudomonas* spp.	-	Onbasli and Aslim (2009)

used as a substrate to produce rhamnolipid biosurfactants (Onbasli and Aslim 2009). Other wastes from the sugar industry, such as soy molasses (Solaiman et al. 2004) and sugarcane molasses (Rodrigues et al. 2006b), have also been used as a carbon source in biosurfactant production.

6.4.2 *Improvement and Optimization of Bioprocessing in the Production of Biosurfactants*

Several factors have been reported to affect the overall process of biosurfactant production and to directly influence the quantity and quality of the final product. These factors include downstream procedures, the availability of suitable economic recovery methods, the media components, and the process conditions (Mukherjee et al. 2006). Environmental hazards and recovery expenses could be cut down significantly if solvents that are low cost, less toxic, and widely available were used for processing biosurfactants.

One study investigated the production of biosurfactant from *Lactococcus lactis 53* and *Streptococcus thermophilus A* to determine the potential use of an alternative fermentative medium (Rodrigues et al. 2006b). The study used a relatively inexpensive and economical alternative to synthetic media, namely, cheese whey and molasses, to produce a probiotic bacteria-derived biosurfactant. The method reduced the medium preparation costs by about 60%–80% and increased the yield of biosurfactant by about 1.2–1.5 times (mass per gram cell dry weight), for both strains. Therefore, this alternative substrate could be used to economically-produce biosurfactants (Rodrigues et al. 2006b).

One optimization study used soybean oil refinery residue as a substrate to produce enhanced tension-active emulsifying biosurfactant derived from *Candida lipolytica* (Rufino et al. 2008). The study found that the optimum production of biosurfactant production by *C. lipolytica* was achieved with a medium containing 1% glutamic acid and 6% soybean oil refinery residue (Rufino et al. 2008).

Another study investigated the production efficiency of surfactin derived from *Bacillus subtilis* BS5 in a mineral salt medium based on varied fermentation conditions and medium components (carbon, nitrogen, and minerals) (Abdel-Mawgoud et al. 2008). A yield of 0.35 g/l biosurfactant was achieved with a medium supplemented with glucose (20 g/l) and $NaNO_3$ (5 g/l). The authors then replaced the glucose with 160 ml/l molasses, and recorded a threefold increase in the biosurfactant yield, to 1.12 g/l, under similar physicochemical conditions, i.e., pH (6.5–6.8) at 30°C.

According to Smyth et al. (2010), the complexity and cost of the overall production process of biosurfactants will affect the purity of the end-product. Only high-purity biosurfactants are safe for human consumption or are safe to use as ingredients in food products. However, the production of these biosurfactants involves a number of complex extraction and purification methods. Cost-effective downstream processing is needed to achieve an optimum production of biosurfactants. The most commonly-used downstream processes, namely adsorption, ultrafiltration, ammonium sulfate precipitation, solvent extraction, acid precipitation, centrifugation, and ion exchange chromatography, are shown in Table 5 (Mukherjee et al. 2006, Singh and Saini 2014).

Table 5. Downstream processes for purification and recovery of biosurfactants (adapted from Mukherjee et al. 2006, Helmy et al. 2011).

Downstream recovery method	Advantages/Disadvantages
Acid precipitation	Low cost, efficient in crude biosurfactant recovery, batch mode, cheap and simple method
Organic solvent extraction	Efficient in crude biosurfactant recovery and partial purification, but more toxic in nature
Solvent extraction (using Methyl tertiary-butyl ether)	Less toxic than conventional solvents, reusable, cheap
Ammonium sulfate precipitation	Effective in isolation of certain type of polymeric biosurfactants
Centrifugation	Reusable, effective in crude biosurfactant recovery, but required to setup the instrument
Membrane ultrafiltration	Fast, one-step recovery, high level of purity, no chemical addition
Adsorption on polystyrene resins	Fast, one-step recovery, high level of purity, reusability
Ion-exchange chromatography	High purity, reusability, fast recovery

6.4.3 Development of Recombinant Strains for Enhanced Biosurfactant Yield

Tailor-made recombinant strains or mutations have been reported to increase biosurfactant yields by several-fold (Table 6). Nevertheless, this approach has still not been properly tested, and research into this area is still in its infancy (Mukherjee et al. 2006). To achieve truly low-cost large-scale biosurfactant production, recombinant or hyperproducer microbial strains might also need to be applied, even with already optimized downstream processes and process conditions. These strains are important to introduce to products that have better commerciality and to economize the production process (Helmy et al. 2011).

Different medium and culture conditions cause *Candida* strains to produce different yields of biosurfactant. One study encultured *C. lipolytica* UCP 0988 for 72 hours and managed to extract 8 g/l of crude biosurfactant, indicating a growth-associated production. Meanwhile, enculturing *Candida sphaerica* for 144 hours culture yielded a biosurfactant extract of 4.5 g/l (Sobrinho et al. 2008). After 144 hours of experiment, *C. sphaerica* produced a biosurfactant yield of 9 g/l (Luna et al. 2012). In yet another study, *C. lipolytica*, with canola oil and glucose as substrates, reported a biosurfactant yield of 8 g/l (Sarubbo et al. 2006).

The most effective strategy to optimize the production of rhamnolipids is the metabolic engineering of non-pathogenic strains. For instance, non-pathogenic strains that have been engineered were able to produce rhamnolipids. Bioengineering also enabled some bacteria to produce rhamnolipids, that could not have produced them otherwise (Tisso et al. 2016). Even a strain that could only produce mono-rhamnolipids could be engineered to produce di-rhamnolipids, or vice versa. Some bioengineering tactics take a different approach, i.e., by introducing genes to promote the production of the biosurfactant (Dobler et al. 2016).

Since sophisticated techniques, skills, and apparatuses are required to improve strains via the recombinant technique, and accompanied with increased research and development costs, there are still issues concerning their application. Nevertheless, this technique will certainly simplify the production process and make commercialization of the biosurfactant more feasible (Helmy et al. 2011).

Table 6. Recombinant strains of microorganisms with improved yield (adapted from Mukherjee et al. 2006).

Mutant or recombinant strain	Increased yield or production	Reference
Pseudomonas aeruginosa PTCC 1637	10 times more production	Tahzibi et al. (2004)
Bacillus licheniformis KGL11	12 times more production	Lin et al. (1998)
Recombinant *Bacillus subtilis* MI 113	8 times more surfactin production	Ohno et al. (1995)
Recombinant *Gordonia amarae*	4 times more production of trehalose lipid biosurfactant	Dogan et al. (2006)

7. Conclusion

The increasing demand for natural food ingredients and green technology, combined with the versatile role of biosurfactants, is a niche area that has increased the market demand for biosurfactants. The use of recombinant or selective strains of microbes with high yield capacities, combined with economical and renewable substrates for biosurfactant production, is gaining increased industrial attention. The existing trends in the market indicate a high demand for various biosurfactants. For example, phospholipids, glycolipids, rhamnolipids, lipopeptides, and sophorolipids are now widely being used in numerous products, namely detergents, paints, textiles, agriculture, makeup, food, and in the pharmacological industry, among others. However, there is limited and expensive availability of biosurfactants in the market, despite their many commercially attractive properties and many advantages over synthetic surfactants, possibly due to their high production costs, complicated down streaming and purification processes, and low yields. Non-economical production is the main factor that is preventing the commercialization of biosurfactants. Therefore, future R&D should focus on optimizing the biosurfactant production process via a simple and inexpensive recovery technique. In this way, biosurfactants could be a potential substitute to their chemical counterparts. Meanwhile, the barriers to entering the food industry mainly hinge on the toxicity and safety of the biosurfactants. Therefore, more studies and *in vivo* tests should be conducted to determine the biosurfactants' interaction and mechanism in a food matrix. Also, not many studies have explored the effect of biosurfactants on the sensory evaluation of food (Mulligan et al. 2014). The extensive use of biosurfactants could potentially eliminate the human and environmental challenges such as pollution associated with synthetic surfactants.

Acknowledgement

The authors acknowledge the financial support from Ministry of Higher Education Malaysia under the Fundamental Research Grant (FRGS-554016 and 4F993) and Universiti Teknologi Malaysia under the Trans-disciplinary Research University Grant (05G90).

References

Abalos, A., A. Pinazo, M.R. Infante, M. Casals, F. Garcia et al. 2001. Physicochemical and antimicrobial properties of new rhamnolipids produced by *Pseudomonas aeruginosa* AT10 from soybean oil refinery wastes. Langmuir. 17: 1367–71.

Abdel-Mawgoud, A., M. Aboulwafa and N. Hassouna. 2008. Optimization of surfactin production by *Bacillus subtilis* isolate BS5. Appl. Biochem. Biotechnol. 150: 305–25.

Adamczak, M. and W. Bednarski. 2000. Influence of medium composition and aeration on the synthesis of biosurfactants produced by *Candida antarctica*. Biotechnology Letters 22: 313–316.

Ángeles, D.R., Á.M. Miguel and E.L. Víctor. 2019. Advances on research in the use of agro-industrial waste in biosurfactant production. World Journal of Microbiology and Biotechnology 35: 155. https://doi.org/10.1007/s11274-019-2729-3.

Araújo, H.W.C., R.F.S. Andrade, D. Montero-Rodríguez, D. Rubio-Ribeaux, C.A. Alves da Silva et al. 2019. Sustainable biosurfactant produced by *Serratia marcescens* UCP 1549 and its suitability for agricultural and marine bioremediation applications. Microbial Cell Factories 18: 1–13. https://doi.org/10.1186/s12934-018-1046-0.

Araújo, S.C.d.S., R.C.B. Silva-Portela, D.C. de Lima, F. MMB, A. WJ, da S. UB et al. 2020. MBSP1 : a biosurfactant protein derived from a metagenomic library with activity in oil degradation. Scientific Reports 10: 1–13. https://doi.org/10.1038/s41598-020-58330-x.

Banat, M.I., S.R. Makkar and S.S. Cameotra. 2000. Potencial commercial applications of microbial surfactants. Appl. Microbiol. Biotechnol. 53: 495–508.

Barros, F., C. Quadros, M. Maróstica Júnior and G. Pastore. 2007. Surfactin: chemical, technological and functional properties for food applications. Quim. Nova. 30: 409–14.

Bednarski, W., M. Adamczak, J. Tomasik and M. Plaszczyk. 2004. Application of oil refinery waste in biosynthesis of glycolipids by yeast. Bioresource Technol. 95: 15–8.

Bezerra, de S.S., J.M. de L. Humberto, D.R. Raquel, L.F.P. Ana, A.S. Leonie et al. 2013. Assessment of toxicity of a biosurfactant from *Candida sphaerica* UCP 0995 cultivated with industrial residues in a bioreactor. Electronic Journal of Biotechnology 16: 4–4.

Bouwman, S., B.D. Lund, M.F. Driesen and G.D. Schmidt. 1982. Growth of thermoresistant *stretococci* and deposition of milk constituents on plates of heat exchangers during long operating times. J. Food Prot. 45: 806–12.

Cameotra, S.S., S.R. Makkar, J. Kaur and S.K. Mehta. 2010. Synthesis of biosurfactants and their advantages to microorganisms and mankind. Adv. Exp. Med. Biol. 672: 261–80. doi:10.1007/978-1-4419-5979-9_20.

Campos, J.M., T.L.M. Stamford, L.A. Sarubbo, J.M. de Luna et al. 2013. Microbial biosurfactants as additives for food industries. Biotechnology Progress 29: 1097–1108.

Campos, J.M., T.L.M. Stamford, L. Sarubbo, M. de L. Asfora, R.D. Rufino et al. 2014. Production of a Bioemulsifier with potential application in the food industry. Appl. Biochem. Biotechnol. 172: 3234–52. https://doi.org/10.1007/s12010-014-0761-1.

Campos, J.M., T.L.M. Stamford, R.D. Rufino, J.M. Luna, T. Christina et al. 2015. Formulation of mayonnaise with the addition of a bioemulsifier isolated from Candida utilis. Toxicology Reports 2: 1164–70.

Campos, J.M., T.L.M. Stamford and L.A. Sarubbo. 2019. Characterization and application of a biosurfactant isolated from Candida utilis in salad dressings. Biodegradation 30: 313–324. https://doi.org/10.1007/s10532-019-09877-8.

Cavalero, D.A. and D.G. Cooper. 2003. The effect of medium composition on the structure and physical state of sophorolipids produced by *Candida bombicola* ATCC 22214. Journal of Biotechnology 103: 31–41. https://doi.org/10.1016/S0168-1656(03)00067-1.

Chen, C., T. Lin, Y. Shieh. 2015. Emulsification and antioxidation of biosurfactant extracts from Chinese medicinal herbs fermentation *in vitro*. Journal of Bioscience and Bioengineering 120(4): 387–95. https://doi.org/10.1016/j.jbiosc.2015.02.010.

Cooper, G.D. and A.D. Cavalero. 2003. The effect of medium composition on the structure and physical state of sophorolipidss produced by *Candida bombicola* ATCC 22214. J. Biotechnol. 103: 31–41.

Cortés-Camargo, S., N. Pérez-Rodríguez, R.P. de Souza-Oliveira, B.E. Barragán-Huerta, J.M. Domínguez et al. 2016. Production of biosurfactants from vine-trimming shoots using the halotolerant strain *Bacillus tequilensis* ZSB10. Ind. Crops Prod. 79: 258–66. https://doi.org/10.1016/j.indcrop.2015.11.003.

Daniel, E.O. 2017. Biosurfactants and surfactants interacting with the membranes and proteins: Same but different? Biochimica et Biophysics Acta 1859. 639–649.

Das, P., S. Mukherjee and R. Sen. 2008. Antimicrobial potential of a lipopeptide biosurfactant derived from a marine *Bacillus circulans*. J. Appl. Microbiol. 104: 1675–84.

De Araujo, Livia, V., A. Fernanda, Lins. Ulysses, M. de M.S.A. Lídia et al. 2011. Rhamnolipid and surfactin inhibit Listeria monocytogenes adhesion. Food Research International 44: 481–88.

de Freitas, F., E.A.V. Jakeline and N. Marcia. 2019. The antibacterial activity of rhamnolipid biosurfactant is pH dependent. Food Research International 116: 737–44.

Desai, J.D. and I.M. Banat. 1997. Microbial production of surfactants and their commercial potential. Microbiology and Molecular Biology Reviews 61: 47–64.

Dobler, L., L.F. Vilela, R.V. Almeida and B.C. Neves. 2016. Rhamnolipids in perspective: gene regulatory pathways, metabolic engineering, production and technological forecasting. New Biotechnology 33: 123–35. doi:10.1016/j.nbt.2015.09.005.

Dogan, I., K.R. Pagilla, D.A. Webster and B.C. Stark. 2006. Expression of Vitreoscilla haemoglobin in *Gordonia amarae* enhances biosurfactant production. J. Ind. Microbiol. Biotechnol. 33: 693–700.

Doost, A.S., J. Van Camp, K. Dewettinck and P. Van der Meeren. 2019. Production of thymol nanoemulsions stabilized using Quillaja Saponin as a biosurfactant: Antioxidant activity enhancement. Food Chemistry 293: 134–43.

do Valle, G., Z. Milene and N. Marcia. 2012. Evaluation of rhamnolipid and surfactin to reduce the adhesion and remove biofilms of individual and mixed cultures of food pathogenic bacteria. Food Control. 25: 441–47.

Drakontis, C.E. and S. Amin. 2020. Biosurfactants: formulations, properties, and applications. Current Opinion in Colloid & Interface Science. doi:10.1016/j.cocis.2020.03.013.

Dubey, V., J. Kirti, A. Asha and S.K. Singh. 2005. Adsorption-desorption process using wood-based activated carbon for recoveryof biosurfactant from fermented distillary wastewater. Biotechnol. Prog. 21: 860–67.

Dusane, D.H., V.S. Pawar, Y.V. Nancharaiah et al. 2011. Anti-biofilm potential of a glycolipid surfactant produced by a tropical marine strain of *Serratia marcescens*. Biofouling 27: 645–54.

Epand, M.R. and J.H. Vogel. 1999. Diversity of antimicrobial peptides and their mechanisme of action. Biochim. Biophys. Acta 1462, pp. 11–28.

Fernandes, P.A.V., I.R. De Arruda, A.F.A.B. Dos Santos and A.A. De Araujo. 2007. Antimicrobial activity of surfactants produced by *Bacillus subtilis* R14 against multidrug-resistant bacteria.Brazilian Journal of Microbiology 38: 704–9.

Fontes, G.C., N.M. Ramos, P.F.F. Amaral, M. Nele, M.A.Z. Coelho et al. 2012. Renewable resources for biosurfactant production by *Yarrowia lipolytica*. Brazilian Journal of Chemical Engineering 29: 483–493.

Galie, S., C. Garcia-Gutierrez, E.M. Miguelez, C.J. Villar, F. Lombo et al. 2018. Biofilms in the food industry: health aspects and control methods. Frontiers in Microbiology 9: 18.

Giri, S.S., E.C. Ryu, V. Sukumaran and S.C. Park. 2019. Microbial Pathogenesis Antioxidant, antibacterial, and anti-adhesive activities of biosurfactants isolated from *Bacillus* strains. Microbial Pthogenesis 132(September 2018): 66–72. https://doi.org/10.1016/j.micpath.2019.04.035.

Haba, E., M.J. Espuny, M. Busquets and A. Manresa. 2000. Screening and production of rhamnolipids by *Pseudomonas aeruginosa* 47T2 NCIB 40044 from waste frying oils. J. Appl. Microbiol. 88: 379–87.

He, Z., Z. Weiwei, Z. Xiaoyu, Z. Haizhen, L. Yingjian, L. Zhaoxin et al. 2017. Influence of surfactin on physical and oxidative stability of microemulsions with docosahexaenoic acid. Colloids and Surfaces B: Biointerfaces 151: 232–39.

Helmy, Q., E. Kardena, N. Funamizu and Wisjnuprapto. 2011. Strategies toward commercial scale of biosurfactant production as potential substitute for it's chemically counterparts. Int. J. Biotechnology 12: 66–86.

Hmidet, N., J. Nawel, O. Manel, P.A. María, N. Moncef et al. 2020. Antioxidant properties of *Enterobacter cloacae C3* lipopeptides *in vitro* and in model food emulsion. Journal of Food Processing and Preservation 44: e14337.

Hwang, Y.H., K.P. Byung, H.L. Jong, S.K. Myoung, B.S. In, C.P. Seung et al. 2008. Evaluation of genetic and developmental toxicity of surfactin C from *Bacillus subtilis* BC1212. Journal of Health Science 54: 101–6.

Irfan, M.M. and S.S. Mahsa. 2014. Rhamnolipids: well-characterized glycolipids with potential broad applicability as biosurfactants. Industrial Biotechnology 10: 285–91.

Iroha, O.K., O.U. Njoku, V.N. Ogugua and V.E. Okpashi. 2015. Characterization of biosurfactant produced from submerged fermentation of fruits baggase of yellow cashew (*Anacardium occidentale*) using Pseudomonas Aeruginosa 9: 473–81.

Ishii, N., T. Kobayashi, K. Matsumiya, M. Ryu, Y. Hirata et al. 2012. Transdermal administration of lactoferrin with sophorolipid. Biochem. Cell Biol. 90: 504–12.

Jain, R.M., K. Mody, N. Joshi, A. Mishra, B. Jha et al. 2013. Effect of unconventional carbon sources on biosurfactant production and its application in bioremediation. Int. J. Biol. Macromol. 62: 52–58. https ://doi.org/10.1016/j.ijbio mac.2013.08.030.

Jemil, N., H.B. Ayed, A. Manresa, M. Nasri, N. Hmidet et al. 2017. Antioxidant properties, antimicrobial and anti-adhesive activities of DCS1 lipopeptides from *Bacillus methylotrophicus*. BMC Microbiology 17: 1–11. https://doi.org/10.1186/s12866-017-1050-2.

Joe, M.M., K. Bradeeba, R. Parthasarathi, P.K. Sivakumaar, P.S. Chauhan et al. 2012. Development of surfactin based nanoemulsion formulation from selected cooking oils : Evaluation for antimicrobial activity against selected food associated microorganisms. Journal of the Taiwan Institute of Chemical Engineers 43(2): 172–180. https://doi.org/10.1016/j.jtice.2011.08.008.

Kadaikunnan, S., S.R. Thankappan, M.K. Jamal, S.A. Naiyf, M. Ramzi et al. 2015. *In-vitro* antibacterial, antifungal, antioxidant and functional properties of *Bacillus amyloliquefaciens*. Annals of Clinical Microbiology and Antimicrobials 14: 9.

Kim, H.S., J.W. Jeon, B.H. Kim, C.Y. Ahn, H.M. Oh et al. 2006. Extracellular production of a glycolipid biosurfactant, mannosylerythritol lipid, by *Candida* sp. SY16 using fed batch fermentation. Appl. Microbiol. Biotechnol. 70: 391–96.

Kiran, G.S., P. Sethu, S. Arya, B.P. Gopal, P. Navya et al. 2017. Production of lipopeptide biosurfactant by a marine *Nesterenkonia* sp. and its application in food industry. Frontiers in Microbiology 8: 1138.

Kralova, I. and J. Sjoblom. 2009. Surfactants used in food industry: a review. J. Dispers. Sci. Technol. 30: 1363–83.

Kralova, I. and J. Sjöblom. 2017. Surfactants Used in Food Industry : A Review Surfactants Used in Food Industry : A Review 2691(April). https://doi.org/10.1080/01932690902735561.

Lin, S.C., K.G. Lin, C.C. Lo and Y.M. Lin. 1998. Enhanced biosurfactant production by a *Bacillus licheniformis* mutant. Enzyme Microb. Technol. 23: 267–73.

Liu, Z.F., G.M. Zeng, H. Zhong, X. Yuan, F.H.Y. Zhong et al. 2012. Effect of dirhamnolipid on the removal of phenol catalyzed by laccasein aquoues solution. World J. Microbiol. Biotechnol. 28: 175–81.

Lobato, A.K.C.L., A.F. Almeida, M.S. Bezerra, L.M.B. Júnior, L.C.L. Santos et al. 2013. Biosurfactant production from industrial residues using microorganisms isolated from oil wells. International Review of Chemical Engineering (I.RE.CH.E.) 5: 310–16.

López, P.A., R.L. Lorena, R.F. Myriam, B.M. Ana, M.C. José et al. 2019. Effect of biosurfactant extract obtained from the corn-milling industry on probiotic bacteria in drinkable yogurt. Journal of the Science of Food and Agriculture 99: 824–30.

Lourenço, L.A., M.M.D. Alberton, L.B.B. Tavares, U. de S.S.M.A. Guelli, R.M. García et al. 2018. Biosurfactant production by Trametes versicolor grown on two-phase olive mill waste in solid-state-fermentation. Environ. Technol. 39: 3066–76. https://doi.org/10.1080/09593 330.2017.13744 71.

Luna, J.M. De, L. Sarubbo and G.M. De. Campos-takaki. 2009. A new biosurfactant produced by *Candida glabrata* UCP 1002 : Characteristics of stability and application in oil recovery. Brazilian Archives of Biology and Technology 52: 785–93.

Luna, J.M., R.D. Rufino, G.M. Campos-Takaki and L.A. Sarubbo. 2012. Properties of the biosurfactant produced by *Candida sphaerica* cultivated in low-cost substrates. Chem. Eng. Trans. 27: 67–72.

Marchant, R. and M.I. Banat. 2012. Biosurfactants: a sustainable replcement for chemical surfactants? Biotechnol. Lett. 34: 1597–1605.

Martínez, T.A., V.R. Rodríguez and H.C.A. Ilizaliturri. 2018. Culture media formulation and growth conditions for biosurfactants production by bacteria. International Journal of Environment Sciences & Natural Resources 10: 555790. DOI: 10.19080/IJESNR.2018.10.555790.

Merghni, A., I. Dallel, E. Noumi, Y. Kadmi, H. Hentati et al. 2017. Antioxidant and antiproliferative potential of biosurfactants isolated from *Lactobacillus casei* and their anti-biofilm effect in oral *Staphylococcus aureus* strains. Microbial Pathogenesis 104: 84–89. https://doi.org/10.1016/j.micpath.2017.01.017.

Mellor, G.E., A.B. Jessica and A.D. Gary. 2011. Evidence for a role of biosurfactants produced by *Pseudomonas fluorescens* in the spoilage of fresh aerobically stored chicken meat. Food Microbiology 28: 1101–4.

Mohan, P.K., N. George and K.Y. Ernest. 2006. Biokinetics of biodegradation of surfactants under aerobic, anoxic and anaerobic conditions. Water Research 40: 533–40.

Mouafo, H.T., A.M.B. Baomog, J.J.B. Adjele, A.T. Sokamte, A. Mbawala et al. 2020. Microbial profile of fresh beef sold in the markets of ngaoundere, cameroon, and antiadhesive activity of a biosurfactant against selected Bacterial Pathogens. Journal of Food Quality 10.

Mukherjee, S., P. Das and R. Sen. 2006. Towards commercial production of microbial surfactants. Trends in Biotechnology 24: 509–15.

Mukherjee, K.A. and K. Das. 2010. Microbial surfactants and their potential applications: an overview. Adv. Exp. Med. Biol. 672: 54–64. doi:10.1007/978-1-4419-5979-9_4.

Mulligan, C.N., S.K. Sharma and A. Mudhoo. 2014. Biosurfactants: Research Trends and Applications, 1st Edition, CRC Press, Boca Raton.

Muthusamy, K., S. Gopalakrishnan, T.K. Ravi and P. Sivachidambaram. 2008. Biosurfactants, properties, commercial production and application. Current Science 94: 736–47.

Nalini, S. and R. Parthasarathi. 2018. Optimization of rhamnolipid biosurfactant production from Serratia rubidaea SNAU02 under solid-state fermentation and its biocontrol ef fi cacy against Fusarium wilt of eggplant. Annals of Agrarian Sciences 16: 108–115. https://doi.org/10.1016/j.aasci.2017.11.002.

Nayarisseri, A., P. Singh and S.K. Singh. 2018. Screening, isolation and characterization of biosurfactant producing *Bacillus subtilis* strain ANSKLAB03. Bioinformation, 14(6): 304–14.

Neta, Nair, do A.S., José, C.S. dos S., Soraya, de O.S et al. 2012. Enzymatic synthesis of sugar esters and their potential as surface-active stabilizers of coconut milk emulsions. Food Hydrocolloids 27: 324–31.

Nitschke, M. and G. Pastore. 2003. Cassava flour wastewater as a substrate for biosurfactant production. Appl. Biochem. Biotechnol. 105: 295–301.

Nitschke, M. and G.M. Pastore. 2004, 2005. Biosurfactant production by *B. subtilis* using cassava-processing effluent. Appl. Biochem. Biotechnol. 112: 163–72.

Nitschke, M. and G. Pastore. 2006. Production and properties of a surfactant obtained from *Bacillus subtilis* grown on cassava wastewater. Bioresource Technol. 97: 336–41.

Nitschke, M. and S. Coasta. 2007. Biosurfcatantsin food industry. Trends Food Sci Tech. 18: 252–9.

Nitsche, M. and S.S.e. Silva. 2016. Recent food applications of microbial surfactants. Critical Reviews in Food Science and Nutrition, 8398. https://doi.org/10.1080/10408398.2016.1208635.

Ocampo, G.Y., G.S. Coutiño, C.B. González and A.W. Villarreal. 2017. Utilization of agroindustrial waste for biosurfactant production by native bacteria from Chiapas. De Gruyter Open. 2: 341–49. DOI 10.1515/opag-2017-0038.

Ohno, A., T. Ano and M. Shoda. 1995. Production of a lipopeptide antibiotic, surfactin, by recombinant *Bacillus subtilis* in solid-state fermentation. Biotechnol. Bioeng. 47: 209–14.

Olasanmi, I.O. and R.W. Thring. 2018. The role of biosurfactants in the continued drive for environmental sustainability. Sustainability 10: 4817. doi:10.3390/su10124817.

Onbasli, D. and B. Aslim. 2009. Biosurfactant production in sugar beet molasses by some *Pseudomonas* sp. Journal of Environmental Biology 30: 161–63.

Otzen, D.E. 2017. Biosurfactants and surfactants interacting with membranes and proteins: Same but different? Biochimica et Biophysica Acta - Biomembranes. 4: 639–49.

Pekin, G., F. Vardar-Sukan and N. Kosaric. 2005. Production of sophorolipids from *Candida bombicola* ATCC 22214 using Turkish corn oil and honey. Eng. Life Sci. 5: 357–62.

Płaza, G. and V. Acha. 2020. Biosurfactants: Eco-friendly and innovative biocides against biocorrosion. International Journal of Molecular Sciences 21: 2152. doi:10.3390/ijms21062152.

Portilla, R.O.M., A.B. Moldes, A.M. Torrado and J.M. Domínguez. 2007. Lactic acid and biosurfactants production from hydrolyzed distilled grape marc. Process Biochem. 42; 1010–20. https ://doi.org/10.1016/j.procb io.2007.03.011.

Prades, J., O. Vogler, R. Alemany, F.M. Gomez, S.R. Funari et al. 2014. Plant pentacyclic triterpenic acids as modulators of lipid membrane physical properties. Biochim. Biophs. Acta 1838, pp. 752–60.

Punrat, T., T. Jiraporn, C.N. Suchada, A. Jirarat, T. Suthep et al. 2020. Production of a sophorolipid biosurfactant by *Wickerhamomyces anomalus* MUE24 and its use for modification of rice flour properties. SCIENCEASIA 46: 11–18.

Ribeiro, B.G. and L.A. Sarubbo. 2020. Biosurfactants : production and application prospects in the food industry. Biotechnology Progress. https://doi.org/10.1002/btpr.3030.

Ribeiro, B.G., J.M.C. Guerra and L.A. Sarubbo. 2020a. Potential food application of a biosurfactant produced by *Saccharomyces cerevisiae* URM 6670. Frontiers in Bioengineering and Biotechnology 8: 1–13. https://doi.org/10.3389/fbioe.2020.00434.

Ribeiro, B.G., B.O. De Veras, S. Aguiar, J. Medeiros, C. Guerra et al. 2020b. Biosurfactant produced by *Candida utilis* UFPEDA1009 with potential application in cookie formulation. Electronic Journal of Biotechnology. https://doi.org/10.1016/j.ejbt.2020.05.001.

Rodrigues, L., M.B. Ibrahim, T. José and O. Rosario. 2006a. Biosurfactants: potential applications in medicine. Journal of Antimicrobial Chemotherapy 57: 609–18.

Rodrigues, L.R., J.A. Teixeira and R. Oliveira. 2006b. Low-cost fermentative medium for biosurfactant production by probiotic bacteria. Biochemical Engineering Journal 32: 135–42.

Ron, Z.E. and E. Rosenberg. 2001. Natural roles of biosurfactants. Environ. Microbiol. 3: 229–236. doi:10.1046/j.1462-2920.2001.00190.x.

Rosenberg, E. and E.Z. Ron. 1999. High- and low-molecular-mass microbial surfactants. Applied Microbiology and Biotechnology 52: 154–62.

Rufino, R.D., L.A. Sarubbo, B.B. Neto and G.M. Campos-Takaki. 2008. Experimental design for the production of tensio-active agent by *Candida lipolytica.* Journal of Industrial Microbiology & Biotechnology 35: 907–914. doi:10.1007/s10295-008-0364-3.

Rufino, R.D., J.M. de Luna, G.M. de Campos Takaki, L.A. Sarubbo et al. 2014. Characterization and properties of the biosurfactant produced by *Candida lipolytica* UCP 0988. Electronic Journal of Biotechnology 17: 34–38. https://doi.org/10.1016/j.ejbt.2013.12.006.

Santos, D.K.F., R.D. Rufino, J.M. Luna, V.A. Santos and L.A. Sarubbo. 2016. Biosurfactants: multifunctional biomolecules of the 21st Century. Int. J. Mol. Sci. 17: 401.

Santos, D.K.F., A.H.M. Resende, D.G. De Almeida, I.M. Banat, L.A. Sarubbo et al. 2017. *Candida lipolytica* UCP0988 biosurfactant : potential as a bioremediation agent and in formulating a commercial related product. Frontiers in Microbiology 8: 1–11. https://doi.org/10.3389/fmicb.2017.00767.

Sarubbo, L.A., J.M. Luna and G.M. Campos-Takaki. 2006. Production and stability studies of the bioemulsifier obtained from a new strain of *Candida glabrata* UCP 1002. Electron. J. Biotechnol. 9. http://dx.doi.org/10.2225/vol9-issue4-fulltext-6.

Satpute, K. Surekha, Banpurkar G. Arun, Dhakephalkar K. Prashant, Banat M. Ibrahim, A. Chopade et al. 2010. Methods for investigating biosurfactants and bioemulsifiers: A review. Critical Reviews in Biotechnology, pp. 1–18, Early Online.

Satpute, S.K., N.S. Mone, P. Das, I.M. Banat, A.G. Banpurkar et al. 2019. Inhibition of pathogenic bacterial biofilms on PDMS based implants by *L. acidophilus* derived biosurfactant. BMC Microbiology 19: 1–15.

Sekhon, K.K., S. Khanna and S.S. Cameotra. 2012. Biosurfactant production and potential correlation with esterase activity. Journal of Petroleum & Environmental Biotechnology 3: 7. DOI: 10.4172/2157-7463.1000133.

Shepherd, R., J. Rockey, Sutherland W. Ian and S. Roller. 1995. Novel bioemulsifiers from microororganisms for use in foods. J. Biotechnol. 40: 2017–17.

Singh, P.B. and H.S. Saini. 2013, 2014. Exploitation of agro-industrial wastes to produce low-cost microbial surfactants. pp. 445–471. *In*: Brar, S.K. et al. (eds.). Biotransformation of Waste Biomass into High Value Biochemicals. DOI 10.1007/978-1-4614-8005-1_18, © Springer, New York 2014.

Singh, P. and B.N. Tiwary. 2016. Isolation and characterization of glycolipid biosurfactant produced by *Pseudomonas otitidis* strain isolated from Chirimiri coal mines, India. Bioresources and Bioprocessing 3: 42.

Singh, P., Y. Patil and V. Rale. 2018. Biosurfactant production: emerging trends and promising strategies. Journal of Applied Microbiology 126: 2–13. doi:10.1111/jam.14057.

Silva, F.S.R. de C., G.D. Almeida, D. Rufino, L.M.J. Raquel, V. Santos et al. 2014. Applications of biosurfactants in the petroleum industry and the remediation of oil spills. Int. J. Mol. Sci. 15: 12523–42.

Silva, S.S., J.W.P. Carvalho, C.P. Aires and M. Nitschke. 2017. Disruption of *Staphylococcus aureus* biofilms using rhamnolipid biosurfactants. Journal of Dairy Science 100(10): 7864–73.

Silva, A.C.S. da, P.N. dos Santos, T.A.L.e Silva, R.F.S. Andrade, G.M. Campos-Takaki et al. 2018. Biosurfactant production by fungi as a sustainable alternative. Arquivos Do Instituto Biológico 85(e0502017): 1–12. https://doi.org/10.1590/1808.

Smyth, T.J.P., A. Perfumo, R. Marchant and I.M. Banat. 2010. Isolation and analysis of low molecular weight microbial glycolipids. Handbook of Hydrocarbon and Lipid Microbiology. Springer, Berlin, pp. 3705–3723.

Soares, da S.R.C.F., D.G. de Almeida, P.P.F. Brasileiro, R.D. Rufino, J.M. de Luna, L.A. Sarubbo et al. 2019. Production, formulation and cost estimation of a commercial biosurfactant. Biodegradation 30: 191–20. https://doi.org/10.1007/s1053 2-018-9830-4.

Sobrinho, H.B.S., R.D. Rufino, J.M. Luna, A.A. Salgueiro, G.M. Campos-Takaki et al. 2008. Utilization of two agroindustrial by-products for the production of a surfactant by *Candida sphaerica* UCP0995. Process Biochem. 43: 912–7. http://dx.doi.org/10.1016/j.procbio.2008.04.013.

Suresh, C.C.R., T. Lohitnath, D.J.M. Kumar and P.T. Kalaichelvan. 2012. Production and characterization of biosurfactant from *Bacillus subtilis* MTCC441 and its evaluation to use as bioemulsifier for food bio-preservative. Advances in Applied Science Research 3: 1827–31.

Tahzibi, A., F. Kamal and M.M. Assadi. 2004. Improved production of rhamnolipids by a *Pseudomonas aeruginosa* mutant. Iran. Biomed. J. 8: 25–31.

Tangsuphoom, N. and N.C. John. 2009. Effect of surface-active stabilizers on the surface properties of coconut milk emulsions. Food Hydrocolloids 23: 1801–09.

Tiso, T., P. Sabelhaus, B. Behrens, A. Wittgens, F. Rosenau et al. 2016. Creating metabolic demand as an engineering strategy in *Pseudomonas Putida*—rhamnolipid synthesis as an example. Metabolic Engineering Communications 3: 234–44. doi:10.1016/j.meteno.2016.08.002.

Tripathi, L., M.S. Twigg, A. Zompra, K. Salek, V.U. Irorere et al. 2019. Biosynthesis of rhamnolipid by a *Marinobacter* species expands the paradigm of biosurfactant synthesis to a new genus of the marine microflora. Microbial Cell Factories 18: 1–12. https://doi.org/10.1186/s12934-019-1216-8.

Trummler, K., F. Effenberger and C. Syldatk. 2003. An integrated microbial/enzymatic process for production of rhamnolipids and l-(+)-rhamnose from rapeseed oil with *Pseudomonas* sp. DSM 2874. Eur. J. Lipid. Sci. Tech. 105: 563–71.

Van Haesendonck, I.P.H. and C.A.V. Emmanuel. 2004. Rhamnolipids in bakery products. International Application Patent (PCT) WO. 2004-040984.

Villarreal, A.W., L.R. Lopez, H.C. Gonzalez, C.B. Gonzalez, G.Y. Ocampo et al. 2016. Characterization of bacteria isolation of bacteria from Pinyon Rhizospehere, producing biosurfactants from agro-industrial waste. Polish Journal of Microbiology 65: 183–89.

Vincent, P., B.F. Valentin, V.A. Eugene, K. Jürgen, W. Rainer et al. 2011. Adsorption of protein−surfactant complexes at the water/oil interface. Langmuir 27: 965–971. DOI: 10.1021/la1040757.

William, C.W. 2010. Describing the mechanism of antimicrobial peptide action with the interfacial activity model. ACS Chem. Biol. 5: 905–917. doi: 10.1021/cb1001558.

Yalcin, E. and K. Cavusoglu. 2010. Structural analysis and antioxidant activity of a biosurfactant obtained from *Bacillus subtilis* RW-I. Turkish J. Biochem-Turk Biyokimya Dergisi. 35: 243–47.

Yan, X., S. Gu, X. Cui, Y. Shi, S. Wen et al. 2019. Microbial Pathogenesis Antimicrobial, anti-adhesive and anti-biofilm potential of biosurfactants isolated from *Pediococcus acidilactici* and *Lactobacillus plantarum* against *Staphylococcus aureus* CMCC26003. Microbial Pthogenesis 127(October 2017): 12–20. https://doi.org/10.1016/j.micpath.2018.11.039.

Zajic, J., C. Panchal and D. Westlake. 1976. Bio-emulsifiers. CRC Crit. Rev. Microbiol. 5: 39–66.

Zouari, R., S. Besbes, S. Ellouze-chaabouni and D. Ghribi-aydi. 2016. Cookies from composite wheat – sesame peels flours : Dough quality and effect of *Bacillus subtilis* SPB1 biosurfactant addition. Food Chemistry 194: 758–769. https://doi.org/10.1016/j.foodchem.2015.08.064.

7

Biosurfactants

Production and Applications in Food

Muzafar Zaman,[1] *Shahnawaz Hassan,*[1] *Sabah Fatima,*[1,*] *Basharat Hamid,*[1] *Shabeena Farooq,*[1] *Ishfaq Qayoom,*[1] *Hina Alim,*[2] *Vibhor Agarwal*[3] and *R Z Sayyed*[4,*]

1. Introduction

Biosurfactants are a group of surface-active compounds produced by different microorganisms (Shekhar et al. 2015) containing both hydrophobic and hydrophilic moieties. These compounds have the ability to accumulate at the interface between air–liquid or liquid–liquid phases, reduce surface tension and stabilize emulsions (Bi et al. 2015). Such properties of biosurfactants have attracted industrial attention in various activities involving detergency, lubrication, emulsification, solubilization or the dispersion of different phases. Biosurfactants offer advantages over synthetic surfactants like, high biodegradability, low toxicity, eco-friendliness and efficacy under various environmental conditions (Radzuan et al. 2017, Ribeiro et al. 2020).

The contribution of biosurfactant production varies from 3–5% to the food and agriculture sector (Banat and Thavasi 2018). Biosurfactants can be produced from different substrates like waste products from agro-industries, such as corn steep liquor, glycerol, and waste frying oil, which reduces their cost of production and enhances their industrial applications (Gudiña et al. 2015). The production of biosurfactants by microorganisms particularly yeasts does not pose any risk of pathogenicity or

[1] Department of Environmental Science, University of Kashmir, Hazratbal, Srinagar 190006, Jammu and Kashmir, India.

[2] Department of Life Sciences, University of Mumbai, Vidyanagari campus, Santacruz (East), Mumbai 400 098, India.

[3] Department of Geology, University of Dayton, Dayton, OH, US, 45469.

[4] Department of Microbiology, PSGVP Mandal's Arts, Science and Commerce College, Shahada 425409, Maharashtra, India.

* Corresponding authors: sabahfatima333@gmail.com; sayyedrz@gmail.com

toxicity, making them suitable for utilization in food formulations. On the basis of their chemical composition biosurfactants are classified into low molecular weight BSs which include phospholipids, lipopeptides, glycolipids neutral lipids and high molecular weight BSs which comprise of polymers. The glycolipids (sophorolipids) produced by *Starmerella bombicola*, lipopeptide (surfactin) from *Bacillus subtilis* and polymeric biosurfactants (emulsan) from *Acinetobacter calcoaceticus* are some of best studied BSs with diverse industrial applications (Jadeja et al. 2019, Jezierska et al. 2019, Jha et al. 2016).

Biosurfactants are used to substitute synthetic surfactants in commonly used products (Louhasakul et al. 2020) like detergents (Fei et al. 2020), cosmetics (Fernández-Peña et al. 2020), domestic cleaning and grooming products (Rebello et al. 2020) and in healthcare as anti-tumor, anti-inflammatory and antimicrobial agents (Saimmai et al. 2020). The condition of the soil in agriculture can be improved by eliminating heavy metals and avoiding the harmful behavior of certain pests which cause dramatic benefit losses (Mnif and Ghribi 2016). The stabilization of gold and silver nanoparticles with the use of BSs has been reported (Rane et al. 2017).

Biosurfactants often play an important role in food formulations. In the food industry BSs can be used as stabilizers in food preparations and inhibit the microbial growth and their adhesion to contact surfaces of food. Pathogenic microorganisms from food processing machinery can be eliminated and neutralized with BSs acting as cleaning agents (Gudiña and Rodrigues 2019). BSs are nowadays purified and incorporated in various formulations by several companies for industrial applications in food, e.g., SyntheZyme LLC and Rhamnolipid Companies Inc. USA, sell rhamnolipids and sophorolipids for food processing purposes. This chapter reflects on the classifications, properties and different industrial applications of BSs in food.

2. Classification

BS can be classified on the basis of chemical structure, molecular weight and ionic charge. On the basis of chemical structure they are classified into glycolipids, phospholipids, lipopolysaccharides, lipoproteins and biopolymers (Najmi et al. 2018). Structures of some important BSs that are used in food processing are represented in Fig. 1.

Glycolipids

Glycolipids are low molecular weight biomolecules (628–826 Da) composed of carbohydrates and lipids. They have critical micellar concentration between 20 and 366 mg/L. Glycolipids are trehalose lipids, rhamnolipids, mannosylerythritol lipids (MELs), and sophorolipids on the basis of the nature of the carbohydrate (Dobler et al. 2017). Glycolipids are produced by different microorganisms like, trehalose lipids by species of *Rhodococcus* (Kuyukina and Ivshina 2019), *Pseudomonas* sp. and *Saccharomyces* sp. produce rhamnolipids (Bahia et al. 2018, Varjani and Upasani 2019) and sophorolipids by species of *Candida* (Campos et al. 2019, Gaur et al. 2019).

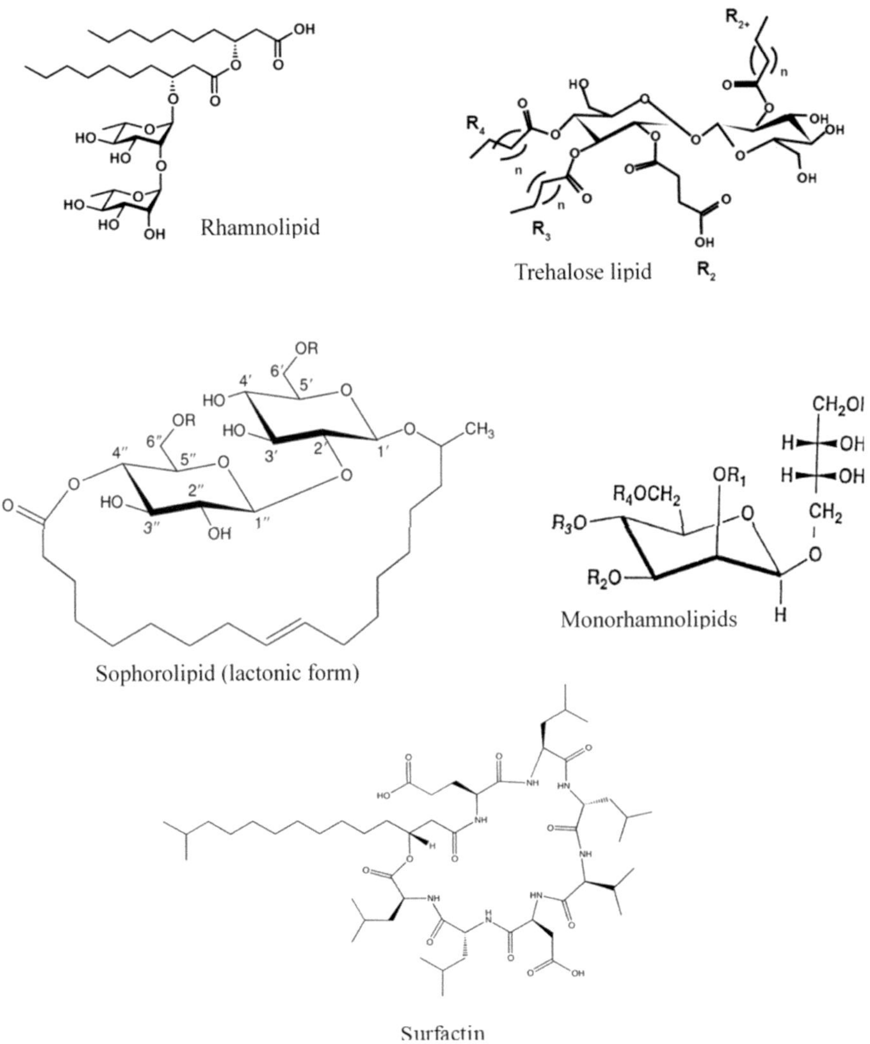

Figure 1. Structure of main biosurfactants.

Rhamnolipids

Rhamnolipids are used in processing of foods and comprise of glycosyl head group linked to the fatty acid tail, usually β-hydroxydecanoic acid. On the basis of a number of rhamnose groups present in structure, they are monorhamnolipids and dirhamnolipids. Large scale production of rhamnolipid has been achieved by species of Burkholderia and Pseudomonas due to the presence of all necessary key enzymes (Dobler et al. 2017).

Lipopeptides

Lipopeptides are higher molecular weight biosurfactants (≥ 1000 Da) and consist of proteins or peptides or lipids at a CMC value of 10 μmol/L or 23 mg/L. They have a surface tension of 72 to 20 mN/m (Wu et al. 2017). The most common BS in the lipopeptide group is surfactin which is produced by *Bacillus subtilis.* Another species *B. licheniformis* produces BS similar in structure to surfactin and stable in unfavorable conditions. Surfactin possess low cytotoxic potential, and even at lower concentrations their efficacy remains unaffected (Meena and Kanwar 2015).

Phospholipids

Phospholipids are a type of BS that contain phosphate groups connected to fatty acid groups (5–12 C atoms) with a CMC ranging from 58.9 mM to 0.007 mM. These biomolecules find various industrial applications in food preparations and have surface tension reduction from72 to 21 mN/m (Sun et al. 2018).

Polymeric BS

Polymerics are higher molecular weight biomolecules (≥ 1000 Da) composed of complex structures of heteropolysaccharides with surface tension reduction from 72 to 30 mN/m (Shekhar et al. 2015). Emulsan, alasan and liposan are some of the best studied biosurfactants used by the food industry.

Particulate BS

Particulate BSs form microemulsions through partition of membrane vesicles and also play a significant role in alkanic absorption by microbial cells. They can be obtained from microbes including *Acinetobacter calcoaceticus*, *Pseudomonas marginalis* and *Cyanobacteria.*

3. Properties

For their feasibility and applicability in the food industry, it is essential to understand the individual properties of BSs like surface activity to comprehend their performance aspects in emulsification, cleansing and foaming. They have the ability to assemble and form micelles which enhance their specificity. A large number of micelles are responsible for lower surface tensions which are influenced by enhanced BS concentration. They impart surface tension reduction which makes them more efficient and effective in terms of their viable applications as compared to chemical surfactants (Sun et al. 2018). Surface and interfacial values measure effectiveness of BSs, whereas CMC values determine their efficiency (Luna et al. 2016). Presence of BSs has been found to increase biodegradability by solubilizing the pollutants (Akintunde et al. 2015). This property has its application in food cleansing processes, e.g., the removal of heavy metals (Arab and Mulligan 2018) from food by the formation of heavy metal-biosufactant precipitates as represented in Fig. 2. BSs with higher molar mass promote the interaction and reduce repulsive forces which prevent

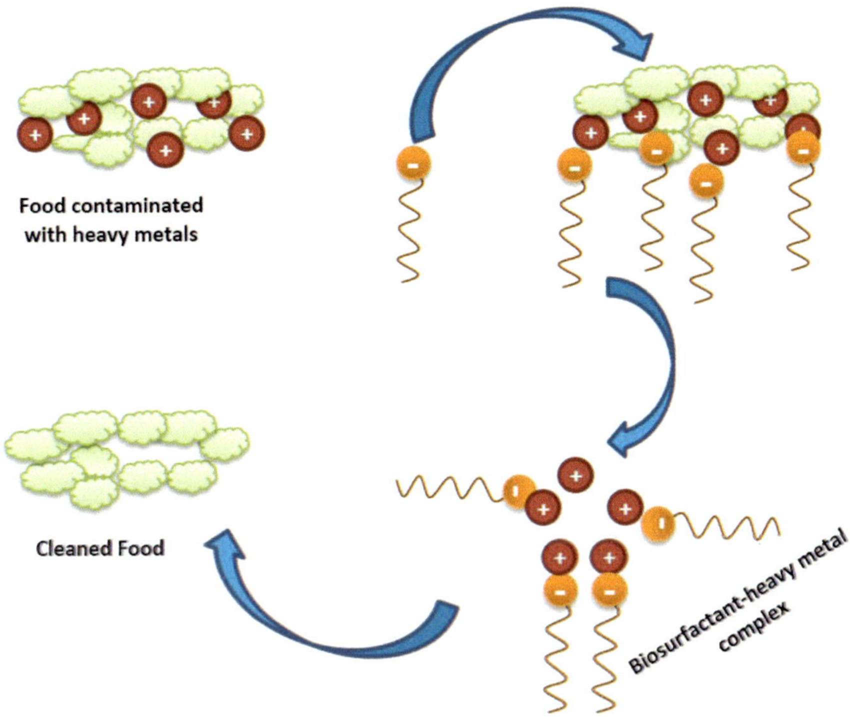

Figure 2. Food cleansing (heavy metals) mechanism of BS.

the formation of emulsions in different types of food formulations comprised of two phases likes oil and water (Fariq and Saeed 2016).

In addition, BSs have other functions like hydrophobic compound solubilization (Liu et al. 2017), and possess antioxidant, antimicrobial and antibiofilm properties (Kaczorek et al. 2018). Due to these properties BSs have progressively become the focus of exciting research for their potential applications. A brief account of different features pertaining to properties of BS is described below.

Biodegradability

The BSs that have a microbial origin can be degraded easily (Santos et al. 2016) while as surfactants synthesized chemically are a matter of concern and prompt environmental harm. Therefore demand and interest for less toxic and biodegradable surfactants as an alternative is rapidly increasing.

Tolerance to different pH, Temperature and Ionic Strength

This is the capacity of BS to function over a large range of environmental conditions. A majority of BSs are not affected by environmental factors like pH and temperature. Glycolipids (sophorolipids, rhamnolipids) remain effective at 120ºC and withstand a wide range of pH (5–10). The activity of bio emulsifier was stable to a wide range of sodium chloride concentrations, pH in the range of 3–7 and temperature (30–100ºC) (Mnif and Ghribi 2016).

Interface and Surface Activity

This is the ability of surface or adhesive forces of liquids to withstand external forces. Some BSs are very helpful in interfacial and surface tension reduction.

Low Toxicity

BS are usually considered as non-toxic biomolecules which makes them appropriate for food uses. Microbial-derived sophorolipids produced by *Candida bombicola* possess less toxicity profile which makes them convenient in food industries (Hoa et al. 2017).

4. Biosurfactant Production

Currently, BSs are formed by the fermentation process. The major challenge however is that the cost of BS production is 10–12 times more than that of synthetic surfactants (Lotfabad et al. 2016). In order to meet the growing demand for BSs, it is imperative to explore cost effective ways of production and yield optimization. The feat of formation of BSs depends upon the way of carrying out the fermentation process, the type of culture media, carbon source and the type of fermenting microorganism. Therefore alterations in conditions of culture have a substantial effect on the structure of BSs depending upon their desired application (Santos et al. 2016). The type of microorganism also influences the structure of the BS. For synthesis of various groups of BSs such as glycolipids, manno-proteins and other complexes having a blend of lipids, carbohydrates and proteins with the same carbon source are utilized. In addition, efficiency can be varied by measuring the conversion rate of substrate into product according to the metabolism used by the microorganism (Campos et al. 2015). High yields, alternative economically feasible substrates and cost efficient bioprocesses are the important challenges for competitive BS development. Different raw materials comprising of oil wastes, plant derived oils, distillery wastes, molasses and starchy substance are found to support the production of BSs (Rodrigues et al. 2017). The utilization of different other substrates such as green coconut, sugarcane bagasse and straw from the carnauba palm as carbon and nitrogen sources to obtain BS with suitable characteristics has also been reported recently.

Different bacteria (Jimoh and Lin 2019), Actinobacteria (Kuyukina et al. 2015), fungi (Silva et al. 2018) and yeasts (Jezierska et al. 2018) have been tested as producers of glycolipid biosurfactants. In general, *Pseudomonas* sp. are known for rhamnolipid BS production. Mannosylerythritol lipid and Sophorolipid BS are primarily produced by strains belonging to *Candida* (Dolman et al. 2017),

Table 1. Different substrates along with different types of BS and the generating strain.

Raw material	**Producer strain**	**Biosurfactant**	**Reference**
Waste canola oil	*Pseudomonas aeruginosa*	Rhamnolipids	(Pérez-Armendárizn et al. 2019)
Agro-industrial wastes	*Pseudomonas aeruginosa AB4*	Rhamnolipids	(Gudiña et al. 2016)
Corncob	*Starmerella bombicola*	Sophorolipids	(Konishi et al. 2015)
Waste cooking oil	*Pseudozyma aphidis*	Mannosylerythritol lipids	(Niu et al. 2019)
Castor oil	*Pseudozyma tsukubaensis*	Mannosylerythritol Lipids	(Niu et al. 2019)
Waste frying oils	*Pseudomonas aeruginosa OG1*	Rhamnolipids	(De Rienzo et al. 2016)
Palm oil soapstocks	*Stenotrophomonas acidaminiphila TW3*	Glycolipids	(Onlamool et al. 2020)

Wickerhamiella (Liu et al. 2016) and *Pseudozyma* (Claus and Van Bogaert 2017) species. Table 1 lists the different substrates used in the processing of different types of BSs and the generating strains.

Biosurfactant-producing Yeasts

Scientific investigations reveal that yeasts received more focus over bacteria, as bacterial BSs have limited use in the cosmetics and the food industry as a result of the pathogenic features of the majority besides, in contrast with yeasts, bacteria produce lower BS concentrations (Ribeiro et al. 2020). Some yeasts are non-toxic or non-pathogenic which allows them to be utilized for a wide range of industrial applications (Nwaguma et al. 2019). The yeasts that have been recently reported for BS production include *Starmerella bombicola* (Wang et al. 2019), *C. utilis* (Campos et al. 2019), *C. lipolytica* (Santos et al. 2017), *Candida sphaerica* (Luna et al. 2016), *Saccharomyces cerevisiae* (Kreling et al. 2020) and *Meyerozyma guilliermondii* (Kapoor et al. 2020). They have the ability to produce compounds with surfactant activities besides emulsification, antioxidant and anti-microbial properties. Yeasts belonging to genus *Saccharomyces* are commonly explored in the food industry. Table 2 lists the different substrates used in production of different types of BSs using yeast generating strains.

4.1 Purification and Extraction

Glycolipid biosurfactants need to be purified from their fermentation broth for usage in agriculture or food additives. In fact, without an appropriate and economical way for product recovery, the manufacturing process remains incomplete. Various techniques for extracting and purifying glycolipids have been developed which are selected according to physical and chemical properties and the possible application of the active surface compounds and their required degree of purity. The presence of the glycolipid BS in the culture medium or extract is determined by colorimetric

Table 2. Different substrates along with different types of BS and the generating yeast strain.

Raw material	Producer yeast strain	Biosurfactant	Reference
Sorghum bagasse and corn fiber	*Candida bombicola*	Sophorolipids	(Samad et al. 2015)
Agricultural by-products	*C. tropicalis*	sophorolipids	(Ojha et al. 2019)
Corn steep liquor and whey	*C. glabrata*	lipopeptides	(Lima et al. 2017)
Sugarcane molasses, waste frying oil and corn steep liquor	*C. bombicola.*	sophorolipids	(Resende et al. 2019)
Olive oil and ammonium nitrate	*S. cerevisiae*	Glycolipids	(Ribeiro et al. 2020)

methods before extraction and purification. These tests are used to identify the glycolipid sugar moieties, with the most widely used anthrone or orcinol test. Thin-layer chromatography (TLC) is an easy means of detecting glycolipid existence (Kubicki et al. 2020) and should be determined before purification. Solvent extraction, acid precipitation and foam fractionation are the most common techniques used for glycolipid extraction. BSs are soluble in organic solvents due to their hydrophobic moieties. Ethanol, methanol, butanol, acetone and ethyl acetate are most commonly used solvents for BS recovery. Acid precipitation and ethyl acetate extraction are typically common methods used for glycolipid recovery from fermentation media (Najmi et al. 2018). Chromatography, ionic exchange, membrane ultrafiltration and adsorption on resins or activated carbon are different strategies developed for purification. Gel filtration and high performance liquid chromatography are used for fractionation and purification of glycolipid BSs. Scientific investigation suggested that a single downstream approach is not enough to recover and purify the substance, instead a multi-stage recovery approach is more successful. Therefore, the substance of desired degree of purity can be obtained in such a multi-stage biosurfactant recovery.

5. Applications of BSs in Foods

To foster feasible and sustainable products stringent microorganism control measures are levied by the food industry to safeguard consumer safety and human well being. The generated end product should be of premier quality and acceptance containing lower contents of artificial and chemically manufactured compounds and greater natural constituents to meet the expectations of a consumer considerably (Garcia-Ortiz et al. 2020). Over the years the demand has ascended to reduce the synthetic additives in food products by utilizing the plant based mixtures (lecithin and gum Arabic) which have been commercially produced and are well recognized (Hasenhuettl 2019). However process formulation modification (microwave cooking and irradiation) profoundly impact the essential properties of these natural ingredients. Therefore it necessitates the formulation of novel additives with stabilizing, emulsifying properties including their production, isolation and identification (Faustino and

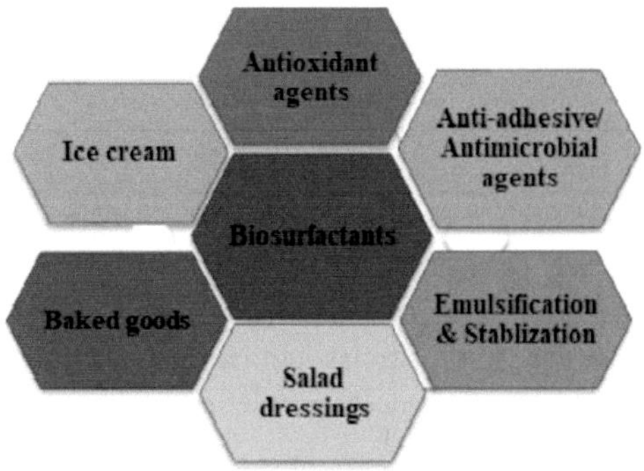

Figure 3. Applications of biosurfactants in food.

Vasco 2019). Figure 3 presents the application of BSs for various food processing purposes.

5.1 Biosurfactants as Emulsifying and Stabilizing Agents

Emulsification plays a major role in the uniformity and texture of various food products. Food emulsions are typically complex mixtures of several modules impacted by parameters like pH or salt content and processing conditions significantly affect their stability. For many centuries numerous compounds have been utilized as emulsifiers in food formulations to aid in formation and stabilization of emulsions to improve their sensory attributes (appearance, consistency, and texture) and enhancing shelf life and large scale storage. Successful emulsion prompts homogenization to increase the stability and formation of lipid droplets to minimize the phase separation to increase the capacity to control the clustering of globules (McClements and Gumus 2016). Food industries use a variety of additives (thickeners, stabilizers and emulsifiers) to enhance the performance which otherwise becomes a difficult entity to achieve due to presence of complex microstructures (Souza et al. 2016, Pascual-Villalobos et al. 2019). Biosurfactants can be used as a replacement or enhancers to diminish the use of emulsifiers to maintain the novel integrity and texture of foods (Pascual-Villalobos et al. 2019). The natural compounds amplify the characteristics of food and increase their tolerance and adaptability to adverse conditions. Bio surfactants are actually denominated as bio emulsifiers with extremely good emulsifying activity reported in systems consisting of two or more phases (Gallo et al. 2019, Campos et al. 2019).

Being used most frequently as food additives, the market value of emulsifiers has shown an increasing trend over the last few years. This rise in market value has opened new windows for emerging emulsifiers such as biosurfactants owing to their excellent surface and emulsifying activities. Different emulsifying properties of biosurfactants such as *Acinetobacter radioresistens* KA53 producing Alasan, Mannoproteins from *Saccharomyces cerevisiae*, Uronic acid bioemulsifiers from *Halomonas eurihalina* and *Klebsiella* species have been reported and investigated (Uzoigwe et al. 2015).

5.2 *Biosurfactants in Salad Dressings*

Investigations carried out by the scientific community have indicated that microbial surfactants are as effective emulsifiers in salad dressings such as mayonnaise maintaining the consistency and texture intact. In order to achieve better stability and consistency biosurfactants from *C. utilis* have been used in salad dressings in combination with metal spatula (Campos et al. 2019). Due to the solubilisation capacity and efficiency under harsh conditions of temperature and extreme pH biosurfactants show promise for this type of application as indicated by the investigations that yeast producing biosurfactants show enhanced stabilizing activity and unveil substantial thermal stability (Ribeiro et al. 2020).

5.3 *Baked Goods*

In recent years the consumer demand for healthier novel food formulations has tremendously increased and as a result the presence flour oriented foods has expanded at a phenomenal rate achieving inordinate prominence (Antoniewska et al. 2018). The addition of essential ingredients and additives for their higher nutritional values have been extensively investigated. The utilization of biosurfactants in baked goods has proved to be an imperative formulation necessarily for improving the product quality. Although only a little exploration has taken place, biosurfactant formulations have enhanced the textile profile of the dough, volume and conservation of baked goods demonstrating its vibrant potential in the industrial sector (Antoniewska et al. 2018). Bio emulsifiers obtained from *B. subtilis* when employed have significantly improved the bread texture profile as well as reduced the microbial proliferation sensitivity in comparison to commercially prepared emulsifiers like glycerol monsteareate (Mnif et al. 2015). Similarly biosurfactants produced from *Nesterenkonia* sp. and *S. cerevisiae* reduce the hardness and chewiness besides improving the cohesion and elasticity resulting in softer products of high quality (Kiran et al. 2017). Studies have indicated that a biosurfactant produced from *C. bombicola* improves the thermal and antioxidant properties in muffin formulations and this biomolecule can be utilized as an alternative for vegetable fat (da Silva Araújo et al. 2020). Such a formulation does not only improve the physical and textural setup but also enhances the nutritional value of muffins by replacing trans-fatty acids.

5.4 *Biosurfactants as Antimicrobial and Anti-adhesive Agents in Foods*

Biosurfactants have also emerged as potent agents for their antimicrobial and anti-adhesive properties and their substantial role in microbial proliferation prevention and control adherence of microorganisms ensuring safe product quality (Kieliszek et al. 2017). Food industries can implement the biosurfactant approach for decreasing microbial adhesion and pathogenic colonization employing mechanism of adsorption as indicated by the investigations. Due to their anti-microbial property, biosurfactants are considered to be safe biological entities. The application of biosurfactants as potential anti-adhesive agents has been utilized in dairy processing products preventing the encrustation problem caused by microorganism adhesion to heat exchangers (Marcelino et al. 2020).

5.5 *Biosurfactants as Antioxidant Agents in Foods*

The antioxidant property is reflected as an imperative characteristic in food industry formulations owing to their capability to lessen the rate of cardiovascular diseases and increase the shelf life of products by impeding or retarding oxidation reactions. There is growing consumer concern for healthier foods and such food formulations with these properties favor higher sales and consumption (Balan et al. 2019). Biosurfactants have been reported to have potential as antioxidant agents, as they are proficient in free radical sequestration. The antioxidant activity of biosurfactants can also be attributed to the presence of unsaturated fatty acids (Kiran et al. 2017). Biosurfactants produced from *C. utilis* have shown higher complex reducing capacities almost greater than 70% as compared to those for radical sequestration. A linear relationship has also been reported between the biosurfactants concentration and antioxidant activity (Ribeiro et al. 2020). Similarly biosurfactants obtained from Lactobacillus has shown the best results (> 77%) at the same concentration (Merghni et al. 2017). Among the most versatile biosurfactants, Mannosylerythritol lipids (MELs) possess strong interfacial and biochemical properties. Their antioxidant activity has been determined using free radical and superoxide anion scavenging tests *in vitro* and these tests have shown that it has highest antioxidant and protective effects in cells and can be utilized as anti-ageing and whitening skin ingredients (Yoo et al. 2019). Biosurfactants obtained from *B. subtilis* RW-1 can be employed as an alternative natural antioxidant. Similarly a Polysaccharide emulsifier from *Klebsiella* is under development in France that has been reported to be a source of rhamnose from which furaneol can be obtained that can act as a flavor precursor (McClements and Gumus 2016).

5.6 *Ice cream*

Although the use of biosurfactants in frozen desserts has not been validated by researchers, however some authors have reported that there can be a possibility that biosurfactants could be important ingredients and emulsifiers in ice cream formulations producing quality end products. This has further resulted in increased control consistency and maintains physical integrity by solubilizing aromatic oils (Anjum et al. 2016, Kieliszek et al. 2017). Biosurfactants form a thin easily penetrable layer that imparts strong resistance to partial coalescence in ice creams. Biosurfactants in combination with globular proteins (hydrophobins) produced by fungi can also be used as stabilizing agents in ice creams as they strongly resist denaturation utilizing adsorption and adhesion agents (Penfold and Thomas 2019). Biosurfactants with low molecular weight have also shown a tremendous potential and promise as stabilizing agents in ice creams owing to their low surface tension and lower balancing time for emulsions. There is also another viable approach in which a combination of both low molecular weight surfactants and mixture of proteins can be employed, however utmost care should be taken such that the proportion should be in synergy without disrupting the adsorption process between the different phases (Zhu et al. 2019).

Table 3. Application of biosurfactants in food industry and their source microorganisms.

Biosurfactants	**Microorganism**	**Application**	**References**
Glycolipids	*Bacillus* sp. *MTCC5877*	Anti-adhesive, antimicrobial, removal of heavy metals from food	(Anjum et al. 2016)
Glycolipipeptides	*Candida utilis*	Emulsifier for salad dressing	(Campos et al. 2019)
Glycolipids	*Saccharomyces cerevisiae URM 6670*	Replacement of egg yolk in bakery	(Ribeiro et al. 2020)
Sophorolipids	*Candida albicans SC5314 and Candida glabrata CBS138*	Emulsifying ability and antibacterial agent	(Gaur et al. 2019)
Sophorolipids	*Candida bombicola ATCC 22214*	Antimicrobial agent	(Yang et al. 2019)
Rhamnolipids	*Pseudomonas aeruginosa*	Emulsification	(Ozdal et al. 2017)
Lipopeptides	*Nesterenkonia* sp.	Emulsifier and emulsion stabilizing agent	(Kiran et al. 2017)
Sophorolipids	*Candida albicans SC5314 and Candida glabrata CBS138*	Emulsifying ability	(Gaur et al. 2019)
Rhamnolipids	*Pseudomonas aeruginosa*	Antimicrobial activity	(Vijayakumar and Saravanan 2015)

5.7 *Technological Challenges in the Food Industry*

The application of biosurfactants in food industry formulations in addition to nutrition value also has resulted in enhancement in physical as well as textural characteristics but still their usage in food industry is still limited and has not been exploited to a greater extent. The possible reason could be the food matrix complexity and different ratio of components subjected to diverse types of processing (Coelho et al. 2020). Thus it is very important to understand food formulation complexity as well as processing techniques and interactions of the ingredients present to efficiently utilize the concept of biosurfactants in food industry without hampering the quality and sustainment of the product. Another imperative obstacle that limits the biosurfactants usage in food industry is high cost of application compromising its competitiveness in relation to the ingredients already used. Therefore it is mandatory to establish a sustainable economic approach not only to reduce the volume of biosurfactants used but also impart greater efficiency in low concentrations (da Silva Araújo et al. 2020).

6. Conclusions

Biosurfactants exhibit various remarkable properties like emulsification, biodegradability and low toxicity which enable their utilization in different food formulations. They have potential food preservation and quality improvement applications. Food valorization through production of glycolipids has emerged as a convenient technology in terms of production cost reduction and byproduct management. Different methods have been formulated for purification and extraction of biosurfactants for their use in high value products. Further, the reduction of cost for bioprocessing of biosurfactants and enhanced yield should be a matter of interest for future study.

References

Akintunde, T., O. Abioye, S. Oyeleke, B. Boboye, U. Ijah et al. 2015. Remediation of iron using rhamnolipid-surfactant produced by *Pseudomonas aeruginosa*. Research Journal of Environmental Sciences 9(4): 169.

Anjum, F., G. Gautam, G. Edgard and S. Negi. 2016. Biosurfactant production through *Bacillus* sp. MTCC 5877 and its multifarious applications in food industry. Bioresource Technology 213: 262–269.

Antoniewska, A., J. Rutkowska, M.M. Pineda and A. Adamska. 2018. Antioxidative, nutritional and sensory properties of muffins with buckwheat flakes and amaranth flour blend partially substituting for wheat flour. LWT 89: 217–223.

Arab, F. and C.N. Mulligan. 2018. An eco-friendly method for heavy metal removal from mine tailings. Environmental Science and Pollution Research 25(16): 16202–16216.

Bahia, F.M., G.C. de Almeida, L.P. de Andrade, C.G. Campos, L.R. Queiroz et al. 2018. Rhamnolipids production from sucrose by engineered *Saccharomyces cerevisiae*. Scientific Reports 8(1): 1–10.

Balan, S.S., P. Mani, C.G. Kumar and S. Jayalakshmi. 2019. Structural characterization and biological evaluation of Staphylosan (dimannooleate), a new glycolipid surfactant produced by a marine *Staphylococcus saprophyticus* SBPS-15. Enzyme and Microbial Technology 120: 1–7.

Banat, I. and R. Thavasi. 2018. Downstream Processing of Microbial Biosurfactants and their Environmental and Industrial Applications (pp. 16): CRC Press.

Bi, J., F. Yang, D. Harbottle, E. Pensini, P. Tchoukov, S. Simon et al. 2015. Interfacial layer properties of a polyaromatic compound and its role in stabilizing water-in-oil emulsions. Langmuir 31(38): 10382–10391.

Campos, J.M., T.L. Stamford, R.D. Rufino, J.M. Luna, T.C.M. Stamford et al. 2015. Formulation of mayonnaise with the addition of a bioemulsifier isolated from *Candida utilis*. Toxicology Reports 2: 1164–1170.

Campos, J., T. Stamford and L. Sarubbo. 2019. Characterization and application of a biosurfactant isolated from *Candida utilis* in salad dressings. Biodegradation 30(4): 313–324.

Claus, S. and I.N. Van Bogaert. 2017. Sophorolipid production by yeasts: a critical review of the literature and suggestions for future research. Applied Microbiology and Biotechnology 101(21): 7811–7821.

Coelho, A.L.S., P.E. Feuser, B.A.M. Carciofi, D. de Oliveira, C.J. de Andrade et al. 2020. Biological activity of mannosylerythritol lipids on the mammalian cells. Applied Microbiology and Biotechnology, pp. 1–11.

da Silva Araújo, S.C., R.C. Silva-Portela, D.C. de Lima, M.M.B. da Fonsêca, W.J. da Silva Araújo et al. 2020. MBSP1: a biosurfactant protein derived from a metagenomic library with activity in oil degradation. Scientific Reports 10(1): 1–13.

Dobler, L., B.R. de Carvalho, W.d.S. Alves, B.C. Neves, D.M.G. Freire et al. 2017. Enhanced rhamnolipid production by *Pseudomonas aeruginosa* overexpressing estA in a simple medium. PloS one 12(8): e0183857.

Dolman, B.M., C. Kaisermann, P.J. Martin and J.B. Winterburn. 2017. Integrated sophorolipid production and gravity separation. Process Biochemistry 54: 162–171.

Dolman, B.M., F. Wang and J.B. Winterburn. 2019. Integrated production and separation of biosurfactants. Process Biochemistry 83: 1–8.

Fariq, A. and A. Saeed. 2016. Production and biomedical applications of probiotic biosurfactants. Current Microbiology 72(4): 489–495.

Faustino, B. and A.B. Vasco. 2019. Early maladaptive schemas and cognitive fusion on the regulation of psychological needs. Journal of Contemporary Psychotherapy, pp. 1–8.

Fei, D., G.W. Zhou, Z.Q. Yu, H.Z. Gang, J.F. Liu, S.Z. Yang et al. 2020. Low-Toxic and nonirritant biosurfactant surfactin and its performances in detergent formulations. Journal of Surfactants and Detergents 23(1): 109–118.

Fernández-Peña, L., E. Guzmán, F. Leonforte, A. Serrano-Pueyo, K. Regulski et al. 2020. Effect of molecular structure of eco-friendly glycolipid biosurfactants on the adsorption of hair-care conditioning polymers. Colloids and Surfaces B: Biointerfaces 185: 110578.

Gallo, S., T. Brochado, L. Brochine, D. Passareli, S. Costa et al. 2019. Effect of biosurfactant added in two different oil source diets on lamb performance and ruminal and blood parameters. Livestock Science 226: 66–72.

Garcia-Ortiz, A., K.S. Arias, M.J. Climent, A. Corma, S. Iborra et al. 2020. Transforming methyl levulinate into biosurfactants and biolubricants by chemoselective reductive etherification with fatty alcohols. ChemSusChem. 13(4): 707–714.

Gaur, V.K., R.K. Regar, N. Dhiman, K. Gautam, J.K. Srivastava et al. 2019. Biosynthesis and characterization of sophorolipid biosurfactant by *Candida* spp.: Application as food emulsifier and antibacterial agent. Bioresource Technology 285: 121314.

Gudiña, E.J., A.I. Rodrigues, E. Alves, M.R. Domingues, J.A. Teixeira et al. 2015. Bioconversion of agro-industrial by-products in rhamnolipids toward applications in enhanced oil recovery and bioremediation. Bioresource Technology 177: 87–93.

Gudiña, E.J., A.I. Rodrigues, V. de Freitas, Z. Azevedo, J.A. Teixeira et al. 2016. Valorization of agro-industrial wastes towards the production of rhamnolipids. Bioresource Technology 212: 144–150.

Gudiña, E.J. and L.R. Rodrigues. 2019. Research and production of biosurfactants for the food industry. Bioprocessing for Biomolecules Production, pp. 125–143.

Hasenhuettl, G.L. 2019. Synthesis and commercial preparation of food emulsifiers. Food Emulsifiers and their Applications (pp. 11–39): Springer.

Hoa, N.L.H. K. Eun-Ki, T.T. Ha, N.D. Duy, H.Q. Khanh et al. 2017. Production and characterization of sophorolipids produced by *Candida bombicola* grown on sugarcane molasses and coconut oil. Asia-Pacific Journal of Science and Technology 22(2).

Jadeja, N.B., P. Moharir and A. Kapley. 2019. Genome sequencing and analysis of strains *Bacillus* sp. AKBS9 and *Acinetobacter* sp. AKBS16 for biosurfactant production and bioremediation. Applied Biochemistry and Biotechnology 187(2): 518–530.

Jezierska, S., S. Claus and I. Van Bogaert. 2018. Yeast glycolipid biosurfactants. Febs Letters 592(8): 1312–1329.

Jezierska, S., S. Claus, R. Ledesma-Amaro and I. Van Bogaert. 2019. Redirecting the lipid metabolism of the yeast *Starmerella bombicola* from glycolipid to fatty acid production. Journal of Industrial Microbiology and Biotechnology 46(12): 1697–1706.

Jha, S.S., S.J. Joshi and G. SJ. 2016. Lipopeptide production by *Bacillus subtilis* R1 and its possible applications. Brazilian Journal of Microbiology 47(4): 955–964.

Jimoh, A.A. and J. Lin. 2019. Biosurfactant: A new frontier for greener technology and environmental sustainability. Ecotoxicology and Environmental Safety 184: 109607.

Kaczorek, E., A. Pacholak, A. Zdarta and W. Smułek. 2018. The impact of biosurfactants on microbial cell properties leading to hydrocarbon bioavailability increase. Colloids and Interfaces 2(3): 35.

Kapoor, D., P. Sharma, M.M.M. Sharma, A. Kumari, R. Kumar et al. 2020. Microbes in pharmaceutical industry. Microbial Diversity, Interventions and Scope (pp. 259–299): Springer.

Kieliszek, M., A.M. Kot, A. Bzducha-Wróbel, S. BŁażejak and I. Gientka. 2017. Biotechnological use of Candida yeasts in the food industry: a review. Fungal Biology Reviews 31(4): 185–198.

Kiran, G.S., S. Priyadharsini, A. Sajayan, G.B. Priyadharsini, N. Poulose et al. 2017. Production of lipopeptide biosurfactant by a marine *Nesterenkonia* sp. and its application in food industry. Frontiers in Microbiology 8: 1138.

Konishi, M., Y. Yoshida and J.-i. Horiuchi. 2015. Efficient production of sophorolipids by *Starmerella bombicola* using a corncob hydrolysate medium. Journal of Bioscience and Bioengineering 119(3): 317–322.

Kreling, N., M. Zaparoli, A. Margarites, M. Friedrich, A. Thomé et al. 2020. Extracellular biosurfactants from yeast and soil–biodiesel interactions during bioremediation. International Journal of Environmental Science and Technology 17(1): 395–408.

Kuyukina, M.S., I.B. Ivshina, T.A. Baeva, O.A. Kochina, S.V. Gein et al. 2015. Trehalolipid biosurfactants from nonpathogenic *Rhodococcus actinobacteria* with diverse immunomodulatory activities. New Biotechnology 32(6): 559–568.

Kuyukina, M.S. and I.B. Ivshina. 2019. Production of trehalolipid biosurfactants by *Rhodococcus*. Biology of Rhodococcus (pp. 271–298): Springer.

Kubicki, S., I. Bator, S. Jankowski, K. Schipper, T. Tiso et al. 2020. A straightforward assay for screening and quantification of biosurfactants in microbial culture supernatants. Frontiers in Bioengineering and Biotechnology 8: 958.

Lima, R.A., R.F. Andrade, D.M. RodrÃguez, H.W. Araújo, V.P. Santos et al. 2017. Production and characterization of biosurfactant isolated from *Candida glabrata* using renewable substrates. African Journal of Microbiology Research 11(6): 237–244.

Liu, X.-g. X.-j. Ma, R.-s. Yao, C.-y. Pan, H.-b. He et al. 2016. Sophorolipids production from rice straw via SO 3 micro-thermal explosion by Wickerhamiella domercqiae var. sophorolipid CGMCC 1576. AMB Express 6(1): 60.

Liu, Y., G. Zeng, H. Zhong, Z. Wang, Z. Liu et al. 2017. Effect of rhamnolipid solubilization on hexadecane bioavailability: enhancement or reduction? Journal of Hazardous Materials 322: 394–401.

Lotfabad, T.B., N. Ebadipour and R. RoostaAzad. 2016. Evaluation of a recycling bioreactor for biosurfactant production by *Pseudomonas aeruginosa* MR01 using soybean oil waste. Journal of Chemical Technology and Biotechnology 91(5): 1368–1377.

Louhasakul, Y., B. Cheirsilp, R. Intasit, S. Maneerat, A. Saimmai et al. 2020. Enhanced valorization of industrial wastes for biodiesel feedstocks and biocatalyst by lipolytic oleaginous yeast and biosurfactant-producing bacteria. International Biodeterioration and Biodegradation 148: 104911.

Luna, J.M., A. Santos Filho, R.D. Rufino and L.A. Sarubbo. 2016. Production of biosurfactant from *Candida bombicola* URM 3718 for environmental applications. Chemical Engineering Transactions 49: 583–588.

Marcelino, P.R.F., F. Gonçalves, I.M. Jimenez, B.C. Carneiro, B.B. Santos et al. 2020. Sustainable production of biosurfactants and their applications. Lignocellulosic Biorefining Technologies, pp. 159–183.

McClements, D.J. and C.E. Gumus. 2016. Natural emulsifiers—Biosurfactants, phospholipids, biopolymers, and colloidal particles: Molecular and physicochemical basis of functional performance. Advances in Colloid and interface Science 234: 3–26.

Meena, K.R. and S.S. Kanwar. 2015. Lipopeptides as the antifungal and antibacterial agents: applications in food safety and therapeutics. BioMed Research International.

Merghni, A., I. Dallel, E. Noumi, Y. Kadmi, H. Hentati et al. 2017. Antioxidant and antiproliferative potential of biosurfactants isolated from *Lactobacillus casei* and their anti-biofilm effect in oral *Staphylococcus aureus* strains. Microbial Pathogenesis 104: 84–89.

Mnif, I., S. Mnif, R. Sahnoun, S. Maktouf, Y. Ayedi et al. 2015. Biodegradation of diesel oil by a novel microbial consortium: comparison between co-inoculation with biosurfactant-producing strain and exogenously added biosurfactants. Environmental Science and Pollution Research 22(19): 14852–14861.

Mnif, I. and D. Ghribi. 2016. Glycolipid biosurfactants: main properties and potential applications in agriculture and food industry. Journal of the Science of Food and Agriculture 96(13): 4310–4320.

Najmi, Z., G. Ebrahimipour, A. Franzetti and I.M. Banat. 2018. *In situ* downstream strategies for cost-effective bio/surfactant recovery. Biotechnology and Applied Biochemistry 65(4): 523–532.

Niu, Y., J. Wu, W. Wang and Q. Chen. 2019. Production and characterization of a new glycolipid, mannosylerythritol lipid, from waste cooking oil biotransformation by *Pseudozyma aphidis* ZJUDM34. Food science and Nutrition 7(3): 937–948.

Nwaguma, I., C. Chikere and G. Okpokwasili. 2019. Isolation and molecular characterization of biosurfactant-producing yeasts from saps of Elaeis guineensis and Raphia africana. Microbiology Research Journal International, pp. 1–12.

Ojha, N., S.K. Mandal and N. Das. 2019. Enhanced degradation of indeno (1, 2, 3-cd) pyrene using *Candida tropicalis* NN4 in presence of iron nanoparticles and produced biosurfactant: a statistical approach. 3 Biotech 9(3): 1–13.

Onlamool, T., A. Saimmai, N. Meeboon and S. Maneerat. 2020. Enhancement of glycolipid production by *Stenotrophomonas acidaminiphila* TW3 cultivated in low cost substrate. Biocatalysis and Agricultural Biotechnology, 101628.

Ozdal, M., S. Gurkok and O.G. Ozdal. 2017. Optimization of rhamnolipid production by *Pseudomonas aeruginosa* OG1 using waste frying oil and chicken feather peptone. 3 Biotech 7(2): 117.

Pascual-Villalobos, M.J., P. Guirao, F.G. Díaz-Baños, M. Cantó-Tejero, G. Villora et al. 2019. Oil in water nanoemulsion formulations of botanical active substances. Nano-Biopesticides Today and Future Perspectives (pp. 223–247): Elsevier.

Penfold, J. and R. Thomas. 2019. Adsorption properties of plant based bio-surfactants: Insights from neutron scattering techniques. Advances in Colloid and interface Science 274: 102041.

Pérez-Armendáriz, B., C. Cal-y-Mayor-Luna, E.G. El-Kassis and L.D. Ortega-Martínez. 2019. Use of waste canola oil as a low-cost substrate for rhamnolipid production using *Pseudomonas aeruginosa*. AMB Express 9(1): 61.

Radzuan, M.N., I.M. Banat and J. Winterburn. 2017. Production and characterization of rhamnolipid using palm oil agricultural refinery waste. Bioresource Technology 225: 99–105.

Rane, A.N., V.V. Baikar, V. Ravi Kumar and R.L. Deopurkar. 2017. Agro-industrial wastes for production of biosurfactant by *Bacillus subtilis* ANR 88 and its application in synthesis of silver and gold nanoparticles. Frontiers in Microbiology 8: 492.

Rebello, S., A. Anoopkumar, R. Sindhu, P. Binod, A. Pandey et al. 2020. Comparative life-cycle analysis of synthetic detergents and biosurfactants—an overview. Refining Biomass Residues for Sustainable Energy and Bioproducts (pp. 511–521): Elsevier.

Resende, A.H.M., J.M. Farias, D.D. Silva, R.D. Rufino, J.M. Luna et al. 2019. Application of biosurfactants and chitosan in toothpaste formulation. Colloids and Surfaces B: Biointerfaces 181: 77–84.

Ribeiro, B.G., J.M. Guerra and L.A. Sarubbo. 2020. Biosurfactants: production and application prospects in the food industry. Biotechnology Progress, e3030.

Ribeiro, B.G., M.M. dos Santosb, I.A. da Silvac, H.M. Meirad, A.M. de Oliveirad et al. 2020. Study of the biosurfactant production by *Saccharomyces cerevisiae* URM 6670 using agroindustrial waste. Chemical Engineering, 79.

Rodrigues, M.S., F.S. Moreira, V.L. Cardoso and M.M. de Resende. 2017. Soy molasses as a fermentation substrate for the production of biosurfactant using *Pseudomonas aeruginosa* ATCC 10145. Environmental Science and Pollution Research 24(22): 18699–18709.

Saimmai, A., W. Riansa-ngawong, S. Maneerat and P. Dikit. 2020. Application of biosurfactants in the medical field. Walailak Journal of Science and Technology (WJST) 17(2): 154–166.

Samad, A., J. Zhang, D. Chen and Y. Liang. 2015. Sophorolipid production from biomass hydrolysates. Applied Biochemistry and Biotechnology 175(4): 2246–2257.

Santos, D.K.F., R.D. Rufino, J.M. Luna, V.A. Santos, L.A. Sarubbo et al. 2016. Biosurfactants: multifunctional biomolecules of the 21st century. International Journal of Molecular Sciences 17(3): 401.

Santos, D.K.F., H.M. Meira, R.D. Rufino, J.M. Luna, L.A. Sarubbo et al. 2017. Biosurfactant production from *Candida lipolytica* in bioreactor and evaluation of its toxicity for application as a bioremediation agent. Process Biochemistry 54: 20–27.

Shekhar, S., A. Sundaramanickam and T. Balasubramanian. 2015. Biosurfactant producing microbes and their potential applications: a review. Critical Reviews in Environmental Science and Technology 45(14): 1522–1554.

Silva, A.C.S.d., P.N.d. Santos, R.F.S. Andrade and G.M. Campos-Takaki. 2018. Biosurfactant production by fungi as a sustainable alternative. Arquivos Do Instituto Biológico, 85.

Souza, P. M., M. Freitas-Silva, T.A.d.L. e Silva, G.K. Silva, M.A. Lima et al. 2016. Factorial design based medium optimization for the improved production of biosurfactant by mucor polymorphosphorus. Int. J. Curr. Microbiol. App. Sci. 5(11): 898–905.

Sun, N., J. Chen, D. Wang and S. Lin. 2018. Advance in food-derived phospholipids: Sources, molecular species and structure as well as their biological activities. Trends in Food Science and Technology 80: 199–211.

Uzoigwe, C., J.G. Burgess, C.J. Ennis and P.K. Rahman. 2015. Bioemulsifiers are not biosurfactants and require different screening approaches. Frontiers in Microbiology 6: 245.

Varjani, S. and V.N. Upasani. 2019. Evaluation of rhamnolipid production by a halotolerant novel strain of *Pseudomonas aeruginosa*. Bioresource Technology 288: 121577.

Vijayakumar, S. and V. Saravanan. 2015. Biosurfactants-types, sources and applications. Research Journal of Microbiology 10(5): 181.

Wang, H., S.L. Roelants, M.H. To, R.D. Patria, G. Kaur, N.S. Lau et al. 2019. *Starmerella bombicola*: recent advances on sophorolipid production and prospects of waste stream utilization. Journal of Chemical Technology and Biotechnology 94(4): 999–1007.

Yang, L., Y. Li, X. Zhang, T. Liu, J. Chen et al. 2019. Metabolic profiling and flux distributions reveal a key role of acetyl-CoA in sophorolipid synthesis by *Candida bombicola*. Biochemical Engineering Journal 145: 74–82.

Yoo, J.W., Y.K. Hwang, B. Sung-Ah, Y.J. Kim, J.H. Lee et al. 2019. Skin whitening composition containing mannosylerythritol lipid: Google Patents.

Zhu, Z., B. Zhang, B. Chen, J. Ling, Q. Cai et al. 2019. Fly ash based robust biocatalyst generation: a sustainable strategy towards enhanced green biosurfactant production and waste utilization. RSC Advances 9(35): 20216–20225.

8

Rhamnolipid Biosurfactants

Production and Applications

Hebatallah H Abo Nahas,[1*] *Gunis Kibar,*[2,3,4] *Gasser M Khairy,*[5] *Sh Husien,*[6] *Hoda Azmy Elkot,*[7] *Ahmed M Abdel-Azeem*[8] and *Essa M Saied*[9,10*]

1. Introduction

Biosurfactants, also named microbial surfactants, are an emerging and assorted class of biomolecules that are being exploited in various fields. They are surface-active compounds produced by microorganisms (bacteria, yeast or fungi). As commonly known, biosurfactants are classified by their molecular weight and chemical composition. According to these combinations, there are low molecular weight biosurfactants, like, glycolipids, lipopeptides, fatty acids and phospholipids; or high molecular weight biosurfactants like polymeric biosurfactants (Vecino et al. 2017). Glycolipids are the most studied microbial biosurfactants and among them Rhamnolipids (RLs), are the glycolipid-type biosurfactants produced mostly by *Pseudomonas aeruginosa* (Sekhon Randhawa and Rahman 2014), and the other *Pseudomonas* species that have been detailed to produce RLs are *P. alcaligenes*,

[1] Zoology Department, Faculty of Science, Suez Canal University, Ismailia 41522, Egypt.
[2] Department of Materials Engineering, Adana Alparslan Turkes Science and Technology University, 01250 Adana, Turkey.
[3] Center for Engineering in Medicine, Massachusetts General Hospital, Harvard Medical School.
[4] Shriners Hospitals for Children, Boston, MA, 02114, USA.
[5] Chemistry Department, Faculty of Science, Suez Canal University, Ismailia 41522, Egypt.
[6] Egyptian Petroleum Research Institute (EPRI), Nasr City, Cairo 11727, Egypt.
[7] Neuroscience and Biotechnology,Faculty of Pharmacy, Masoura University, Egypt.
[8] Botany and Microbiology Department, Faculty of Science, Suez Canal University, Ismailia 41522, Egypt.
[9] Chemistry Department, Faculty of Science, Suez Canal University, Ismailia 41522, Egypt.
[10] Institute for Chemistry, Humboldt Universität zu Berlin, Brook-Taylor-Str. 2, 12489 Berlin, Germany.
* Corresponding authors: hebahasssan350@yahoo.com; saiedess@hu-berlin.de

P. chlororaphis, *P. fluorescens*, *P. nitroreducens* and *P. putida* (Bharali et al. 2018). RLs are known dominating commercial biosurfactants appropriate for different industries and bioremediations. They are composed of a hydrophilic head formed by one or two rhamnose molecules, known as mono-RL and di-RL respectively, and a hydrophobictail. The tail comprises up to three molecules of hydroxyl fatty acids of changing chain length from 8 to 14 off which β-hydroxy decanoic acid dominates (Abdel-Mawgoud et al. 2009).

Monteiro et al. (2005) revealed that RLs reduce the surface tension of water from 72 to 30 mN/m with a critical micelle concentration (CMC) of 5–200 mg.L^{-1} and show an emulsification index of over 70%. Being of microbial root, RLs with high biodegradability possess an extra advantage over the synthetic surfactants in soil washing and bioremediation processes (Silva et al. 2010). RLs possess the capability to enhance the microbial degradation of chlorinated hydrocarbons, heavy metals, PAHs and petroleum hydrocarbons from contaminated soil and water (Banat et al. 2000). A biosurfactant produced by *P. aeruginosa* SR17 was utilized to evaluate its proficiency in the improvement of bioremediation of oil contaminated soil. It was observed that the degradation of total petroleum hydrocarbon (TPH) along with the addition of RL at 1.5 g was 86.1% and 80.5% in two soil samples containing 6800 ppm and 8500 ppm of TPH respectively (Patowary et al. 2018).

Recently (2010–2016), 53 novel rhamnolipid homologs or congeners had also been discovered and identified. These compounds exhibited multi-physicological functions with multiple applications. *P. aeruginosa* has long been considered as the primary species that produces rhamnolipids. However, several other species which do not belong to the Pseudomonadacaeae family in the taxonomical classification are also recorded to produce rhamnolipids such as the pathogens *Burkholderia pseudomallei*, *Burkholderia mallei* and the nonpathogenic *Burkholderia thailandensis* (Toribio et al. 2010). *Acinetobacter calcoaceticus*, *Enterobacter* sp., and *Pantoea* sp. (Hošková et al. 2015). Clearly, the extended rhamnolipid-producing microorganisms will facilitate researchers to screen industrially safe and nonpathogenic alternatives to human pathogen *P. aeruginosa*. However, it is worth noting that when a novel bacterial species is recorded to synthesize rhamnolipids, in most cases only one isolate can be chosen for further studies, and metabolic pathways and engineering were not adequately elucidated. Thus, until now, only few microorganisms can be selected as real rhamnolipid producers. It became clear that wide range approaches will be required to be developed, based on the presence of the rhlA, rhlB, or rhl Chomologs in the biosynthetic genes of bacteria or many characterizations of rhamnolipid surfactants in all probability (Irorere et al. 2017) setup several standards for reporting new recommended rhamnolipid-producing strains. For example, genes responsible for rhamnolipid production should be completely characterized based on molecular and bioinformatics tools. The absolute quantitative yields and molecular structures of these compounds should be determined (Chen et al. 2017). RLs are not only efficient surfactants but also show great antimicrobial activity against many other pathogens and disrupt host defenses during infections (Nitschke et al. 2005). RLs have been considered in several applications which include fungicides, bactericides and wound healing (Aparna et al. 2012). The advantages of RLs over those chemically synthesized have been described extensively. Because of their latent advantages, biosurfactants are broadly used in many industries such as agriculture, chemical, cosmetics, food production, and pharmaceutics. Lately, new companies have been founded and are trying to establish a market in spite of the high cost of these products (Sekhon Randhawa and Rahman 2014). Nevertheless, the process is still conducted on only a slight scale and these manufacturers are mostly associated with universities or research centers (Rikalović et al. 2015). A growing number of studies and blatant registrations associated with rhamnolipid production are focused on overcoming the high cost of rhamnolipid synthesis. Essentially, they seek to find more desirable substrates for the process (Dobler et al. 2020) due to a high percentage of costs derived from the materials used such as carbon and nitrogen sources for the microorganism, and from the downstream processes (Sekhon Randhawa and Rahman 2014). Other investigations have focused on specific components of the process, ignoring other features that should be analyzed together, to considerably decrease the costs. Replacement of conventional substrates by industrial wastematerials is a valuable strategy to decrease the high prices of RLs. Studies in this area deal mostly with wastes from oil refineries or protein-based industries (Dobler et al. 2020). The *above-mentioned inconsistency regarding the actual efficiency of biosurfactants in* bioremediation and biomedicine is the driving force behind this manuscript, which is focused on providing a critical overview of recent advances in RL biosurfactants related studies. The aim of this chapter is to plot the development in the field of biosurfactant-mediated bioremediation and biomedicine covering techniques where such compounds have found considerable efficacy.

2. Swift Production of Rhamnolipid Biosurfactants

Rhamnolipids were distinguished by their excellent physico-chemical properties and ability to achieve high fermentation yields that made them potential candidates for study by a lot of researchers. They are a class of glycocidic biosurfactants which consist of rhamnose that is known as a sugar moiety "hydrophilic part" and β-hydroxylated fatty acid chains as a hydrophobic part (Abdel-Mawgoud et al. 2010, Chong and Li 2017). Rhamnolipids may contain one or more rhamnose molecules where rhamnolipids with one rhamnose sugar molecule are called mono-rhamnolipids, while the ones containing two rhamnose molecules are known as Di-rhamnolipids. Additionally, they have been reported as molecules with a high surface activity, which can reduce the surface tension from nearly ~72 mN/m to ~30 mN/m (Janek et al. 2013). By contrast to other chemical surfactants rhamnolipids, are highly biodegradable and biocompatible molecules giving them priority for application in many industries such as; cosmetics, food, pesticides, pharmaceuticals, bioremediation, petroleum recovery, detergents (Chen et al. 2010, Mańko et al. 2014). However, because of the huge industrial market demand for rhamnolipids, they are considered not compatible as chemical surfactants due to their high production costs, and their limited yield and applicability especially when produced on a large-scale (Dhanarajan and Sen 2014, Gudiña et al. 2016, Henkel et al. 2012, Hošková et al. 2015). In consequence, improving rhamnolipids by—scaling up the production process to prepare large compatible and marketable quantities generated a great interest lately (Jiang et al. 2020, Nitschke et al. 2005). Therefore, understanding the biosynthesis and genetic molecule regulation system, and optimization of fermentation strategies of rhamnolipids production helps improving its yield and productivity. As a result, alternative low cost production strategies can be applied in the rhamnolipids biosynthesis process (Lovaglio et al. 2015, Shao et al. 2017). In the following sections, key enzymes in rhamnolipids production, optimization fermentation strategies, and strategies for low cost production will be discussed as follows:

3. Key Enzymes in Rhamnolipid Synthesis

Rhamnolipids are a biosurfactant compounds that are produced from different microorganisms such as bacteria, fungi, and yeast, and the gram-negative bacteria *Pseudomonas aeruginosa* that are considered the main producer species for them (Gong et al. 2015, Kim et al. 2015, Rahman et al. 2010). Other species such as *Acinetobacter calcoaceticus* (Rooney et al. 2009), *Enterobacter* spp. (Hošková et al. 2013), *Burkholderia* spp. (Hošková et al. 2015), and *Nocardiopsis* sp. (Roy et al. 2015) were successfully applied for rhamnolipids production under suitable conditions. However, *Pseudomonas aeruginosa* has been used as a model strain to understand the main genes and enzymes that are responsible for rhamnolipids biosynthesis (Abdel-Mawgoud et al. 2014). Figure 1 presents a diagram that shows the key enzymes that help in rhamnolipids biosynthesis process and Table 1 has a summary of each enzyme name and role. The key enzymes will be explained through the elucidation of the rhamnolipids biosynthesis mechanism. For instance, the process of producing rhamnolipids in *Pseudomonas aeruginosa* involves several steps where the precursors for rhamnolipid synthesis are the sugar (dTDP-L-rhamnose) "hydrophilic part and hydrophobic moieties such 3-3-hydroxyalkanoyloxy alkanoic acid (HAA)".

Rhamnolipids biosynthesis process is stated in a three step process as follows.

3.1 *Sugar Moiety Biosynthesis*

Rhamnose is a deoxy-hexose sugar and its activated form L-rhamnose that was derived from a glucose scaffold through four steps by the influence of the four enzymes gene that are known as rmlBDAC. At first glucose-6-phosphate converted into glucose-1-phosphate by AlgC enzyme. After that, four repetitive enzymes was responsible for the convertion of glucose-6-phosphate into dTDP-L-rhamnose. The first enzyme is RmlA, which promoted the transfer of thymidyldiphosphate nucleotide (TDP) into glucose-1-phosphate by a nucleophilic attack of the oxygen from the hexose to the thymidyl triphosphate nucleotide (TTP), to form dTDP-D-glucose and two molecules of inorganic phosphate. Whereas, the second enzyme RmlB catalyzed the formation of dTDP-4-keto-6-deoxy-D-glucose by C4-OH group of the D-glucose oxidation and followed by H2O molecule removal (Allard et al. 2001). The third pathway enzyme catalyzed the epimerization of C3 and C5 in dTDP-4-keto-6-deoxy-D-glucose, forming dTDP-4-keto- 6-deoxy-L-mannose (Graninger et al. 1999). This molecule leads to the formation of dTDP-Lrhamnose by the act of the fourth enzyme RmlD (Lovaglio et al. 2015, Müller and Hausmann 2011, Olvera et al. 1999, Rahim et al. 2000).

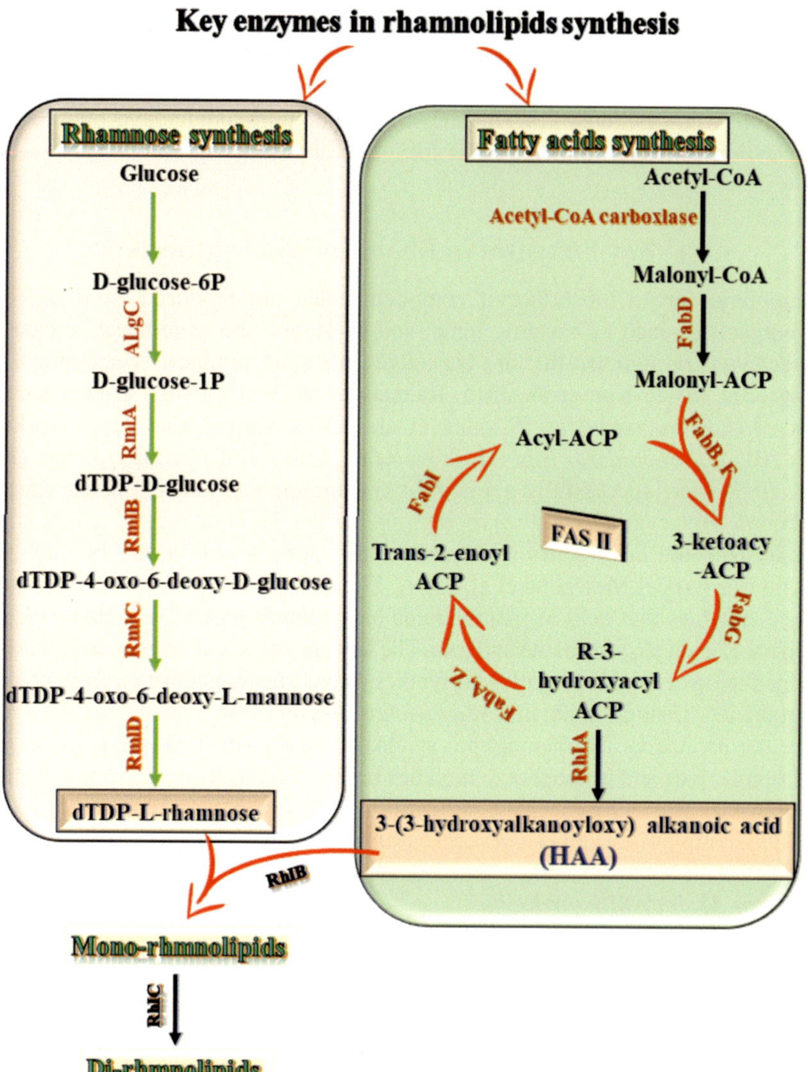

Figure 1. Diagram that explains the key Enzymes involved in the rhamnolipids biosynthesis process (Abdel-Mawgoud et al. 2014).

3.2 Lipid Moiety, 3-3-hydroxyalkanoyloxy Alkanoic Acid Biosyththesis

The enzymatic machinery for the FASII of the *Pseudomonas aeruginosa* is similar to the one described in *Esherchia coli* bacteria that consisted of two stages known as initiation and elongation stages (Abdel-Mawgoud et al. 2011). At first, in the initiation stage acetyle-CoA carboxylase enzyme catalyzes the synthesis of Malonyl CoA from Acetyle-CoA and Malonyl-ACP formation is catalyzed by FabD. Whereas

Table 1. Key enzymes name, function involved in the rhamnolipids biosynthesis process.

Enzyme	Name	Role	References
ALgC	Phosopho mannomutase	Involved in the synthesis of the HAAs, the fatty acid dimers, from two 3-hydroxyfatty acid precursors.	Pham et al. (2004)
RmlA	glucose 1-phosphate thymidytransferase	Promotes the transfer of thymidyldiphosphate nucleotide (TDP) to glucose-1-phosphate to form dTDP-D-glucose.	Giraud et al. (1992)
RmlB	dTDP-D-glucose-4,6-dehydratase	Catalyze the Oxidation process that occurs to convert dTDP-D-glucose to dTDP-4-keto-6-deoxy-D-glucose.	Lindhout et al. (2009)
RmlC	dTDP-4-dehydrorhamnose-3,5-epimerase-4,6-dehydratase	Catalyzes the epimerization of C3 and C5 in dTDP-4-keto-6-deoxy-D-glucose, which results in dTDP-4-keto-6-deoxy-L-mannose formation.	Rahim et al. (2000)
RmlD	dTDP-4-dehydrorhamnose reductase	Helps in the final formation of dTDP-L-rhamnose.	Graninger et al. (1999)
Acetyl-CoA carboxylase	-	First synthesizes Acetyl-CoA into Malonyl-CoA.	Kutchma et al. (1999)
FabD	malonyl-CoA-ACP transacylase	Derives Malonyl-ACP from Malonyl-CoA.	
FabB, F	beta-ketoacyl-ACP reductase	Catalyzes the condensation between Malonyl-ACP and Acyl-ACP, FabB for Saturated fattyacids, and FabF for unsaturated.	Hoang and Schweizer (1997)
FabG	NADBH-dependant enoyl-ACP reductase	Helps catalyze beta-katonyl into beta hydroxylacyl-ACP.	Campos-García et al. (1998)
FabA, Z	beta-hydroxylacyl-ACP dehydratase	Catalyzes β-hydroxyacyl-ACP dehydration to form trans-2-enoyl-ACP.	Hoang and Schweizer (1997)
FabI	NADH dependant enoyl-ACP reductase	Reduces trans-2-enoyl-ACP to Acyl-ACP.	Hoang and Schweizer (1997)
RhlA	3-(3hydroxyalkonyloxy) alkanoate synthetase	involved in the synthesis of the HAAs, the fatty acid dimers, from two 3-hydroxy fatty acid precursors.	Deziel et al. (2003)
RhlB	rhamnosytransferase I	Uses dTDP-L-rhamnose and an HAA molecule as precursors to produce mono-rhamnolipids. (The enzyme catalyzing the second reaction).	Burger et al. (1966)
RhlC	rhamnosytransferase II	Mono-RLs link with another molecule from dTDP-L-rhamnose, to produce di-rhamnolipids with its help (the enzyme catalyzing the third reaction).	Rahim et al. (2001)

in the elongation stage, four chemical reactions were stated as follows: The first reaction step is the condensation of Malonyl-ACP with the acyle-CoA by the aid of FabB enzyme for saturated fatty acids and FabF enzyme for unsaturated fatty acids to form 3-keto-acyle. The second step is the reduction of 3-ketoacyle into 3-hydroxyacyl-ACP that were catalyzed by FabG. After that, the third reaction in which FabA and FabZ catalyzed the dehydration of 3-hydroxyacyl-ACP to transform to trans-2-enoyl-ACP. The last reaction step is the reduction of trans-2-enoyl-ACP in the presence of NADH and catalyzed by FabI (Hoang and Schweizer 1997).

3.3 *Three Key Enzymes Responsible for Rhamnolipids Final form Formation*

Most of all previous enzymes that were demonstrated in the rhamnolipids biosynthesis process were stated in many types of bacteria except the three key enzymes RhlA, RhlB and RhlC, which were observed only in *Pseudomonas aeruginosa* and *Burkholderia* sp. bacterial strain (Dobler et al. 2016, Dubeau et al. 2009, Olvera et al. 1999). RhlA is responsible for the formation of 3-3-hydroxyalkanoyloxy alkanoic acid (HAAs) fatty acids dimer from R-3-hydroxyacyl-ACP (Deziel et al. 2003). Nevertheless, RhlB uses dTDP-L-rhamnose and HAAs to form rhamnolipids that yield Mono-RLs. The third reaction is catalyzed by RhlC, which helps the combination between Mono-RLs together with dTDP-L-rhamnose to form Di-RLs (Zhu and Rock 2008). Furthermore, these three enzymes are important enzymes that control the production of rhamnolipids. Accordingly, it was proposed that, other bacterial forms especially non-pathogenic forms could synthesize rhamnolipids by transferring the three enzyme genes into them (Chong and Li 2017).

3.4 *Optimization Fermentation Strategies for Enhancing Rhamnolipid Yield*

Rhamnolipids production through the fermentation process by using limited growth conditions and using the resting cell in the production process resulted in increasing production levels because of the limitations of one or more component of the medium. Different studies recorded increasing production of rhamnolipids when its culture reached the stationary phase due to the limitations in ions and nitrogen (Chayabutra et al. 2001). Furthermore, rhamnolipids production is achieved by using free and immobilized cells where after the cultivation process the microorganism is separated from the culture medium under optimum conditions and the wet biomass is used for the secondary metabolites production (Ramana and Karanth 1989). For instance, the production yield of *Pseudomonas* sp. DSM2874 was improved from 0.16 to 0.23 in the resting cell application and from 0.61 to 3.30 by nitrogen nutrient limitation in the medium. Moreover, two new rhamnolipids types RL3 and RL4 were detected under these conditions (Syldatk et al. 1985). In summary, the quantity of rhamnolipids produced is significantly influenced by the source of Carbon, nitrogen, and the medium ions; environmental conditions such as pH, temperature, and aeration; in addition to the cultivation strategies (Nitschke et al. 2005).

3.4.1 Nutrition Factors

The amount of produced rhamnolipids was affected by different nutritional factors such as Carbon and Nitrogen sources, and multivariate ions and they are discussed as follows:

3.4.1.1 Carbon Source

Carbon sources are important in the rhamnolipids production process where they influence the biosynthesis by either repression or inductions. Oil substrate carbon sources; vegetable oils, and n-alkanes were the carbon substrates that achieved higher rhamnolipid yields than water-soluble carbon sources; glucose, mannitol, methanol, and glycerol (Dobler et al. 2020, Sim et al. 1997). Additionally, hydrophobic substrates such as long chain alcohols, corn oil and lard yielded more rhamnolipid production from a ratio of nearly 100:165 compared to that of hydrophilic substrates such as succinic acid and glucose in a ratio of 12:36 (Mata-Sandoval et al. 2001, Shatila et al. 2020).

3.4.1.2 Nitrogen Source

The Nitrogen source had a key role in the rhamnolipids biosynthesis production. It was reported that, nitrate is the best nitrogen source for rhamnolipids surfactant production by growing *P. aeruginosa* 44T1 on olive oil. By cultivation of *P. aeruginosa* under a nitrogen-limited condition, a correlation between biosurfactant production and glutamine-synthethase activity can be observed. Previously, it was reported that at a C/N ratio of 16/1:18/1 under nitrogen limitation conditions a maximum rhamnolipids production was achieved, whereas, no surfactant production was observed at C/N ratio of 11/1 "no-Nitrogen limitations" (Gong et al. 2020, Guerra-Santos et al. 1986, Wu et al. 2019).

3.4.1.3 Multivariate Ions

Multivariate cations limitation leads to a high biosurfactant production yield. For instance, limiting the concentrations of Mg, Ca, Na, K, and trace element salts influences the production of rhamnolipids by *P. aeruginosa* DSM 2659. Moreover, limitations of phosphate iron element leads to rhamnolipids overproduction and this was obvious through growing of *P. aeruginosa* ATCC 10145 under phosphorus limitation conditions that achieved the most effective rhamnolipids production 4- to 5-fold higher specific productivity than the conventional Nitrogen limitation (Guerra-Santos et al. 1984, Sharma et al. 2020).

3.4.2 Environmental Conditions

In order to obtain large quantities from rhamnolipid biosurfactants, it is important to optimize the bioprocess environmental conditions such as pH, temperature, aeration and agitation speed (Adamczak and odzimierz Bednarski 2000, Desai and Banat 1997, Zinjarde and Pant 2002). It was reported that the maximum production of rhmanolibids by *P. aeruginosa* was observed at a pH range from 6 to 6.5 and the production decreased sharply by increasing the pH above 7. While, the best demonstrated temperature that achieved a high rhamnolipid production yield was 37ºC (Syldatk et al. 1985). At last, the aeration and agitation rates facilitate oxygen transfer from the gaseous phase into the liquid phase, therefore they are considered as significant factors for a high rhamnolipid production. The best rhamnolipid production of 45.5 g/L^{-1} was obtained by applying an air flow of 1 vvm (Guerra-Santos et al. 1984, Sharma et al. 2019, Wu et al. 2019).

3.4.3 *Cultivation Strategies*

Rhamnolipids are secondary metabolites produced in the stationary phase. Hence, most of the cultivation strategies aim to induce rhamnolipid biosurfactants production by limiting at least one of the medium components. There are different cultivation technologies that have been reported for the rhamnolipids over production process and they are explained as follows:

3.4.3.1 Batch and Fed-Batch Strategies

It is the best effective strategy that helps achieving a high rhamnolipids bioproductivity. This can be explained by the optimal low concentrations of all substrates which can be applied and the specific growth rate which can be controlled by the feeding process. By comparing this strategy with the continuous one, a limited contamination risk can be noticed, although this strategy had not been adopted with rhamnolipids production yet. However, the final rhamnolipids concentrations reported by these continuous strategies was reported in the range of 6–95 g/l concentration that were in the same order of magnitude as those reported in the fed batch cultivation where it recorded a concentration range of 5–112 g/l rhamnolipids (Hajfarajollah et al. 2019, Kumar and Das 2018).

By using fish oil as a carbon source and urea as a nitrogen source for fed-batch *P. aeruginosa* BYK-2 cultivation leads to high production yields of 0.75 g/g and a final concentration of 17g/l can be achieved after 216 hours of cultivation time in contrast to the batch strategy that gave a specific yield YP/S of 0.68 g/g (Lee et al. 2004).

3.4.3.2 Resting Cells Cultivation

Resting cells cultivation process of *Pseudomonas* sp. DSM 2874 is a two-step process where, the biomass firstly produced and harvested is followed by a resting cells suspension in a buffer solution. Rhamnolipids production was influenced by the addition of the carbon source (Trummler et al. 2003). It was reported that rhamnolipids production yield improved from 0.16 to 0.23 g/g and YP/X from 0.61 to 3.30 g/g in a comparison form between resting cell conditions and growth-limiting conditions especially with nitrogen limitation (Suhaila et al. 2019, Syldatk et al. 1985).

3.4.3.3 Continuous and Semicontinuous Cultivations

Foaming problem is one of the problems that face rhamnolipids fermentation strategies, Therefore, semicontinuous cultivation methods have been displayed and applied with integrated continuous products removal by flotation. The best conditions for semicontinuous cultivation of rhamnolipids production was stated for *Pseudomonas* sp. DSM 2874 through the combination of glycerol carbon source and calcium alginate-immobilized cells (Guerra-Santos et al. 1984). Immobilized biocatalyst can be used in the cultivation process several times after appropriate regeneration of the cells (Siemann and Wagner 1993).

On the other hand, using the continuous cultivation process for rhamnolipids production proved it to be highly efficient in their over production. For instance, many continuous cultivation processes were reported for *P. aeruginosa* DSM 2659 production by using a dextrose carbon source under excess carbon and phosphate amounts in addition to nitrogen and iron limitations (de Oliveira et al. 2020).

3.4.3.4 Solid State Fermentation

As a result of the foaming process that is considered to be a serious obstacle through the production of rhmnolipids in bioreactors that work under areated conditions in a liquid medium (Rodríguez et al. 2020). Hence, application of solid-state fermentation was proposed as a solution for the foam formation problem that caused a low rhamnolipids yield productivity. RL production by *P. aeruginosa* UFPEDA 614 was optimized and grown on a solid medium mixed with a solution that contained glycerol where the yield recorded was 46 g/L concentration of rhamnolipids (Neto et al. 2008).

3.5 *Strategies to reduce Rhamnolipids Production Cost*

Promising and potential applications of rhamnolipids encouraged many researchers to investigate various production techniques in order to reduce their production costs. In general, high costs of rhamnolipids production occur due to a return to the use of expensive substrates in the fermentation process and the high purification costs that limit their application in the industrial sector (Henkel et al. 2012). Therefore, there are many suggested strategies in order to reduce the process costs such as;

The production conditions were optimized (Borges et al. 2015) using different production processes (Nalini and Parthasarathi 2014), Genetic engineering strategies and screening for new natural producer strains (Lovaglio et al. 2015, Roy et al. 2015). Many trials have been implemented to isolate the rhamnolipids producer gene in order to facilitate its transfer to other organisms other than the pathogenic one *Pseudomonas aeruginosa.* This strategy will help reducing rhamnolipids production costs especially in large-scale production (Abouseoud et al. 2008, Cruz 2020, Twigg et al. 2018). Identification of low cost substrates as raw materials for rhamnolipids production have been suggested, lately in order to reduce the production costs. Many hydrophobic carbon sources and edible vegetable oils were used as carbon substrates for rhamnolipids production. Feed stock contributes 30–80% of the total rhamnolipids fermentation costs. Therefore, using low cost substrates such as agro-industrial by-products, agriculture and cooking waste oils, and glucose-based agriculture wastes prove to be efficient in fermentation cost reduction (Ramírez et al. 2015). Recovery and purification processes of rhamnolipids constitute 60 to 80% of the total production process costs, as a result looking for new low cost purification strategies is an optimistic solution for reducing the total process cost Even though, there are various traditional methods for rhamnolipids production such as acid precipitation, organic solvents extraction by "ethyl acetate, chloroform, and methanol", and different chromatographic methods. However, the previous methods had many disadvantages as its high costs, generated a large quantity of solvent as a toxic waste after the purification process, and resulted in biosurfactant activity loss. Therefore, researchers have developed new non-conventional methods for the purification process such as the ISPR method "in situ product removal" technique (Decesaro et al. 2020). Using of ISPR methods, leads to continuous recovery of the end-products from the fermentation broth and in consequence reduces end-product inhibitory effects (Baker and Chen 2010). These previous methods were used in the recovery of small molecules such as; pharmaceutical and food ingredients, fuels, and industrial chemicals (Woodley et al. 2008). Foam fractionation and ultrafiltration were considered one of the most used ISPR methods for effective application especially with large-scale production processes (Chen et al. 2006, Müller et al. 2012).

4. Microbial Diversity of Rhamnolipids Producers

Rhamnolipids, containing L-rhamnose and β-hydroxy greasy corrosive moieties, have appeared as a class of glycolipid biosurfactants during the last few years, mainly within industries due to their distinctive chemical features (Dobler et al. 2016). *Pseudomonas aeruginosa* are observed as the best manufacturers of rhamnolipids, but due to their pathogenic nature, much consideration has been given for rhamnolipid production from nonpathogenic strains. On the contrary, recombinant DNA innovation is mostly explored for the large-scale production of rhamnolipids (Dobler et al. 2016). Gene regulation in case of rhamnolipid production by *Pseudomonas aeruginosa* has been found to be complex in nature. This complexity has been a major challenge for manufacturing-scale production of rhamnolipids. On the other hand, rhamnolipid manufacture through recombinant DNA technology can diminish the risk regarding utilization of *P. aeruginosa* (Dobler et al. 2016).

Pseudomonas aeruginosa provide two forms of rhamnolipids, i.e., mono- or di-rhamnolipid forms (Soberón-Chávez et al. 2005). Away from *Pseudomonas aeruginosa*, a few rhamnolipid manufacturing Pseudomonas strains have been separated such as *Pseudomonas fluorescens*, *Pseudomonas putida*, and *Pseudomonas chlororaphis* (Gunther et al. 2005, Gunther et al. 2006, Martinez-Toledo et al. 2006, Sharma et al. 2007, Vasileva-Tonkova et al. 2006). *Burkholderia* species are moreover powerful rhamnolipid manufacturers, and detailed species are *Burkholderia thailandensis*, *Burkholderia mallei*, and Burkholderia pseudomallei (Dubeau et al. 2009, Häußler et al. 1998, Toribio et al. 2010).

4.1 *Fungal Rhamnolipids verses Bacterial Rhamnolipids*

It has been found that biosurfactants are produced by microbes (Bouassida et al. 2018), and yeasts, but rarely by filamentous fungi (Silva et al. 2014). Investigating studies on filamentous fungi are rare. It has been found that the few filamentous fungi explored have appeared potential deliverers of biosurfactants, with higher yields when compared to those from yeasts, but mainly when compared to those from bacteria. Hence (Yang et al. 2012) proposed that the discharge of a high production of biosurfactants by filamentous fungi has been credited to cell wall stiffness.

Utilizing filamentous fungi meets the requests of society for concerns regarding the environment to be taken under consideration. This is typically reflected in the new legislation seeking for items obtained via microbial, toxicity-free options, instead of depending on commercial surfactants synthesized from petroleum. They are flexible alternatives, and includes the feasible utilization of agrarian by-products in the rummage around for modern surfactants with wide appropriateness and a lower cost (Silva et al. 2018).

Table 2 shows the studies performed with yeasts, the genera and species most examined, as well as the synthesized class of biosurfactants. It is shown that most yeasts of the class Candida are the most examined among diverse sorts of biosurfactants that have the potential to be produced economically. It is conceivable that the broader utilization of yeast to produce these substances with incredible success is related primarily to its GRAS status (considered as safe). Microorganisms that have a GRAS status don't present a hazard of actuating harmfulness and pathogenic reactions. In expansion, the flexibility of the metabolic pathways of yeasts, particularly *Candida lipolytica*, and *Yarrowia lipolytica*, are considered to have unconventional metabolism,which is highlighted in the appearance of their development in a wide variety of substrates with diverse chemical natures to produce biosurfactants.

It ought to be shown that filamentous fungi (Table 3) are less abused than yeasts, due to their slower growth. However, they are great producers of biosurfactants, as well as emulsifiers, with steady emulsions, and have an excellent capacity to decrease stress. In addition, they promote the scattering of hydrophobic compounds, which enables them to be applied in numerous sectors (Silva et al. 2014).

4.2 *Rhamnolipid Biosurfactants from Untapped Marine Resources*

Marine fungi and bacteria acquire particular physiological and metabolic characters and are a potential source of novel biomolecules, as biosurfactants. Some of the biosurfactant compounds synthesized by marine microorganisms display anti-adhesive, antimicrobial and anti-biofilm activity against multi-drug resistant pathogens. However, marine biosurfactants have not been broadly investigated, basically due to the challenges related with their isolation and isolates growth. Culture-independent techniques (metagenomics) constitute a promising application to investigate the genetic resources of otherwise inaccessible marine microorganisms without the requirement of culturing them, and can contribute to reveal novel biosurfactants with potential biological activities (Gudiña et al. 2016).

Gudiña et al. (2016) reported that the strain *Streptomyces* sp. ISP2-49E, was isolated from marine sediment recovered from Galveston Bay (Texas). This isolate produced the rhamnolipid biosurfactant L-rhamnosyl-L-rhamnosyl-β-hydroxydecanoyl-β-hydroxydecanoate (Rha-Rha-C10-C10), being the first record of a rhamnolipid-producing *Streptomyces* strain. However, the basic rhamnolipid producers in a marine environment are *P. aeruginosa* strains, an opportunistic human pathogen. Therefore, the use of substitute non-pathogenic rhamnolipid producers such as *Streptomyces* sp. can contribute to the safe utilization of rhamnolipids as therapeutic agents. This can be accomplished using either non-pathogenic natural

Table 2. Yeast producing biosurfactant.

Biosurfactant	**Microorganism**
1-Glycolipids	*Candida bogoriensis* *Candida bombicola* *Candida glabrata* *Candida ishiwadae* *Candida batistae* *Candida sphaerica UCP0995* *Wickerhamomyces anomalus CCMA 0358* *Pseudozima fusifornata*
2-Lipopeptides	*Candida glabrata*
3-Complex carbohydrates/proteins/lipids	*Candida lipolytica UCP0988* *Candida lipolytica IA1055* *Candida tropicalis* *Candida valida* *Candida boleticota* *Candida ingens* *Candida utilis* *Yarrowia lipolytica NCIM 3589* *Debaryomyces polymorphus*
4-Complex Carbohydrates /proteins	*Candida lipolytica ATCC8662* *Candida utilis* *Candida ingens* *Saccharomyces cerevisiae* *Rhodotorula glutinis* *Kluyveromyces marxianus*
5-Sophorolipids	*Pseudozyma aphidis* *Candida bombicola* *Torulopsis petrophilum* *Candida (Torulopsis) apicola* *Candida bogoriensis* *Candida antarctica* *Pseudozyma rugulosa* *Candida* sp. *SY16* *Kurtzmanomyces* sp.

Table 3. Types of bioemulsifier or biosurfactants produced by filamentous fungi.

Biosurfactant	**Microorganism**
Glycopeptide	*Aspergillus niger*
Lipopeptide	*Penicillium chrysogenum* SNP5
Complex Carbohydrate/protein/lipid	*Cunninghamella echinulata*
Not identified	*Fusarium* sp. *Fusarium*, *Penicillium* and *Trichoderma* *Aspergillus* spp. *Aspergillus fumigatus* *Aspergillus niger* *Rhizopus arrhizus*

rhamnolipid-producing strains, or engineered non-pathogenic hosts expressing the genes required for the synthesis of rhamnolipids.

On other hand halotolerant extremophiles are required to generate biosurfactants with great resilience for the bioremediation of oil spills (Řezanka et al. 2011) recorded new RL-producing extremophilic bacteria like, the thermophilic strains, *T. aquatic, Thermus* sp. and *Meiothermus ruber.* RL derivatives were characterized as mono-rhamnolipid and di-rhamnolipid homologues possessing one or two 3-hydroxy-fatty acids, saturated, monounsaturated or diunsaturated, even- or odd-chains, up to unusually long chains with 24 carbon atoms (Řezanka et al. 2011). Prospecting the marine environment would yield new halotolerant biosurfactants, however the marine environment records are limited. Recently some investigations proposed marine microbes as significant sources of biosurfactants involving RL-derivatives. Glycolipid derivatives may envisage RL biosurfactants from marine ancestors. For example, the rhlB gene product (GenBank Accession Numbers: FJ372668 and FJ372667) was amplified from marine Actinobacteria producing a biosurfactant, which was characterized with glycolipid moieties. It was found that the biosurfactant produced by marine Actinobacteria was steady up to 5% NaCl. A glycolipoprotein biosurfactant was recovered from *Oceanobacillus* sp. BRI sampled from Antarctica. This biosurfactant was detailed to possibly emulsify diverse forms of petroleum hydrocarbons. Investigation of hydrothermal vents and extreme marine habitats would reveal new biosurfactant derivatives active at stresses like, extreme pressure, salinity and temperature. Particularly, the literature reported that biosurfactants with high-tolerance capacities have not been investigated. A recent report on the isolation of cold-tolerant glycolipid bio surfactants supports the possibility that such high-tolerance biosurfactants could be gotten from harsh habitats (Kiran et al. 2016).

Tripathi et al. (2019) reported that *Marinobacter* sp. MCTG107b produces a glycolipid biosurfactant. The observed products possessed *m/z* values that corresponded to values for known rhamnolipids, demonstrating that the biosurfactant synthesized by strain MCTG107b was a mixture of rhamnolipid congeners. An assortment of isolated rhamnolipid congeners were present in purified cell-free supernatant extracts from culture samples of the *Marinobacter* strain. These congeners involved mono- and di-rhamnolipids; however, there was an overwhelming inclination toward the synthesis of di-rhamnolipid (95.39% of total rhamnolipid abundance). The congener with the highest relative abundance (52.45%) possessed a *m/z* value of 651.73. This value correlated with α-l-rhamnopyranosyl-α-l-rhamnopyranosyl-β-hydroxydecanoyl-β-hydroxydecanoate (Rha-Rha-C10-C10) with a molecular weight of 650.79 Da. The next most abundantly synthesized congeners were Rha-Rha-C10-C10CH3 (23.07%), Rha-Rha-C10 (5.13%), Rha-Rha-C10-C12 (5.01%), Rha-Rha-C10-C12CH3 (3.26%) and Rha-C14:2 (3.18%). This study extends the paradigm of rhamnolipid biosynthesis to a new genus of bacterium from the marine habitats. Rhamnolipids produced from *Marinobacter* have prospects for industrial application due to their potentialialty of being synthesized from cheap, renewable feed stocks with essentially decreased pathogenicity compared to *P. aeruginosa* strains.

5. Inimitable Applications of Rhamnolipids

Throughout the years rhamnolipids are becoming widely pertinent in various industries and are posturing a genuine threat to the synthetic surfactants. Before venturing into the recent production economics of rhamnolipids it is imperative to assess the main applications of rhamnolipids that make them an obvious choice among other biosurfactants. A list of major applications of rhamnolipids that cater to the extensive range of industrial demands includes bioremediation and enhanced oil recovery (EOR). Rhamnolipids display outstanding emulsification properties, proficiently remove crude oil from contaminated soil and ease bioremediation of oil spills (Costa et al. 2010). Rhamnolipids are already used for soil remediation for improving soil quality and are now additionally getting explored for plant pathogen elimination, for assisting the absorption of fertilizers and nutrients through roots and as biopesticides (Sachdev and Cameotra 2013). Rhamnolipids as detergents and cleaners are natural emulsifiers and surface active agents leading to their extensive usage in detergent compositions, laundry products, shampoos and soaps (Sekhon Randhawa and Rahman 2014). Pharmaceuticals and therapeutics: Rhamnolipids display low toxicity, surface active properties and antimicrobial activities against numerous microbes (*Bacillus cereus, Micrococcus luteus, Staphylococcus aureus, Listeria monocytogenes*) thereby showing promising applications in pharmaceuticals and therapeutics (Magalhães and Nitschke 2013).

6. Conclusion: Prospectives for the Industrial Production of Rhamnolipids

We have gained a fortune of knowledge on rhamnolipidic surfactants of microbial route. Still, even though over 60 years have passed since their first description (Jarvis and Johnson 1949), RLs have not yet been significantly employed in the industry. Certainly, there is still a long way to go before achieving extensive bulk bioproduction of RLs, for both technical and economical aims. Currently, the economic competitiveness of RLs against synthetic surfactants is mostly determined by the low productivity of the bio processes employed. However, this is beginning to change, as environmental compatibility becomes an increasingly imperative factor for the selection of industrial chemicals. Major improvements can be expected if more productive strains can be found and if a better understanding of the underlying regulation can be attained. In view of the complex quorumsensing-regulated induction of RL production in *P. aeruginosa*, further optimization will almost surely be reliant on a more accurate understanding of the mechanisms of regulation. It is expected that significant insights on the regulation and biosynthesis of RLs will be gained from the current systems biological approaches. There are good chances of success in the near future if a more integrated biotechnological approach is efficiently adopted for strain and process development. Additionally, the use of new heterologous RL-producing hosts will help to widen the product spectrum and make it possible to produce single RL congeners.

References

Abalos, A., A. Pinazo, M. Infante, M. Casals, F. Garcia et al. 2001. Physicochemical and antimicrobial properties of new rhamnolipids produced by *Pseudomonas aeruginosa* AT10 from soybean oil refinery wastes. Langmuir 17(5): 1367–1371.

Abdel-Mawgoud, A.M., M.M. Aboulwafa and N.A.-H. Hassouna. 2009. Characterization of rhamnolipid produced by *Pseudomonas aeruginosa* isolate Bs20. Applied Biochemistry and Biotechnology 157(2): 329–345.

Abdel-Mawgoud, A.M., F. Lépine and E. Déziel. 2010. Rhamnolipids: diversity of structures, microbial origins and roles. Applied Microbiology and Biotechnology 86(5): 1323–1336.

Abdel-Mawgoud, A.M., R. Hausmann, F. Lépine, M.M. Müller, E. Déziel et al. 2011. Rhamnolipids: detection, analysis, biosynthesis, genetic regulation, and bioengineering of production. pp. 13–55. *In*: Soberón-Chávez, G. (ed.). Biosurfactants: From Genes to Applications. Springer Berlin Heidelberg. https://doi.org/10.1007/978-3-642-14490-5_2.

Abdel-Mawgoud, A.M., F. Lépine and E. Déziel. 2014. A stereospecific pathway diverts β-oxidation intermediates to the biosynthesis of rhamnolipid biosurfactants. Chemistry & Biology 21(1): 156–164.

Abouseoud, M., A. Yataghene, A. Amrane and R. Maachi. 2008. Biosurfactant production by free and alginate entrapped cells of *Pseudomonas fluorescens*. Journal of Industrial Microbiology & Biotechnology 35(11): 1303–1308.

Adamczak, M. and W. Odzimierz Bednarski. 2000. Influence of medium composition and aeration on the synthesis of biosurfactants produced by *Candida antarctica*. Biotechnology Letters 22(4): 313–316.

Allard, S.T., M.-F. Giraud, C. Whitfield, M. Graninger, P. Messner et al. 2001. The crystal structure of dTDP-D-Glucose 4, 6-dehydratase (RmlB) from Salmonella enterica serovar Typhimurium, the second enzyme in the dTDP-l-rhamnose pathway. Journal of Molecular Biology 307(1): 283–295.

Aparna, A., G. Srinikethan and H. Smitha. 2012. Production and characterization of biosurfactant produced by a novel *Pseudomonas* sp. 2B. Colloids and Surfaces B: Biointerfaces 95: 23–29.

Banat, I.M., R.S. Makkar and S.S. Cameotra. 2000. Potential commercial applications of microbial surfactants. Applied Microbiology and Biotechnology 53(5): 495–508.

Bharali, P., S. Das, A. Ray, S. Pradeep Singh, U. Bora et al. 2018. Biocompatibility natural effect of rhamnolipids in bioremediation process on different biological systems at the site of contamination. Bioremediation Journal 22(3-4): 91–102.

Borges, W., A. Moura, U. Coutinho Filho, V. Cardoso, M. Resende et al. 2015. Optimization of the operating conditions for rhamnolipid production using slaughterhouse-generated industrial float as substrate. Brazilian Journal of Chemical Engineering 32(2): 357–365.

Burger, M., L. Glaser and R. Burton. 1966. [78] Formation of rhamnolipids of *Pseudomonas aeruginosa*. pp. 441–445. *In*: Methods in Enzymology (Vol. 8). Elsevier.

Campos-García, J., A.D. Caro, R. Nájera, R.M. Miller-Maier, R.A. Al-Tahhan et al. 1998. The *Pseudomonas aeruginosa* rhlG gene encodes an NADPH-dependent β-ketoacyl reductase which is specifically involved in rhamnolipid synthesis. Journal of Bacteriology 180(17): 4442–4451.

Chayabutra, C., J. Wu and L.K. Ju. 2001. Rhamnolipid production by *Pseudomonas aeruginosa* under denitrification: effects of limiting nutrients and carbon substrates. Biotechnology and Bioengineering 72(1): 25–33.

Chen, C.Y., S.C. Baker and R.C. Darton. 2006. Continuous production of biosurfactant with foam fractionation. Journal of Chemical Technology & Biotechnology: International Research in Process, Environmental & Clean Technology 81(12): 1915–1922.

Chen, M.L., J. Penfold, R.K. Thomas, T.J.P. Smyth, A. Perfumo et al. 2010. Solution self-assembly and adsorption at the air–water interface of the monorhamnose and dirhamnose rhamnolipids and their mixtures. Langmuir 26(23): 18281–18292. https://doi.org/10.1021/la1031812.

Chong, H. and Q. Li. 2017. Microbial production of rhamnolipids: opportunities, challenges and strategies. Microbial Cell Factories 16(1): 137. https://doi.org/10.1186/s12934-017-0753-2.

Costa, S.G.V.A.O., M. Nitschke, F. Lépine, E. Déziel, J. Contiero et al. 2010. Structure, properties and applications of rhamnolipids produced by *Pseudomonas aeruginosa* L2-1 from cassava wastewater. Process Biochemistry 45(9): 1511–1516. https://doi.org/https://doi.org/10.1016/j.procbio.2010.05.033.

Cruz, R.L. 2020. RhlR Quorum Sensing and Social Dynamics in Cystic Fibrosis-Adapted Isolates of *Pseudomonas aeruginosa.* https://doi.org/10.1007/s10532-018-9833-1.

Dahrazma, B. and C.N. Mulligan. 2007. Investigation of the removal of heavy metals from sediments using rhamnolipid in a continuous flow configuration. Chemosphere 69(5): 705–711. https://doi.org/https://doi.org/10.1016/j.chemosphere.2007.05.037.

de Oliveira, S.P., N.A. Rodrigues, P.A. Casciatori-Frassatto and F.P. Casciatori. 2020. Solid-liquid extraction of cellulases from fungal solid-state cultivation in a packed bed bioreactor. Korean Journal of Chemical Engineering 37(9): 1530–1540.

Decesaro, A., T.S. Machado, Â.C. Cappellaro, A. Rempel, A.C. Margarites et al. 2020. Biosurfactants production using permeate from whey ultrafiltration and bioproduct recovery by membrane separation process. Journal of Surfactants and Detergents 23(3): 539–551. https://doi.org/https://doi.org/10.1002/jsde.12399.

Desai, J.D. and I.M. Banat. 1997. Microbial production of surfactants and their commercial potential. Microbiology and Molecular Biology Reviews 61(1): 47–64.

Deziel, E., F. Lepine, S. Milot and R. Villemur. 2003. rhlA is required for the production of a novel biosurfactant promoting swarming motility in *Pseudomonas aeruginosa*: 3-(3-hydroxyalkanoyloxy) alkanoic acids (HAAs), the precursors of rhamnolipids. Microbiology 149(8): 2005–2013.

Dhanarajan, G. and R. Sen. 2014. Cost analysis of biosurfactant production from a scientist's perspective. Biosurfactants, 159, 153.

Dobler, L., L.F. Vilela, R.V. Almeida and B.C. Neves. 2016. Rhamnolipids in perspective: gene regulatory pathways, metabolic engineering, production and technological forecasting. New Biotechnology 33(1): 123–135. https://doi.org/https://doi.org/10.1016/j.nbt.2015.09.005.

Dobler, L., H.C. Ferraz, L.V. Araujo de Castilho, L.S. Sangenito, I.P. Pasqualino et al. 2020. Environmentally friendly rhamnolipid production for petroleum remediation. Chemosphere 252: 126349. https://doi.org/https://doi.org/10.1016/j.chemosphere.2020.126349.

Dubeau, D., E. Déziel, D.E. Woods and F. Lépine. 2009. Burkholderia thailandensis harbors two identical rhl gene clusters responsible for the biosynthesis of rhamnolipids. BMC Microbiology 9(1): 263.

Giraud, M.-F., G. Leonard, R. Rahim, C. Creuzenet, J. Lam et al. 2000. The purification, crystallization and preliminary structural characterization of glucose-1-phosphate thymidylyltransferase (RmlA), the first enzyme of the dTDP-L-rhamnose synthesis pathway from *Pseudomonas aeruginosa*. Acta Crystallographica Section D: Biological Crystallography 56(11): 1501–1504.

Gong, Z., Y. Peng and Q. Wang. 2015. Rhamnolipid production, characterization and fermentation scale-up by *Pseudomonas aeruginosa* with plant oils. Biotechnology Letters 37(10): 2033–2038.

Gong, Z., Q. He, C. Che, J. Liu, G. Yang et al. 2020. Optimization and scale-up of the production of rhamnolipid by *Pseudomonas aeruginosa* in solid-state fermentation using high-density polyurethane foam as an inert support. Bioprocess and Biosystems Engineering 43(3): 385–392. https://doi.org/10.1007/s00449-019-02234-2.

Graninger, M., B. Nidetzky, D.E. Heinrichs, C. Whitfield, P. Messner et al. 1999. Characterization of dTDP-4-dehydrorhamnose 3, 5-epimerase and dTDP-4-dehydrorhamnose reductase, required for dTDP-L-rhamnose biosynthesis in *Salmonella enterica* serovar Typhimurium LT2. Journal of Biological Chemistry 274(35): 25069–25077.

Gudiña, E.J., J.A. Teixeira and L.R. Rodrigues. 2016. Biosurfactants produced by marine microorganisms with therapeutic applications. Marine Drugs 14(2): 38.

Gudiña, E.J., A.I. Rodrigues, V. de Freitas, Z. Azevedo, J.A. Teixeira et al. 2016. Valorization of agro-industrial wastes towards the production of rhamnolipids. Bioresource Technology 212: 144–150. https://doi.org/https://doi.org/10.1016/j.biortech.2016.04.027.

Guerra-Santos, L., O. Käppeli and A. Fiechter. 1984. *Pseudomonas aeruginosa* biosurfactant production in continuous culture with glucose as carbon source. Applied and Environmental Microbiology 48(2): 301–305.

Guerra-Santos, L.H., O. Käppeli and A. Fiechter. 1986. Dependence of *Pseudomonas aeruginosa* continuous culture biosurfactant production on nutritional and environmental factors. Applied Microbiology and Biotechnology 24(6): 443–448.

Gunther, N.W., A. Nuñez, W. Fett and D.K.Y. Solaiman. 2005. Production of Rhamnolipids by <em>Pseudomonas chlororaphis</em>, a Nonpathogenic Bacterium. Applied and Environmental Microbiology 71(5): 2288. https://doi.org/10.1128/AEM.71.5.2288-2293.2005.

Gunther, N.W., A. Nuñez, L. Fortis and D.K.Y. Solaiman. 2006. Proteomic based investigation of rhamnolipid production by *Pseudomonas chlororaphis* strain NRRL B-30761. Journal of Industrial Microbiology and Biotechnology 33(11): 914–920. https://doi.org/10.1007/s10295-006-0169-1.

Hajfarajollah, H., B. Mokhtarani, A. Tohidi, S. Bazsefidpar, K.A. Noghabi et al. 2019. Overproduction of lipopeptide biosurfactant by *Aneurinibacillus thermoaerophilus* HAK01 in various fed-batch modes under thermophilic conditions. RSC Advances 9(52): 30419–30427.

Häußler, S., M. Nimtz, T. Domke, V. Wray, I. Steinmetz et al. 1998. Purification and Characterization of a Cytotoxic Exolipid of <em>Burkholderia pseudomallei</em>. Infection and Immunity 66(4): 1588. https://doi.org/10.1128/IAI.66.4.1588-1593.1998.

Henkel, M., M.M. Müller, J.H. Kügler, R.B. Lovaglio, J. Contiero et al. 2012. Rhamnolipids as biosurfactants from renewable resources: Concepts for next-generation rhamnolipid production. Process Biochemistry 47(8): 1207–1219. https://doi.org/https://doi.org/10.1016/j.procbio.2012.04.018.

Hoang, T.T. and H.P. Schweizer. 1997. Fatty acid biosynthesis in *Pseudomonas aeruginosa*: cloning and characterization of the fabAB operon encoding beta-hydroxyacyl-acyl carrier protein dehydratase (FabA) and beta-ketoacyl-acyl carrier protein synthase I (FabB). Journal of Bacteriology 179(17): 5326. https://doi.org/10.1128/jb.179.17.5326-5332.1997.

Hošková, M., O. Schreiberová, R. Ježdík, J. Chudoba, J. Masák et al. 2013. Characterization of rhamnolipids produced by non-pathogenic Acinetobacter and Enterobacter bacteria. Bioresource Technology 130: 510–516. https://doi.org/https://doi.org/10.1016/j.biortech.2012.12.085.

Hošková, M., R. Ježdík, O. Schreiberová, J. Chudoba, M. Šír et al. 2015. Structural and physiochemical characterization of rhamnolipids produced by *Acinetobacter calcoaceticus*, *Enterobacter asburiae* and *Pseudomonas aeruginosa* in single strain and mixed cultures. Journal of Biotechnology 193: 45–51. https://doi.org/https://doi.org/10.1016/j.jbiotec.2014.11.014.

Irorere, V.U., L. Tripathi, R. Marchant, S. McClean, I.M. Banat et al. 2017. Microbial rhamnolipid production: a critical re-evaluation of published data and suggested future publication criteria. Applied Microbiology and Biotechnology 101(10): 3941–3951.

Janek, T., M. Łukaszewicz and A. Krasowska. 2013. Identification and characterization of biosurfactants produced by the Arctic bacterium *Pseudomonas putida* BD2. Colloids and Surfaces B: Biointerfaces 110: 379–386. https://doi.org/https://doi.org/10.1016/j.colsurfb.2013.05.008.

Jarvis, F.G. and M.J. Johnson. 1949. A Glyco-lipide produced by *Pseudomonas aeruginosa*. Journal of the American Chemical Society 71(12): 4124–4126. https://doi.org/10.1021/ja01180a073.

Jiang, J., Y. Zu, X. Li, Q. Meng, X. Long et al. 2020. Recent progress towards industrial rhamnolipids fermentation: Process optimization and foam control. Bioresource Technology 298: 122394.

Kim, L.H., Y. Jung, H.-W. Yu, K.-J. Chae, I.S. Kim et al. 2015. Physicochemical interactions between rhamnolipids and *Pseudomonas aeruginosa* Biofilm Layers. Environmental Science & Technology 49(6): 3718–3726. https://doi.org/10.1021/es505803c.

Kiran, G.S., A.S. Ninawe, A.N. Lipton, V. Pandian, J. Selvin et al. 2016. Rhamnolipid biosurfactants: evolutionary implications, applications and future prospects from untapped marine resource. Critical Reviews in Biotechnology 36(3): 399–415.

Kumar, R. and A.J. Das. 2018. Rhamnolipid Biosurfactant: Recent Trends in Production and Application. Springer.

Kutchma, A.J., T.T. Hoang and H.P. Schweizer. 1999. Characterization of a *Pseudomonas aeruginosa* fatty acid biosynthetic gene cluster: purification of acyl carrier protein (ACP) and malonyl-coenzyme A: ACP transacylase (FabD). Journal of Bacteriology 181(17): 5498–5504.

Lee, K.M., S.-H. Hwang, S.D. Ha, J.-H. Jang, D.-J. Lim et al. 2004. Rhamnolipid production in batch and fed-batch fermentation using *Pseudomonas aeruginosa* BYK-2 KCTC 18012P. Biotechnology and Bioprocess Engineering 9(4): 267–273.

Lindhout, T., P.C. Lau, D. Brewer and J.S. Lam. 2009. Truncation in the core oligosaccharide of lipopolysaccharide affects flagella-mediated motility in *Pseudomonas aeruginosa* PAO1 via modulation of cell surface attachment. Microbiology 155(10): 3449–3460.

Lovaglio, R., V. Silva, H. Ferreira, R. Hausmann, J. Contiero et al. 2015. Rhamnolipids know-how: looking for strategies for its industrial dissemination. Biotechnology Advances 33(8): 1715–1726.

Magalhães, L. and M. Nitschke. 2013. Antimicrobial activity of rhamnolipids against Listeria monocytogenes and their synergistic interaction with nisin. Food Control 29(1): 138–142.

Mańko, D., A. Zdziennicka and B. Jańczuk. 2014. Thermodynamic properties of rhamnolipid micellization and adsorption. Colloids and Surfaces B: Biointerfaces 119: 22–29. https://doi.org/https://doi.org/10.1016/j.colsurfb.2014.04.020.

Martinez-Toledo, A., E. Rios-Leal, R. Vazquez-Duhalt, M.d.C. González-Chávez, J. Esparza-Garcia et al. 2006. Role of phenanthrene in rhamnolipid production by *P. putida* in different media. Environmental Technology 27(2): 137–142.

Mata-Sandoval, J.C., J. Karns and A. Torrents. 2001. Effect of nutritional and environmental conditions on the production and composition of rhamnolipids by *P. aeruginosa* UG2. Microbiological Research 155(4): 249–256.

Monteiro, L., R.d.L.R. Mariano and A.M. Souto-Maior. 2005. Antagonism of *Bacillus* spp. against Xanthomonas campestris pv. campestris. Brazilian Archives of Biology and Technology 48: 23–29. http://www.scielo.br/scielo.php?script=sci_arttext&pid=S1516-89132005000100004&nrm=iso.

Müller, M.M. and R. Hausmann. 2011. Regulatory and metabolic network of rhamnolipid biosynthesis: traditional and advanced engineering towards biotechnological production. Applied Microbiology and Biotechnology 91(2): 251–264.

Müller, M.M., J.H. Kügler, M. Henkel, M. Gerlitzki, B. Hörmann et al. 2012. Rhamnolipids—next generation surfactants? Journal of Biotechnology 162(4): 366–380.

Nalini, S. and R. Parthasarathi. 2014. Production and characterization of rhamnolipids produced by *Serratia rubidaea* SNAU02 under solid-state fermentation and its application as biocontrol agent. Bioresource Technology 173: 231–238.

Neto, D.C., J.A. Meira, J.M. de Araújo, D.A. Mitchell, N. Krieger et al. 2008. Optimization of the production of rhamnolipids by *Pseudomonas aeruginosa* UFPEDA 614 in solid-state culture. Applied Microbiology and Biotechnology 81(3): 441.

Nitschke, M., S.G. Costa and J. Contiero. 2005. Rhamnolipid surfactants: an update on the general aspects of these remarkable biomolecules. Biotechnology Progress 21(6): 1593–1600.

Olvera, C., J.B. Goldberg, R. Sánchez and G. Soberón-Chávez. 1999. The *Pseudomonas aeruginosa* algC gene product participates in rhamnolipid biosynthesis. FEMS Microbiology Letters 179(1): 85–90.

Patowary, R., K. Patowary, M.C. Kalita and S. Deka. 2018. Application of biosurfactant for enhancement of bioremediation process of crude oil contaminated soil. International Biodeterioration & Biodegradation 129: 50–60.

Pham, T.H., J.S. Webb and B.H. Rehm. 2004. The role of polyhydroxyalkanoate biosynthesis by *Pseudomonas aeruginosa* in rhamnolipid and alginate production as well as stress tolerance and biofilm formation. Microbiology 150(10): 3405–3413.

Rahim, R., L.L. Burrows, M.A. Monteiro, M.B. Perry, J.S. Lam et al. 2000. Involvement of the rml locus in core oligosaccharide and O polysaccharide assembly in *Pseudomonas aeruginosa*. Microbiology 146(11): 2803–2814.

Rahim, R., U.A. Ochsner, C. Olvera, M. Graninger, P. Messner et al. 2001. Cloning and functional characterization of the *Pseudomonas aeruginosa* rhlC gene that encodes rhamnosyltransferase 2, an enzyme responsible for di-rhamnolipid biosynthesis. Molecular Microbiology 40(3): 708–718.

Ramana, K.V. and N. Karanth. 1989. Factors affecting biosurfactant production using *Pseudomonas aeruginosa* CFTR-6 under submerged conditions. Journal of Chemical Technology & Biotechnology 45(4): 249–257.

Rahman, P.K., G. Pasirayi, V. Auger and Z. Ali. 2010. Production of rhamnolipid biosurfactants by *Pseudomonas aeruginosa* DS10-129 in a microfluidic bioreactor. Biotechnology and Applied Biochemistry 55(1): 45–52.

Ramírez, I.M., K. Tsaousi, M. Rudden, R. Marchant, E.J. Alameda et al. 2015. Rhamnolipid and surfactin production from olive oil mill waste as sole carbon source. Bioresource Technology 198: 231–236.

Řezanka, T., J. Lukavský, L. Nedbalová and K. Sigler. 2011. Effect of nitrogen and phosphorus starvation on the polyunsaturated triacylglycerol composition, including positional isomer distribution, in the alga Trachydiscus minutus. Phytochemistry 72(18): 2342–2351.

Rikalović, M.G., M.M. Vrvić and I.M. Karadžić. 2015. Rhamnolipid biosurfactant from *Pseudomonas aeruginosa*: from discovery to application in contemporary technology. Journal of the Serbian Chemical Society 80(3): 279–304.

Rodríguez, A., T. Gea, A. Sánchez and X. Font. 2020. Agro-wastes and inert materials as supports for the production of biosurfactants by solid-state fermentation. Waste and Biomass Valorization, pp. 1–14.

Rooney, A.P., N.P.J. Price, K.J. Ray and T.-M. Kuo. 2009. Isolation and characterization of rhamnolipid-producing bacterial strains from a biodiesel facility. FEMS Microbiology Letters 295(1): 82–87. https://doi.org/10.1111/j.1574-6968.2009.01581.x.

Roy, S., S. Chandni, I. Das, L. Karthik, G. Kumar et al. 2015. Aquatic model for engine oil degradation by rhamnolipid producing Nocardiopsis VITSISB. 3 Biotech 5(2): 153–164.

Sachdev, D.P. and S.S. Cameotra. 2013. Biosurfactants in agriculture. Applied Microbiology and Biotechnology 97(3): 1005–1016.

Sekhon Randhawa, K.K. and P.K.S.M. Rahman. 2014. Rhamnolipid biosurfactants—past, present, and future scenario of global market [Opinion]. Frontiers in Microbiology 5(454). https://doi.org/10.3389/fmicb.2014.00454.

Sharma, A., R. Jansen, M. Nimtz, B.N. Johri, V. Wray et al. 2007. Rhamnolipids from the rhizosphere bacterium *Pseudomonas* sp. GRP3 that reduces damping-off disease in chilli and tomato nurseries. Journal of Natural Products 70(6): 941–947. https://doi.org/10.1021/np0700016.

Sharma, S., P. Datta, B. Kumar, P. Tiwari, L.M. Pandey et al. 2019. Production of novel rhamnolipids via biodegradation of waste cooking oil using *Pseudomonas aeruginosa* MTCC7815. Biodegradation 30(4): 301–312. https://doi.org/10.1007/s10532-019-09874-x.

Sharma, R., H. Krishna and K.S.M.S. Raghavarao. 2020. Metal ion–enhanced quantification of chloramphenicol in milk using imipramine hydrochloride as diazo-coupling agent. Food Analytical Methods. https://doi.org/10.1007/s12161-020-01837-w.

Shatila, F., M.M. Diallo, U. Şahar, G. Ozdemir, H.T. Yalçın et al. 2020. The effect of carbon, nitrogen and iron ions on mono-rhamnolipid production and rhamnolipid synthesis gene expression by *Pseudomonas aeruginosa* ATCC 15442. Archives of Microbiology 202(6): 1407–1417. https://doi.org/10.1007/s00203-020-01857-4.

Siemann, M. and F. Wagner. 1993. Prospects and limits for the production of biosurfactants using immobilized biocatalysts. Surfactant Science Series, pp. 99–99.

Silva, R.D.C.F., D.G. Almeida, R.D. Rufino, J.M. Luna, V.A. Santos et al. 2014. Applications of biosurfactants in the petroleum industry and the remediation of oil spills. International Journal of Molecular Sciences 15(7): 12523–12542.

Silva, A.C.S.d., P.N.d. Santos, T.A.L.e. Silva, R.F.S. Andrade, G.M. Campos-Takaki et al. 2018. Biosurfactant production by fungi as a sustainable alternative. Arquivos do Instituto Biológico, 85. http://www.scielo.br/scielo.php?script=sci_arttext&pid=S1808-16572018000100602&nrm=iso.

Sim, L., O. Ward and Z. Li. 1997. Production and characterisation of a biosurfactant isolated from *Pseudomonas aeruginosa* UW-1. Journal of Industrial Microbiology and Biotechnology 19(4): 232–238.

Soberón-Chávez, G., F. Lépine and E. Déziel. 2005. Production of rhamnolipids by *Pseudomonas aeruginosa*. Applied Microbiology and Biotechnology 68(6): 718–725. https://doi.org/10.1007/s00253-005-0150-3.

Suh, S.-J., K. Invally and L.-K. Ju. 2019. Rhamnolipids: pathways, productivities, and potential. In Biobased Surfactants (pp. 169–203). Elsevier.

Suhaila, Y.N., A. Hasdianty, N. Maegala, A. Aqlima, A.H. Hazwan et al. 2019. Biotransformation using resting cells of Rhodococcus UKMP-5M for phenol degradation. Biocatalysis and Agricultural Biotechnology 21: 101309.

Syldatk, C., S. Lang, F. Wagner, V. Wray, L. Witte et al. 1985. Chemical and physical characterization of four interfacial-active rhamnolipids from *Pseudomonas* spec. DSM 2874 grown on n-alkanes. Zeitschrift für Naturforschung C 40(1-2): 51–60.

Toribio, J., A.E. Escalante and G. Soberón-Chávez. 2010. Rhamnolipids: Production in bacteria other than *Pseudomonas aeruginosa*. European Journal of Lipid Science and Technology 112(10): 1082–1087. https://doi.org/https://doi.org/10.1002/ejlt.200900256.

Tripathi, L., M.S. Twigg, A. Zompra, K. Salek, V.U. Irorere et al. 2019. Biosynthesis of rhamnolipid by a *Marinobacter* species expands the paradigm of biosurfactant synthesis to a new genus of the marine microflora. Microbial Cell Factories 18(1): 164.

Trummler, K., F. Effenberger and C. Syldatk. 2003. An integrated microbial/enzymatic process for production of rhamnolipids and L-(+)-rhamnose from rapeseed oil with *Pseudomonas* sp. DSM 2874. European Journal of Lipid Science and Technology 105(10): 563–571.

Twigg, M.S., L. Tripathi, A. Zompra, K. Salek, V. Irorere et al. 2018. Identification and characterisation of short chain rhamnolipid production in a previously uninvestigated, non-pathogenic marine pseudomonad. Applied Microbiology and Biotechnology 102(19): 8537–8549.

Vecino, X., J. Cruz, A. Moldes and L. Rodrigues. 2017. Biosurfactants in cosmetic formulations: trends and challenges. Critical Reviews in Biotechnology 37(7): 911–923.

Woodley, J.M., M. Bisschops, A.J. Straathof and M. Ottens. 2008. Future directions for *in-situ* product removal (ISPR). Journal of Chemical Technology & Biotechnology: International Research in Process. Environmental & Clean Technology 83(2): 121–123.

Wu, L.-m., L. Lai, Q. Lu, P. Mei, Y.-q. Wang et al. 2019a. Comparative studies on the surface/interface properties and aggregation behavior of mono-rhamnolipid and di-rhamnolipid. Colloids and Surfaces B: Biointerfaces 181: 593–601. https://doi.org/https://doi.org/10.1016/j.colsurfb.2019.06.012.

Wu, T., J. Jiang, N. He, M. Jin, K. Ma et al. 2019b. High-performance production of biosurfactant rhamnolipid with nitrogen feeding. Journal of Surfactants and Detergents 22(2): 395–402. https://doi.org/10.1002/jsde.12256.

Yang, H., M. Tong and H. Kim. 2012) Influence of bentonite particles on representative gram negative and gram positive bacterial deposition in porous media. Environmental Science & Technology 46(21): 11627–11634.

Zhong, H., X. Yang, F. Tan, M.L. Brusseau, L. Yang et al. 2016. Aggregate-based sub-CMC solubilization of n-alkanes by monorhamnolipid biosurfactant. New Journal of Chemistry 40(3): 2028–2035.

Zhu, K. and C.O. Rock. 2008. RhlA converts β-hydroxyacyl-acyl carrier protein intermediates in fatty acid synthesis to the β-hydroxydecanoyl-β-hydroxydecanoate component of rhamnolipids in *Pseudomonas aeruginosa*. Journal of Bacteriology 190(9): 3147–3154.

Zinjarde, S.S. and A. Pant. 2002. Emulsifier from a tropical marine yeast, Yarrowia lipolytica NCIM 3589. Journal of Basic Microbiology: An International Journal on Biochemistry, Physiology, Genetics, Morphology, and Ecology of Microorganisms 42(1): 67–73.

Zulianello, L., C. Canard, T. Köhler, D. Caille, J.-S. Lacroix et al. 2006. Rhamnolipids are virulence factors that promote early infiltration of primary human airway epithelia by *Pseudomonas aeruginosa*. Infection and Immunity 74(6): 3134–3147.

9

Current Perspectives of Microbial Lipopeptides with Their Advanced Applications for Sustainable Agriculture

Urja Pandya[1] and *Meenu Saraf*[2,*]

1. Introduction

Soil borne plant diseases occur in a wide diversity of plants such as vegetables and fruits, ornamental plants, shrubs and trees. The causal agents such as fungi, oomycetes, nematodes, viruses and few parasitic plants are responsible for soil borne plant diseases. Soil borne fungi such as *Rhizoctonia*, *Fusarium*, *Macrophomina*, *Sclerotiana*, *Sclerotium*, *Gaeumannomyces graminis* including oomycetes *Pythium* and *Phytophthora* are the major causal biological agents of significant soil borne plant diseases (Mathivanan et al. 2006, Jayaprakashvel and Mathivanan 2011, Saraf et al. 2014, Jayaprakashvel et al. 2019, Jeffery et al. 2010).

The major excessive applications of chemical based pesticides in the current decade is responsible not only for contamination of land and water resources but it can also affect the ecological imbalance of soil microorganisms, thus promoting the development of resistant pathogenic microbes giving rise to threats to human health as well as the environment. The steady hazards with respect to health and environment stimulate an alternative solution in the form of lipopeptides produced by many PGPRs and can be used for many agricultural applications (Mandal et al. 2013, Hafeez et al. 2019, Saraf et al. 2014). PGPRs are well studied and applied in managing soil-borne fungal diseases in various agricultural plants as they reduce

[1] Department of Microbiology, Gujarat Vidyapith, Sadra, Gandhinagar 382320, Gujarat, India.

[2] Department of Microbiology and Biotechnology, School of Sciences, Gujarat University, Ahmedabad 380009, Gujarat, India.

* Corresponding author: sarafmeenu@gmail.com

diseases by acting as biocontrol agents (Shaikh and Sayyed 2015, Pandya and Saraf 2014a, Gogoi et al. 2020).

Lipopeptides are among the most popular and interesting class of microbial surfactants. Lipopeptides are amphipathic molecules appraised as multifunctional materials of the 21st century because of their numerous properties of practical utility in various fields like industry, environment and agriculture (Santos et al. 2016). Most of them are generally produced by a variety of microbes especially the members of *Pseudomonas* and *Bacillus* species (Kloepper et al. 2004). Broadly, there are several major types of lipopeptides (LPs): polymyxins, daptomycin, surfactin, iturin and fengycin (Meena and Kanwar 2015, Patel et al. 2015). Total 263 different types of lipopeptides (LPs) were identified by 11 microbial genera. Within these genera, *Pseudomonas* sp. (78) and *Bacillus* sp. (98) classified in 11 and 5 lipopeptides families as reported by Coutte et al. (2017).

2. Soil Borne Phytopathogens

Koike et al. (2003) defined soil borne phytopathogens that cause plant diseases via inoculums that come to the plant through the soil. These pathogens are well established as the cause for many devastating diseases leading to around 25–100% crop loss every year. They may complete their entire life cycle within the soil or may spend part of it on the phyllosphere (Prashar et al. 2013). Soil borne plant pathogens may include a diverse group of organisms including bacteria, fungi, virus and nematodes hence fungal pathogens are the most critical amongst them that cause a large number of diseases (Agrios 2004, Koike et al. 2003, Prashar et al. 2013).

Soil borne pathogens survive either as soil inhabitants (for long periods) or as soil transients (for short periods) to be retained in the soil. They are allocated in soil depending upon the history of cropping, agricultural practices and other attributes. The pathogens are generally present in the top 10 inches of the soil profile, in the vertical plane, whereas field inoculums are collected from susceptible crops that grow in the horizontal plane. Some factors like soil type, pH, texture, moisture, temperature and different nutrient levels also affect the distribution of soil pathogens (Gogoi et al. 2020).

Some of pathogens showed very limited and highly specific host ranges where as others worked as generalists to a lesser or greater degree causing diseases across many host taxa. The intensity of pathogenesis varies from those which destroy crops through killing in the processes of colonization and reproduction to those which are only mildly aggressive and possibly almost commensalists (Dixon and Tilston 2010). Fungal pathogen inoculums survive in soil as forms of microsclerotia, sclerotia, chlamydospores or oospores. Singh et al. (2019) reported soil borne pathogens initiate specific symptoms when shown compatible association. Various symptoms included decaying and flaccid roots, withering and dropping of plants and flaccidity, yellowing, stunting, bark cracking, twig or branch dieing back (Horst 2001, Patil and Solanki 2016, Musheer et al. 2020). Most important soil-borne fungi include *Fusarium* sp., *Phytophthora* sp., *Pythium* sp., *Rhizoctonia solani*, *Sclerotinia* sp., *Slerotium rolfsii*, *Thielaviopsis basicola* and *Verticillium dahlia* (Jeffery et al. 2010, Prashar et al. 2013, Mathivanan et al. 2006, Jayaprakashvel and Mathivanan 2011, Jayaprakashvel et al. 2019).

3. Management Practices to Control various Pathogens

Biotic factors are very important for disease development. A successful interaction occurs between the virulent pathogen and a susceptible host under favorable conditions for a sufficient period of time (Katan 2017). These conditions can be resolved by integrated disease management approaches (IDM). There are various approaches such as cultural practices, chemical techniques, host plant resistance and practices like fumigation, steam treatment, and solarization of soils to some extent to control disease development (Prashar et al. 2013, Kumari and Katoch 2020). Chellemi et al. (2016) reported four steps in soil borne disease management like (i) to prevent introduction and establishment of pathogens to newer cultivating areas (ii) to reduce phytopathogens population below injury level as per economic point of view (iii) to improve natural suppressiveness of soil and (iv) to maintain least manipulation of natural biological and physical properties of soil.

3.1 *Cultural Methods*

These methods include flooding, deep plowing, crop rotation, soil solarization, biofumigation, nutrient management, tillage and others. The disadvantages are time consuming, require skills and deep knowledge, in-effective towards closed related species and difficulties faced in assessing the successfulness (Katan 2010, Jayaprakashvel and Mathivanan 2011, Juroszek and von Tiedemann 2011, Pandey et al. 2018, Hill 2019).

3.2 *Chemical Methods (Use of Chemical Fungicides)*

Chemical fungicides have long been used as agents in reducing plant diseases. However, some chemicals like metam-sodium and carbofuran generally used as soil fumigants which cause environmental pollution as well as toxic effects on human health and also show resistance among pathogens. These chemical fungicides are only efficient for a short time duration in the growing season (Baysal et al. 2008, Zheng et al. 2013). The disadvantages of chemical fungicides include persistence residues, higher cost, damage to environment, non-target effects and more (Jayaprakashvel and Mathivanan 2011).

3.3 *Biological Control*

The situation of environmental losses, reduction in soil microflora and fauna, excessive use of chemical pesticides as well as resistance capacity towards pathogens, leads to a new approach like use of biological control which offers ecofriendly solutions for plant disease management (Chandrashekhar et al. 2012, Pandya and Saraf 2013, Kumari and Kaotch 2020). Many antagonistic PGPR strains like *Agrobacterium*, *Arthrobacter*, *Azotobacter*, *Azospirillum*, *Bacillus*, *Burkholderia*, *Caulobacter*, *Chromobacterium*, *Erwinia*, *Flavobacterium*, *Micrococcus*, *Pseudomonas*, and *Serratia* have been found to control phytopathogens by various mechanisms and widely used in disease management at a global level (Pankhurst and Lynch 2005, Gouda et al. 2018).

3.4 *Host Plant Resistance*

This approach is the most economical and effective for disease management. However, available high yielding cultivars do not have genetic resistance against soil borne pathogens. Breeding programs are also applied but they are expensive and need a longer time for success (Jayaprakashvel and Mathivanan 2011). Some efforts are being made for new sources of resistance in wild relatives of cultivated pulse crops, mapping of resistance genes/quantitative trait loci (QTL) and identifying genetic markers linked with identified resistant(R) genes/QTLs for application of marker assisted selection (MAS) in breeding programs (Winter and Kahl 1995, Kumari and Kaotch 2020).

4. Lipopeptides of PGPR as a Remedy for Disease Management

Plant growth-promoting rhizobacteria (PGPR) are important soil microbial communities which reside on the roots and/or associated with plant roots (Kloepper and Schroth 1978, Ahemad and Kibret 2014, Saraf et al. 2014). PGPR is an umbrella that especially defines those bacteria in particular that substantially impact plant growth attributes, yield and disease resistance through different modes of actions by making a very strong and competitive interconnection with plant root systems (Prasad et al. 2019, Khatoon et al. 2020). PGPR produce a wide array of secondary metabolites such as siderophores, antibiotics, volatile metabolites, and other allelochemicals against soil borne pathogens (Saraf et al. 2014, Jayaprakashvel et al. 2019).

PGPRs excreted several classes of antibiotics such as phenazines, phloroglucinol, pyolutrorin, cyclic lipopeptides and volatile HCN (Hass and Defago 2005). The mechanisms of these antibiotics are partly understood however, their main effects including some processes like (i) inhibition of cell wall synthesis (ii) the arrest of ribosomal RNA formation (iii) deformation of cellular membranes and (iv) inhibition of protein biosynthesis (Maksimov et al. 2011, Kenawy et al. 2019). Antibiosis is mostly considered as one of the most studied mechanisms of PGPR for combating soil borne pathogens. Antibiotics are made of wide and heterogeneous groups of low molecular weight organic compounds that are produced by a wide variety of PGPRs. They are harmful to the growth or metabolic activities of other pathogens at low concentrations (Fravel 1988, Thomashow 1996, Jayaprakashvel et al. 2019).

Lipopeptides (LPs) are very small molecules that are synthesized by various microorganisms as forms of secondary metabolites (Zhao et al. 2017). Antimicrobial properties of lipopeptides are produced non-ribosomally in many bacteria and fungi during cultivation on various carbon sources (Markovitzki et al. 2006, Koh et al. 2017). These LPs can be structurally characterized by a hydrophilic peptide (cyclic or linear form) and a hydrophobic fatty acyl chain of an amphiphilic nature. Most of the native lipopeptides have complex cyclic structures (Koh et al. 2017).

4.1 *LPs of Pseudomonas* sp.

Many plant associated as well as biocontrol strains of *Pseudomonas* spp. and their genes have been reported for production of LPs viz. massetolide A, viscosin, etc. (De Bruijn et al. 2007, 2008, Gross et al. 2007, Loper et al. 2012, Mercado-Blanco,

2015). The general mechanism of synthesis of cyclic LPs occurs via non-ribosomal peptide synthases. Generally they have a chemical structure having a lipid tail (with variability in length and composition) which enhances their capacity to insert membranes thereby disturbing their function and integrity (Mercado-Blanco 2015).

Nielsen et al. (2000) reported fluorescent pseudomonads isolated from sugar beet rhizospheres producing cyclic LPs. Tensin (one type of LPs) produced by *P. fluorescens* 96.578 showed antagonistic activity against *R. solani* (Nielsen et al. 2000, Pathma et al. 2011). *P. fluorescens* DR54 produced cyclic LPs viz. viscosinamide reported for antifungal as well as biosurfactant properties under *in vitro* conditions (Nielsen et al. 2000, 2002, Thrane et al. 2000). Amphisin is produced by *Pseudomonas* sp. DSS73 having lipoundecapeptide synthesized from the non-ribosomal synthesis (Sorensen et al. 2001). This amphisim showed more antifungal activity as compared to tensin and viscosinamide in the case of fluorescent pseudomonad peptide antibiotics (Nielsen et al. 2002, Pathma et al. 2011).

Koch et al. (2002) studied amphisin produced in the stationary phase and amsY gene codes for amphisin synthetase which were controlled by two component regulatory systems like GacA/GacS. Andersen et al. (2003) reported *P. fluorescens* DSS73 produced amphisin which is responsible for the growth control of *P. ultimum* and *R. solani*. Other CLps like Massetolide also produced by various *Pseudomonas* strains and their genes were found to be involved in the massetolide A synthesis in *P. fluorescens* strain SS101 (de Bruijn et al. 2008, Pathma et al. 2011). Tran et al. (2007) observed inoculation of tomato plants with massetolide producing strains increased the tomato plant leave's resistance against *Phytophthora infestans.*

Olorunleke et al. (2015) reported total eight different structural groups of cyclic LPs from *Pseudomonas* sp. having different lengths and composition of the oligopeptide and fatty acid tail. The antimicrobial activity of LPs is related to their ability to perturb biological membranes (Raaijmakers et al. 2006). Nielsen et al. (2005) reported tolaasin and syringpmycin groups of LPs. Orfamide is also a type of LP which is produced by *P. protegens* showing insecticidal activity (Loper et al. 2016). In fluorescent pseudomonads, two component GacS/GacA regulatory mechanism systems showed a special effect of regulation of LPs synthesis (Raaijmakers et al. 2010). The possible roles of new regulatory genes required for biosynthesis of LPs in different species and strains of fluorescent pseudomonads has been studied by Song et al. (2015). Recently, Oni et al. (2020) characterized two LPs viz. Pseudodesmin and Viscosinamide from *Pseudomonas* sp. COR52 and A2W4.9 showing hyphal distortion and/or disintegration of *Rhizoctonia solani* AG2-2 and *Pythium myriotylum* CMR1.

4.2 LPs of *Bacillus* sp.

Many species of *Bacillus* synthesized a large number of lipopeptides. In the study of *Bacillus subtilis*, it was found to produce lipopeptides in a non ribosomal manner by multi domain enzymes viz. Non-ribosomal peptide synthetases (NRPSs) and polyketide synthetases (PKSs) (Finking and Marahiel 2004, Kumar and Johri 2012). Iturin family lipopeptides have 7 different types such as iturin A & C, bacillomycin D, F, L & LC and mycosubtilin with variants. They are heptapeptides with a β-amino fatty acid chain, comprised of 14–17 carbons exhibiting strong antifungal activity against a wide range of yeasts and fungi (Duitman et al. 1999, Tsuge et al. 2001, Moyne et al. 2004, Ongena and Jacques 2008). *B. subtilis* RB14 produced a lipopeptide, i.e., iturin A having operon more than 38 kb long and 4 reading frames (*itu* D, *itu* A, *itu* B and *itu* C). The itu D gene encodes a putative malonyl coenzyme A transacylase. The second gene *itu* A, codes a 449-kDa protein similar to fatty acid synthetase, aminoacid transferase, and peptide synthetase. The third and fourth genes, *itu* B and *itu* C encode 609 and 297 kDa peptide synthetases (Tsuge et al. 2001).

Surfactin is the most studied LP family. The surfactin structures are heptapeptides with an LLDLLDL chiral sequence linked by a β-hydroxy fatty acid (comprised of 13–15 'C' atoms) to form a cyclic lactone ring structure. Surfactin is synthesized by three NRPSs, *Srf* A-C and the enzyme thioesterase/acyltransferase wherein *Srf* D is known to initiate the process (Peypoux et al. 1999, Steller et al. 2004). *B. amyliliquefaciens* FZB42 produced surfactin showed antibacterial activity and formation of biofilm which was reported as powerful biocontrol agent (Chen et al. 2009).

The last group is Fengycin including fengycin A, fengycin B, plipastatin A, and plipastatin B. Fengycin is a bioactive molecule that contains a peptide chain of 10 amino acids linked to a β-hydroxy fatty acid chain that can vary from C-14 to C-17 carbon atoms with a lactone ring (Akpa et al. 2001, Meena and Kanwar 2015). Five open reading frames, namely, *fen* C, *fen* D, *fen* E, *fen* A, and *fen* B, are responsible for the synthesis of fengycin that are located in one operon with a molecular size of 37 kb (Lin et al. 1999). Both iturin and fengycins showed an antagonistic effect against *P. fusca* infecting melon leaves (Romero et al. 2007).

Wang et al. (2013) reported antifungal iturin and bacillomycin D types of LPs from *Bacillus amyloliquefaciens* W19 through HPLC-ESI-MS analysis as well as GC-MS analysis. Similarly, Bacillus *amyloliquefaciens* SH-B10 isolated from deep sea sediment identified fengycin A and a new derivative compound by tandem Q-TOF mass spectroscopy and HPLC analysis by Chen et al. (2010).

Pandya et al. (2017) also reported iturin homologues (m/z 1020–1120), surfactin (m/z 1008.7 and m/z 1022.7), fengycin A and fengycin B (m/z 1400–1550) through MALDI-TOF–MS and LC–ESI–MS/MS analysis from antifungal *B. subtilis* MBCU5 and also performed PCR analysis for confirmation of Iturin A synthetase (KJ531680) and Surfactin synthetase (KJ601726) involved in the biosynthesis of LPs.

5. Biological Activities of LPs in the Agriculture Field

A wide range of biocidals such as bactericidal, fungicidal, insecticidal effects from LPs are of great interest to many researchers (Fira et al. 2018, Sidorova et al. 2018, Maksimov et al. 2020). Etchegaray et al. (2008) studied *B. subitlis* OG produced iturin and surfactin and showed antibacterial activity against *Xanthomonas campestris* pv. *Campestris* and *Xanthomonas axonopodis pv. Citri*, which cause bacteriosis and cancer of citrus plants, respectively. Many researchers reported that LPs of *Pseudomonas* sp. are also very important for motility which may help in colonization with plants and inhibition of soil borne plant pathogens (Andersen et al. 2003, Roongsawang et al. 2003, de Bruijn et al. 2007). A few examples of lipopeptides with suitable antimicrobial action were reported in Table 1.

Pathak and Keharia (2014) identified iturin and surfactin from fungal antagonist *B. subtilis* K1 which was isolated from a banyan tree by using LC-ESI-MS/MS

Table 1. Antimicrobial action of LPs against soil borne pathogens.

Types of Lipopeptides	**Target Fungal Pathogens**	**References**
Fengycin, iturin and surfactin	*Sclerotinia sclerotiorum*	Alvarez et al. (2012)
Fengycin	*Botrytis cinerea, Sphaerotheca fuliginea*	Zhang et al. (2013)
Fengycin, iturin and surfactin	*Pectobacterium carotovorum, Xanthomonas campestris, Podosphaera fusca*	Zeriouh et al. (2014)
Iturin, fengycin and surfactin	*G. graminis var. tritici*	Yang et al. (2018)
Cyclic lipopeptides	*Pythium myriotylum*	Oni et al. (2019)
Tensin	*Rhizoctonia solani*	Nielsen et al. (2000)
Massetolide A	*Phytophthora infestans*	Tran et al. (2007)
Viscosin	*Rhizoctonia solani*	Yang et al. (2014)
Iturin A	*C. gloeosporioides*	Yan et al. (2020)
Surfactin	*Mucor* sp. and *Aspergillus niger*	Meena et al. (2020)

analysis. A new class of lipopeptides named orfamide synthesized from *P. protegens* controlled growth of *Cochliobolus miyabeanus*; this fungus was responsible for brown pot spot diseases on rice and such LPs triggered induced systematic resistance as well in rice plants (Ma et al. 2017). Liu et al. (2019) tested the antimicrobial activity of *B. velezensis* HC6 against *Aspergillus*, *Fusarium*, and aspathogenic bacteria as well (especially *Listeria monocytogenes*).

Three types of antimicrobial lipopeptides viz. iturin, surfactin and fengycin were identified in this strain by high-performance liquid chromatography and MALDI-TOF mass spectrometry. Also this strain reduced production of aflatoxij and ochratoxin when applied in maize plants. Romero et al. (2007) observed that the antagonistic effect of *Bacillus subtilis* against *Podosphaera fusca* is due to the fengycin and iturn types of LPs. The effective control of fire blight disease by *Bacillus amyloliquefaciens* is due to the production of difficidin and bacilysin families of polyketides (Chen et al. 2009). Chowdhury et al. (2015) identified a total of ten gene clusters regarding LPs viz. polyketides, bacilysin, plantazolicin and amylocyclicin in *Bacillus amyloliquefaciens* through genetic (knock out mutagenesis) and chemical techniques (mass spectroscopy).

Bacillus LP was reported to be efficient in wheat against *Zymoseptoria tritici* (Mejri et al. 2017), *Gaeumannomyces graminis var. tritici* (Zhang et al. 2017, Yang et al. 2018), and *Fusarium graminearum* (Gong et al. 2015). This type of LP also reported to be efficient in maize (Chan et al. 2009). Mnif et al. (2015) studied *Bacillus* biosurfactants controlled the growth of *Fusarium solani* in potatoes plants. *B. subtilis* CPA-8 was reported as effective for control of brown rot of stone fruit caused by *Monilinia laxa* and *M. fructicola* (Yánez-Mendizábal et al. 2011). A further study by Yánez-Mendizábal et al. (2012) studied that among all three families of LPs, fengycin was the most active inhibiting agent however the mutant strain was unable to produce fengycin.

Cao et al. (2012) isolated *B. subtilis* from a cucumber field and it produced fengycin shown to be an effective agent against diseases. The fengycin producing strain *B. subtilis* D1/2 reduced fusarium head blight through growth inhibition of

Fusarium graminearum (Chen et al. 2009). Rebib et al. (2012) observed *B. subtilis* SR146 isolated from saline soil inhibits foot root disease caused by *Fusarium culmorum* in wheat plants and further studies explained this strain produced fengycin significantly inhibiting *F. culmorum* spore germination. Many strains of *B. subtilis* isolated from the rhizosphere of avocado trees produced different types of LPs which were most effective against the growth of *F. oxysporum f.* sp. *radicis-lycopersici* and *Rosellinia necatrix* (Cazorla et al. 2007).

In another research study, *B. subtilis* KS03 produced iturin A compound reducing the anthracnose disease in fruits and vegetables caused by *Gloeosporium gloeosporioides* (Cho et al. 2003). Another iturin family of bacillomycin was the main component working as an antagonist against *Podosphaera fusca* which is a causal agent of cucurbit powdery mildew, however fengycin was unable to inhibit the growth of *Podosphaera fusca*. Zeriouh et al. (2011) observed iturin was found to be an antibacterial agent against growth inhibition of Xanthomonas campestris pv. cucurbitae and *Pectobacterium carotovorum* subsp. carotovorum, in curcurbits plants. Elkahoui et al. (2012) studied the potential growth inhibition of *Rhizoctonia solani* through *B. subitilis* LPs both *in vitro* and *in vivo* experiments on potatoes.

Pandya and Saraf (2015) isolated and identified iturin A and surfactin from *Bacillus sonorensis* MBCU2 through MALDI-TOF-MS and liquid chromatography coupled with ESI-MS/MS which play a significant role for suppression of charcoal rot disease in *Arachis hypogaea.* L. Nielsen et al. (1998) studied viscosinamide produced from *P. fluorescens* DR54 and observed an increased plant emergence of sugar beet against *P. ultimmum* infection. Similarly, *P. fluorescens* SS 101 significantly produced massetolide. A reduced infection of tomato leaves by *P. infestans* (Tran et al. 2007). Pedras et al. (2003) isolated pseudophomins A and B from *P. fluorescens* BRG100 showed antifungal activities against *Phoma lingam* (Tode ex Fr.) Desm. and *S. sclerotiorum*, the causal agents of black-leg and white mold diseases respectively.

6. Conclusion

Nature provides a wide structural diversity of various cyclic lipopeptides. According to literature studies based on research work, several strains of *Pseudomonas* and *Bacillus* sp. were isolated and identified by modern techniques such as MALDI-TOFF analysis, ESI-MS/MS analysis and HPLC analysis. Lipopeptides are very useful molecules as both their fatty acid and peptide moieties could be modified for different applications. Particularly, they are potential candidates to be used in pharmaceutical, agriculture, plants and food industries, where antimicrobial properties are applied. Bacillus strains produced excellent types of LPs (iturin, surfactin and fengycin) which are strong antimicrobial agents in direct and indirect ways by affecting and killing soil borne fungal pathogens. These properties make them an excellent option for biocontrol systems for sustainable agricultural production.

References

Agrios, G. 2004. Plant pathology, 5th edn. Elsevier, London.

Ahemad, M. and M. Kibret. 2014. Mechanisms and applications of plant growth promoting rhizobacteria: current perspective. J. King Saud Uni. Sci. 26: 1–20.

Akpa, E., P. Jacques, B. Wathelet, M. Paquot, R. Fuchs et al. 2001. Influence of culture conditions on lipopeptide production by *Bacillus subtilis*. Appl. Biochem. Biotech. 91: 551–561.

Alvarez, F., M. Castro, A. Principe, G. Borioli, S. Fischer et al. 2012. The plant-associated *Bacillus amyloliquefaciens* strains MEP 218 and ARP 23 capable of producing the cyclic lipopeptides iturin or surfactin and fengycin are effective in biocontrol of sclerotinia stem rot disease. J. Appl. Microbiol. 112: 159–174. https://doi.org/10.1111/j.1365-2672.2011.05182.x.

Andersen, J.B., B. Koch, T.H. Nielsen, D. Sorensen, M. Hansen et al. 2003. Surface motility in *Pseudomonas* sp. DSS73 is required for efficient biological containment of the root-pathogenic microfungi *Rhizoctonia solani* and *Pythium ultimum*. Microbiol. 149: 37–46.

Baysal, O., M. Caliskan and O. Yesilova. 2008. An inhibitory effect of a new *Bacillus subtilis* strain (EU07) against *Fusarium oxysporum f.* sp. *radicis-lycopersici*. Physiol. Mol. Plant Pathol. 73: 25–32.

Cao, Y., Z.H. Zhang, N. Ling et al. 2011. *Bacillus Subtilis* SQR 9 can control *Fusarium* Wilt in cucumber by colonizing plant Rroots. Biol. Fertil. Soils 47(5): 495–506.

Cazorla, F.M., D. Romero, A. Perez-Garcia et al. 2007. Isolation and characterization of antagonistic *Bacillus subtilis* strains from the avocado rhizoplane displaying biocontrol activity. J. Appl. Microbiol. 103: 1950–1959.

Chandrashekara, K., C. Chandrashekara, M. Chakravathi and S. Manivannan. 2012. Biological control of plant disease. pp. 147–166. *In*: Singh, V.K., Y. Singh and A. Singh (eds.). Eco-Friendly Innovative Approaches in Plant Disease Management: International Book Distributors, Uttarakhand, India.

Chellemi, D.O., A. Gamliel, J. Katan and K.V. Subbarao. 2016. Development and deployment of system based approaches for the management of soil borne plant pathogens. Phytopathol. 106: 216–225.

Chen, L., N. Wang, X. Wang, J. Hu, S. Wang et al. 2010. Characterization of two anti-fungal lipopeptides produced by *Bacillus amyloliquefaciens* SH-B10. Biores. Technol. 101: 8822–8827.

Chen, X.H., R. Scholz, M. Borriss, H. Junge, G. Mogel et al. 2009. Difficidin and bacilysin produced by plant associated *Bacillus amyloliquefaciens* are efficient in controlling fire blight disease. J. Biotechnol. 140: 3844.

Cho, S.J., S.H. Lee, B.J. Cha et al. 2003. Detection and characterization of the Gloeosporium gloeosporioides growth inhibitory compound iturin A from *Bacillus subtilis* strain KS03. FEMS Microbiol. Lett. 223: 47–51.

Chowdhury, S.P., A. Hartmann, X. Gao and R. Borriss. 2015. Biocontrol mechanism by root-associated *Bacillus Amyloliquefaciens* FZB42-a Review. Front. Microbiol. 6: 780.

Coutte, F., D. Lecouturier, K. Dimitrov, J.S. Guez, F. Delvigne et al. 2017. Microbial lipopeptide production and purification bioprocesses, current progress and future challenges. Biotechnol. J. 12(7): 1600566.

De Bruijin, I., M.J.D. de Kock, M. Yang, P. de Waard, T.A. van Beek et al. 2007. Genome based discovery, structure prediction and functional analysis of cyclic lipopeptide antibiotics in *Pseudomonas* species. Mol. Microbiol. 63: 417–442.

De Bruijn, I., M.J. De. Kock, P. De Waard, T.A. Van Beek, J.M. Raaijmakers et al. 2008. Massetolide A biosynthesis in *Pseudomonas fluorescens*. J. Bacteriol. 190: 2777–2789.

Dixon, G.R. and E.L. Tilton. 2010. Soil borne pathogens and their interactions with the soil environment. Soil Microbiol. Sust. Crop Prod., pp. 197–271.

Duitman, E.H., L.W. Hamoen, M. Rembold, G. Venema, H. Seitz, W. Saenger, F. Bernhard, R. Reinhardt, M. Schmidt, C. Ullrich, T. Stein, F. Leenders and J. Vater. 1999. *Bacillus subtilis* ATCC6633: a multifunctional hybrid between a peptide synthetase, an amino transferase, and a fatty acid synthase. P. Natl.Acad.Sci. USA. 96: 13294–13299.

Etchegaray, A., C.D. Bueno, I.S. de Melo et al. 2008. Effect of a highly concentrated lipopeptide extract of *Bacillus subtilis* on fungal and bacterial cells. Arch. Microbiol. 190: 611–622.

Elkahoui, S., N. Djebali, O. Tabbene et al. 2012. Evaluation of antifungal activity from *Bacillus strains* against *Rhizoctoniasolani*. African J. Biotechnol. 11(18): 4196–4201.

Finking, R. and M.A. Marahiel. 2004. Biosynthesis of non-ribosomal peptides. Ann. Rev. Microbiol. 58: 453–488.

Fira, D., I. Dimkic, T. Beric, J. Lozo and S. Stankovic. 2018. Biological control of plant pathogens by *Bacillus* species. J. Biotech. 10 (285): 44–55.

Fravel, D.R. 1988. Role of antibiosis in the biocontrol of plant diseases. Annu. Rev. Phytopathol. 26: 75–91.

Gogoi, P., P. Kakoti, J. Saikia, R.K. Sarma, A. Yadav et al. 2020. Plant growth promoting rhizobacteria in management of soil borne fungal pathogens. pp. 1–14. *In*: Singh, B.P. et al. (eds.). Management of Fungal Pathogens In Pulses. Springer Nature Switzerland.

Gong, A.D., H. Li, Q.S. Yuan et al. 2015. Antagonistic mechanism of Iturin A and Plipastatin A from *Bacillus amyloliquefaciens* S76-3 from Wheat Spikes against *Fusarium graminearum*. PLoS One. 10(2): e0116871.

Gouda, S., R.G. Kerry, G. Das, S. Paramithiostis, H.S. Shine et al. 2018. Revitalization of plant growth promoting rhizobacteria for sustainable development in agriculture. Microbiol. Res. 206: 131–140.

Gross, H., V.O. Stockwell, M.D. Henkels, B. Nowak-Thompson, J.E. Loper et al. 2007. The genomisotopic approach: a systematic method to isolate products of orphan biosynthetic gene clusters. Chem. Biol. 14: 53–63.

Haas, D. and G. Defago. 2005. Biological control of soil-borne pathogens by fluorescent pseudomonads. Nat. Rev. Microbiol. 3: 307–319.

Hafeez, F.Y., Z. Naureen and A. Sarwar. 2019. Surfactin: An emerging biocontrol tool for agriculture sustaninability. pp. 203–214. *In*: Kumar, V.S. and S. Meena (eds.). Plant Growth Promoting Rhizobacteria for Agricultural Sustanibility, Springer Nature, Singapore.

Hill, S.B. 2019. Pest control-cultural control of insects cultural methods of pest, primarily insect, control; EAP Publication-58.

Horst, R.K. 2001. Plant diseases and their pathogens. *In*: Westcott's plant disease handbook. Springer, Boston. https://doi.org/10.1007/978-1-4757-3376-1_3.

Jayaprakashvel, M. and N. Mathivanan. 2011. Management of plant diseases by microbial metabolites. *In*: Maheshwari, D.K. (ed.). Bacteria in Agrobiology: Plant Nutrient Management. Springer-Verlag, Berlin/Heidelberg. https://doi.org/10.1007/978-3-642-21061-7_10.

Jayaprakashvel, M., C. Chitra and N. Mathivanan. 2019. Metabolites of plant growth promoting rhizobacteria for the management of soil borne pathogenic fungi in crops. pp. 293–315. *In*: Singh, H.B. et al. (eds.). Secondary Metabolites of Plant Growth Promoting Rhizomicroorganisms. Springer Nature Singapore.

Jeffery, S., C. Gardi, A. Jones, L. Montanarella, L. Marmo et al. 2010. European atlas of soil biodiversity. European Commission Publications Office of the European Union, Luxembourg, pp. 17–48.

Juroszek, P. and A. von Tiedemann. 2011. Potential strategies and future requirements for plant disease management under a changing climate. Plant Pathol. 60: 100–12. https://doi. org/10.1111/j.1365-3059.2010.02410.x.

Katan, J. 2010. Cultural approaches for disease management: present status and future prospects. J. Plant Pathol. 92(4): S4.7–S4.9.

Kenawy, A., D.J. Dailin, G.A. Abo-Zaid, R.A. Malek, K.K. Ambehabati et al. 2019. Biosynthesis of antibiotics by PGPR and their roles in biocontrol of plant diseases. pp. 1–35. *In*: Sayyed, R.Z. (ed.). Plant growth promoting rhizobacteria for sustainable stress management for sustainability. Springer Nature Singapore Pte Ltd..

Khatoon, Z., S. Huang, M. Rafique, Ali. Fakhar, M.A. Kamran et al. 2020. Unlocking the potential of plant growth promoting rhizobacteria on soil health and the sustainability of agricultural systems. J. Environ. Manag. 273: 111–118.

Kloepper, J.W. and M.N. Schroth. 1978. Plant growth promoting rhizobacteria on radishes. pp. 879–882. *In*: Proceedings of the fourth international conference on plant pathogen bacteria, vol. 2. INRA, Gilbert-Clarey, Tours.

Kloepper, J.W., C.M. Ryu and S. Zhang. 2004. Induced systemic resistance and promotion of plant growth by *Bacillus* sp. Phytopathol. 94: 1259–1266.

Koch, B., T.H. Nielsen, D. Sorensen, J.B. Andersen, C. Christophersen et al. 2002. Lipopeptide production in *Pseudomonas* sp. strain DSS73 is regulated by components of sugar beet exudates via the Gac two component regulatory system. Appl. Environ. Microbiol. 68: 4509–4516.

Koh, Jun-Jie., S. Lin, R.W. Beuerman and S. Liu. 2017. Recent advances in synthetic lipopeptides as anti-microbial agents: designs and synthetic approaches. Amino Acids 49: 1653–1677.

Koike, S.T., K.V. Subbarao, R.M. Davis and T.A. Turini. 2003. Vegetable diseases caused by soilborne pathogens. ANR publication 8099, the Regents of University of California, Division of Agriculture and Natural Resources: 1–13.

Kumar, A. and B.N. Johri. 2012. Antimicrobial lipopeptides of *Bacillus*: Natural weapons for biocontrol of plant pathogens. pp. 91–111. *In*: Satyanarayana, T. et al. (eds.). Microorganisms in Sustainable Agriculture and Biotechnology. Springer.

Kumari, N. and S. Katoch. 2020. Wilt and root rot complex pulse crops: their detection and integrated management. pp. 93–119. *In*: Singh, B.P. et al. (eds.). Management of Fungal Pathogens in Pulses, Fungal Biology, Springer Nature Switzerland.

Lin. T., C. Chen, L. Chang, G.S. Tschen, S. Liu et al. 1999. Functional and transcriptional analyses of a fengycin synthetase gene, fenc, from *Bacillus subtilis*. J. Bact. 181: 5060–5067.

Liu, Y., K. Teng, T. Wang, E. Dong, M. Zhang et al. 2019. Antimicrobial *Bacillus velezensis* HC6: production of three kinds of lipopeptides and biocontrol in maize. J. Appl. Microbiol. 128: 242–254.

Leoper, J.E., M.D. Henkels, L.I. Rangel, M.H. Olcott, F.L. Walker et al. 2016. Rhizoxin, orfamide A, and chitinase production contribute to the toxicity of *Pseudomonas protegens* strain Pf-5 to *Drosophila melanogaster*. Environ. Microbiol. 18: 3509–3521.

Loper, J.E., K.A. Hassan, D.V. Mavrodi, E.W. Davis, C.K. Lim et al. 2012. Comparative genomics of plant-associated *Pseudomonas* sp.: Insights into diversity and inheritance of traits involved in multitrophic interactions. PLoS Genet 8(7): e1002784.

Ma, Z., M. Ongena and M. Hofte. 2017. The cyclic lipopeptide orfamide induces systemic resistance in rice to *Cochliobolus miyabeanus* but not to *Magnaporthe oryzae*. Plant Cell Rep. 36(11): 1731–1746.

Makovitzki, A., D. Avrahami and Y. Shai. 2006. Ultrashort antibacterial and antifungal lipopeptides. Proc. Natl. Acad. Sci. USA 103(43): 15997–16002.

Maksimov, I.V., R.R. Abizgildina and L.I. Pusenkova. 2011. Plant growth promoting rhizobacteria as an alternative to chemical crop protectors from pathogens (Review). Appl. Biochem. Microbiol. 47: 333–345.

Mandal, S.M., A.E.A.D Barbosa and O.L. Franco. 2013. Lipopeptides in microbial infection control: scope and reality for industry. Biotechnol. Adv. 31: 338–345.

Mathivanan, N., K. Manibhushanrao and K. Murugesan. 2006. Biological control of plant pathogens. pp. 275–323. *In*: Anand, N. (ed.). Recent Trends in Botanical Research. University of Madras, Chennai.

Meena, K.R. and S.S. Kanwar. 2015. Lipopeptides as the antifungal and antibacterial agents: applications in food safety and therapeutics. Biomed. Res. Int. 473050.

Meena, K.R., A. Sharma, R. Kumar and S.S. Kanwar. 2020. Two factor at a time approach by response surface methodology to aggrandize the *Bacillus subtilis* KLP2015 surfactin lipopeptide to use as antifungal agent. J. King Saud University-Sci. 32(10): 337–348.

Maksimov, I.V., B.P. Singh, E.A. Cherepanova, G.F. Burkhanova and R.M. Khairulli. 2020. Prospects and applications of lipopeptide-producing bacteria for plant protection (Review). Appl. Biochem. Microbiol. 56: 15–28.

Mercado-Blanco, J. 2015. *Pseudomonas* strains that exert biocontrol of plant pathogens. pp. 121–172. *In*: Ramos, J.-L. et al. (eds.). *Pseudomonas*. Springer Science+Business Media Dordrecht.

Meijri, S., A. Siah, F. Coutte et al. 2017. Biocontrol of the wheat pathogen *Zymoseptoriatritici* using cyclic lipopeptides from *Bacillus subtilis*. Environ. Sci. Pollut. Res. Int. 25(30): 29822–29833.

Mnif, I., I. Hammami, M.A. Triki, M.C. Azabou et al. 2015. Antifungal efficiency of a lipopeptide biosurfactant derived from *Bacillus subtilis* SPB1 versus the phytopathogenic fungus, *Fusarium solani*. Environ. Sci. Pollut. Res. Int. 22(22): 18137–47.

Moyne, A.L., T.E. Cleveland and S. Tuzun. 2004. Molecular characterization and analysis of the operon encoding the antifungal lipopeptide bacillomycin D. FEMS Microbiol. Lett. 234: 43–49.

Musheer, N., S. Ashraf, A. Chaudhary, M. Kumar and S. Saeed et al. 2020. Role of microbiotic factors against the soil borne phytopathogens. pp. 251–280. Solanki, M.R. (eds.). Phytobiomes: Current Insights and Future Vistas. Springer Nature Singapore.

Nielsen, M.N., J. Sorensen, J. Fels and H.C. Pedersen. 1998. Secondary Metabolite- and Endochitinase-dependent antagonism toward plant-pathogenic microfungi of *Pseudomonas fluorescens* isolates from sugar beet Rhizosphere. Appl. Environ. Microbiol. 64(10): 3563-3569.

Nielsen, T.H., C. Thrane, C. Christophersen, U. Anthoni, J. Sorensen et al. 2000. Structure, production characteristics and fungal antagonism of tensin—a new antifungal cyclic lipopeptide from *Pseudomonas fluorescens* strain 96.578. J. Appl. Microbiol. 89: 992–1001.

Nielsen, T.H., D. Sorensen, C. Tobiasen, J.B. Andersen, C. Christophersen et al. 2002. Antibiotic and biosurfactant properties of cyclic lipopeptides produced by fluorescent *Pseudomonas* sp. from the sugar beet rhizosphere. Appl. Environ. Microbiol. 68: 3416–3423.

Nielsen, T.H., O. Nybroe, B. Koch, M. Hansen and J. Sorensen. 2005. Genes involved in cyclic lipopeptide production are important for seed and straw colonization by *Pseudomonas* sp. strain DSS73. Appl. Environ. Microbiol. 71 (7): 4112–4116.

Olorunleke, F.E., N.P. Kieu and M. Hofte. 2015. Recent advances in *Pseudomonas biocontrol*. pp. 167–198. *In*: Murillo, J., B.A. Vinatzer, R.W. Jackson, D.L. Arnold (eds.). Bacterial-Plant Interactions: Advance Research and Future Trends. Caister Academic Press, Norfolk.

Ongena, M. and P. Jacques. 2008. *Bacillus* lipopeptides: Versatile weapons for plant disease biocontrol. Trends Microbiol. 16: 115–125.

Oni, F.E., O.F. Olorunleke and M. Hofte. 2019. Phenazines and cyclic lipopeptides produced by *Pseudomonas* sp. CMR12a are involved in the biological control of *Pythium myriotylum* on cocoyam (*Xanthosoma sagittifolium*). Biol. Control. 129: 109–114.

Oni, F.E., N. Geudens, A. Adiobo, O.O. Omoboye, E.A. Enow et al. 2020. Biosynthesis and antimicrobial activity of pseudosmin and viscosinamide cyclic lipopeptides produced by *Pseudomonads* associated with the cocoyam rhizosphere. Microrog. 8: 1079.

Pandey, A.K., R.R. Burlakoti, L. Kenyon and R.M. Nair. 2018. Perspectives and challenges for sustainable management of fungal diseases of mungbean [*Vigna radiata* (L.) R. Wilczek var. radiata]: a review. Front Environ. Sci. 6: 53. https://doi.org/10.3389/fenvs.2018.00053.

Pandya, U. and M. Saraf. 2013. Integrate diseases management in groundnut for sustainable productivity. pp. 351–377. *In*: Maheshwari, D.K., M. Saraf and A. Aeron (eds.). Bacteria in Agrobiology: Crop Productivity. Springer-Verlag Berlin, Heidelberg.

Pandya, U. and M. Saraf. 2014(a). *In vitro* evaluation of PGPR strains for their biocontrol potential against fungal pathogens. pp. 293–305. *In*: Kharwar, R.N., R.S. Upadhyay, N.K. Dubey and Richa Raghuwanshi (eds.). Microbial Diversity and Biotechnology in Food Security, Springer India.

Pandya, U. and M. Saraf. 2015. Isolation and identification of allelochemicals produced by *B. sonorensis* for suppression of charcoal rot of *Arachis hypogaea* L. J. Basic Microbiol. 54: 1–10.

Pandya, U., S. Prakash, K. Shende, U. Dhuldhaj, M. Saraf et al. 2017. Multifarious allelochemicals exhibiting actifungal activity from *Bacillus subtilis* MBCU5. 3Biotech 7: 175.

Pankhurst, C.E. and J.M. Lynch. 2005. Biocontrol of soil-borne plant diseases. *In*: Hillel, D. (ed.). Encyclopedia of Soils in the Environment. Elsevier, Amsterdam.

Patel, S., S. Ahmed and J.S. Eswari. 2015. Therapeutic cyclic lipopeptides mining from microbes: latest strides and hurdles. World J. Microbiol. Biotechnol. 31(8): 1177–1193. doi:10.1007/s11274-015-1880-8.

Pathak, K.V. and H. Keharia. 2014. Identification of surfactins and iturins produced by potent fungal antagonist, *Bacillus subtilis* K1 isolated from aerial roots of banyan (*Ficus benghalensis*) tree using mass spectrometry. 3 Biotech. 4: 283–295.

Pathma, J., R.K. Kennedy and N. Sakthivel. 2011. Mechanisms of fluorescent pseudomonads that mediate biological control of phytopathogens and plant growth promotion of crop plants. pp. 77–105. *In*: Maheshwari, D.K. (ed.). Bacteria in Agrobiology: Plant Growth Responses. Springer-Verlag Berlin Heidelberg.

Patil, H.J. and M.K. Solanki. 2016. Molecular prospecting: advancement in diagnosis and control of *Rhizoctonia solani* diseases in plants. Springer International Publishing, pp. 165–185.

Pedras, M.S.C., N. Ismail, J.W. Quail and S.M. Boyetchko. 2003. Structure, chemistry, and biological activity of pseudophomins A and B, new cyclic lipodepsipeptides isolated from the biocontrol bacterium *Pseudomonas fluorescens*. Phytochemi. 62(7): 1105–1114.

Peypoux, E., J.M. Momatin and J. Wallach.1999. Recent trends in the biochemistry of surfactin. Appl. Microbiol. Biotechnol. 51: 553–563.

Prasad, M., R. Srinivasan, M. Chaudhary, M. CHaudhary, L.K. Jat et al. 2019. Plant growth promoting rhizobacteria (PGPR) for sustainable agriculture: perspectives and challenges. pp. 129–157. *In*: PGPR Amelioration in Sustainable Agriculture. Woodhead Publishing. https://doi.org/10.1016/B978-0-12-815879-1.00007-0.

Prashar, P., N. Kapoor and S. Sachdeva. 2013. Biocontrol of plant pathogens using plant growth promoting bacteria. pp. 319–360. *In*: Lichtfouse, E. (ed.). Sustainable Agriculture Reviews, Sustainable Agriculture Reviews 12.

Raaijmakers, J.M., I. de Bruijn and M.J.D. de kock. 2006. Cyclic lipopeptide production by plant-associated *Pseudomonas* sp.: diversity, activity, biosynthesis, and regulation. Mol. Plant. Microbe Interact. 19(7): 699–710.

Raaijmakers, J.M., I. De Bruijn, O. Nybroe and M. Ongena. 2010. Natural functions of lipopeptides from *Bacillus* and *Pseudomonas*: more than surfactants and antibiotics. FEMS Microbiol. Rev. 34(6): 1037–1062.

Rebib, H., H. Abdeljabbar et al. 2012. Biological control of *Fusarium* foot rot of wheat using fengycin-producing *Bacillus subtilis* isolated from salty soil. African J. Biotech. 11(34): 8464–8475.

Romero, D., A. de Vicente, R.H. Rakotoaly, S.E. Dufour, S.E. Veening et al. 2007. The iturin and fengycin families of lipopeptides are key factors in antagonism of *Bacillus subtilis* toward *Podosphaera fusca*. Mol. Plant-Microbe Interact. 20(4): 430–440.

Roongsawang, N., K. Hase, M. Haruki, T. Imanaka, M. Morikawa et al. 2003. Cloning and characterization of the gene cluster encoding arthrofactin synthetase from *Pseudomonas* sp. MIS38. Chem. Biol. 10: 869–880.

Santos, D.K.F., R.D. Rufino, J.M. Luna, V.A. Santos, L.A. Sarubbao et al. 2016. Biosurfactants: multifunctional biomolecules of the 21st century. Int. J. Mol. Sci. 17(3): 401.

Saraf, M., U. Pandya and A. Thakkar. 2014. Role of allelochemicals in plant growth promoting rhizobacteria for biocontrol of phytopathogens. Microbiol. Res. 169: 18–29.

Shaika, S.S. and R.Z. Sayyed. 2015. Role of plant growth-promoting rhizobacteria and their formulation in biocontrol of plant diseases. Springer, pp. 337–351.

Sidorova, T.M., A.M. Asaturova and A.I. Homyak. 2018. *Bacillus subtilis* and their role in the control of phytopathogenic microorganisms. Agri. Biol. 53(1): 29–37.

Singh, P., Q.Q. Song, R. Singh, H.B. Li, M. Solanki et al. 2019. Proteomic analysis of the resistance mechanisms in sugarcane during Sporisorium scitamineum infection. Int. J. Mol. Sci. 20: 569.

Sonh, C., G. Sundqvist, E. Malm, I. de Bruijin, A. Kumar et al. 2015. Lipopeptide biosynthesis in *Pseudomonas fluorescens* is regulated by the protease complex ClpAP. BMC Microbiol. 15(1): 29.

Sorensen D., T.H. Nielsen, C. Christophersen, J. Sorensen, M. Gajhede et al. 2001. Cyclic lipoundecapeptide amphisin from *Pseudomonas* sp. strain DSS73. Acta Crystallogr. C. 57: 1123–1124.

Steller, S., A. Sokoll, C. Wilde, F. Berhard, P. Franke and J. Vater. 2004. Initiation of surfactin biosynthesis and the role of the SrfD thioesterase protein. Biochem. 43: 11331–11343.

Thomashow, L.S. 1996. Biological control of plant root pathogens. Curr. Opin. Biotechnol. 7: 343–347.

Thrane, C., M.N. Nielsen, T.H. Nielsen, S. Olsson, J. Sorensen et al. 2000. Viscosinamide-producing *Pseudomonas fluorescens* DR54 exerts a biocontrol effect on *Pythium ultimum* in sugar beat rhizosphere. FEMS Microbiol. Ecol. 33: 139–146.

Tran, H., A. Ficke, T. Asiimwe, M. Hofte and J.M. Raaijmakers et al. 2007. Role of the cyclic lipopeptide massetolide A in biological control of *Phytophthora infestans* and in colonization of tomato plants by *Pseudomonas fluorescens*. New Phytol. 175: 731–742.

Tsuge, K., T. Akiyama and M. Shoda. 2001. Cloning, sequencing, and characterization of the iturinA operon. J. Bacteriol. 183: 6265–6273.

Wang, B., J. Yuan, J. Zhang, Z. Shen, M. Zhang et al. 2013. Effects of novel bioorganic fertilizer produced by *Bacillus amyloquefaciens* W19 on antagonism of Fusarium wilt of banana. Biol. Fertil. Soils. 49: 435–446.

Winter, P. and G. Kahl. 1995. Molecular marker technologies for plant improvement. World J. Microbiol. Biotechnol. 11: 438–448.

Yan, F., C. Li, X. Ye, Y. Lian, Y. Wu et al. 2020. Antifungal activity of lipopeptides from *Bacillus amyloliquefaciens* MG3 against *Colletotrichum gloeosporioides* in loquat fruits. Biol. Cont. 146: 104–281.

Yánez-Mendizábal, V., I. Vinas et al. 2011. Potential of a new strain of *Bacillus subtilis* CPA-8 to control the major postharvest diseases of fruit. Biocont Sci. Technol. 21(4): 409–426.

Yánez-Mendizábal, V., H. Zeriouh, I. Viñas et al. 2012. Biological control of peach brown rot (*Monilinia* spp.) by *Bacillus subtilis* CPA-8 is based on production of fengycin-like lipopeptides. Eur. J. Plant Pathol. 132(4): 609–619.

Yang, L., X. Han, F. Zhang et al. 2018. Screening *Bacillus* species as biological control agents of *Gaeumannomyces graminis* var. Tritici on wheat. Biol. Cont. 118: 1–9.

Yang, M.M., S.S. Wen, D.V. Mavrodi, D.V. Wettstein, L.S. Thomashow et al. 2014. Biological Control of Wheat Root Diseases by the CLP-Producing Strain *Pseudomonas fluorescens* HC1-07. Phytopathol. 104(3): 248–256.

Zeriouh, H., D. Romero, L. García-Gutiérrez, F.M. Cazorla et al. 2011. The iturin-like lipopeptides are essential components in the biological control arsenal of *Bacillus subtilis* against bacterial diseases of cucurbits. Mol. Plant Microbe Interact. 24: 1540–1552.

Zeriouh, H., A. de Vicente, A. Perez-Garcia and D. Romero. 2014. Surfactin triggers biofilm formation of *Bacillus subtilis* in melon phylloplane and contributes to the biocontrol activity. Environ. Microbiol. 16: 2196–2211.

Zhang, X., B. Li, Y. Wang et al. 2013. Lipopeptides, a novel protein, and volatile compounds contribute to the antifungal activity of the biocontrol agent *Bacillus atrophaeus* CAB-1. Appl. Microbiol. Biotechnol. 97: 9525–9534.

Zhang, Q.X., Y. Zhang, H.H. Shan, Y.H. Tong, X.J. Chen and F.Q. Liu. 2017. Isolation and identification of antifungal peptides from *Bacillus amyloliquefaciens* W10. Environ. Sci. Pollut. Res. Int. 24: 25000–25009.

Zhao, H., D. Shao, C. Jiang, J. Shi, Q.I. Li et al. 2017. Biological activity of lipopeptides from *Bacillus*. Appl. Microbiol. Biotechnol. 101(15): 5951–5960.

Zheng, Yu., F. Chen and M. Wang. 2013. Use of *Bacillus*-based biocontrol agents for promoting plant growth and health. pp. 243–258. *In*: Maheshwari, D.K. (ed.). Bacteria in Agrobiology: Disease Management, Springer-Verlag Berlin Heidelberg.

10

Biosurfactants

Applications for Sustainable Agriculture

*Manju Sharma** and *S M Paul Khurana*

1. Introduction

Agriculture is a major global trade that is essential for the existence of life on earth. Feeding the ever increasing population alongwith shrinking arable land is a pressing call to search for genuine options for increasing productivity. It is well known that irreversible damage has been caused to agriculture using synthetic pesticides, herbicides and fertilizers. Now the time has come when the farmers are ready to rejuvenate conventional farming systems using local inputs hence the scientists are making innovations for sustainable agriculture. A blend of both may be the game changer in the direction of sustainable agriculture.

Increasing reliance on plant biochemicals of microorganism origin seems the only ray of hope to improve agriculture significantly. All over the world, use of biopesticides, biofertilizers, and biosurfactants is increasing due to their eco-friendly nature. Choosing inexpensive raw materials to grow microorganisms to produce biomolecules may also reduce the production cost by 50% (Nitschke et al. 2004). In this chapter our focus will be on applications of biosurfactants for sustainable agriculture.

Biosurfactants are considered as a group of surface-active molecules of microbial origin, amphiphilic in nature having a wide-range of applications due to their unique properties viz. specificity, low toxicity and relative ease of preparation (Mukherjee and Das 2010). Their unique functional properties make them important for diverse industrial use including mining, metallurgy (mainly bioleaching), organic chemicals,

Amity Intsititute of Biotechnology, Amity University Haryana, Gurugram, Maneser, Haryanna-122413, India.

* Corresponding author: msharma@ggn.amity.edu

petrochemicals, petroleum, agrochemicals, fertilizers, foods, beverages, cosmetics, pharmaceuticals and many more. The use of biosurfactants as emulsifiers as well as demulsifiers, and their vast utility as detergents, foaming agents, spreading agents, functional food ingredients and wetting agents is common (Priyam and Singh 2018). Their potential quality of reducing interfacial surface tension had been reported to be worth utilizing for bioremediation and recovery of heavy crude oil and other oils respectively (Volkering et al. 1998).

In agriculture their multitudinous applications make them a popular and ecofriendly choice for imperative crop plant care as follows:

1. Against plant pathogens viz. root diseases, soil-borne and root-borne plant pathogens
2. As biostimulants for Plant Growth Promotions
3. For improving agricultural soil quality by soil remediation
4. Effective weed management
5. Improving the bioavailability of nutrients for plants and biodegradation of soluble organic carbon sources in soil
6. For management of the mycotoxins

The entire world is looking towards biosurfactants as a replacement for harmful synthetic surfactants used in pesticide industries, which require commercial production of biosurfactants on a large scale.

2. Synthetic Surfactants vs. Biosurfactants

Biosurfactants have many qualities as compared to chemical surfactants. In comparison to their chemically synthesized counterparts, biosurfactants are preferred for commercial applications due to a broad substrate availability for their biosynthesis (Fig. 1). A combination of hydrocarbons, carbohydrates and/or lipids, with each other can be used as a carbon source for growing microorganisms (Kosaric 2001) but many a times can be used singly as a cheap source of carbon. Biosurfactants are not only specific to substrates but also specific to bacteria or fungi form which they originate.

Their biodegradable nature makes them easy to use without threatening the surroundings (Mulligan 2005, Ying 2006). Biosurfactants are safer due to their low toxicity (Desai and Banat 1997, Mnif and Ghribi 2016). Their biocompatible nature gives them a status of molecules of preference (Rosenberg and Ron 1999). They also have better foaming properties and higher selectivity (Edwards et al. 2003, Mulligan 2005).

The stepwise life cycle of synthetic surfactants viz. production, formulation, use phase, and discharge phase may impact the environment negatively. The process involved in the production and formulation phases may release or pass off the amounts of chemicals as well as byproducts which may lead to bio-accumulation in the environment (Mungray and Kumar 2008, 2009) with long term adverse effects (Banat et al. 2000) due to leaching of synthetic chemicals/compounds in water and soil.

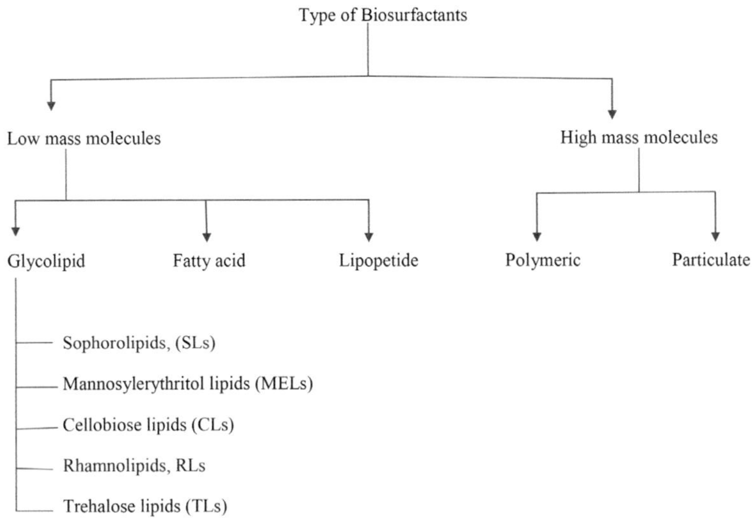

Figure 1. Types of Biosurfactants.

Even microbial biosurfactants have been rated high over plant-based surfactants attributed to their rapid production, multi-functional properties and scale-up capacity. Lecithins, saponins and soy proteins have significant emulsification properties but their production cost cannot be afforded by industry. Besides, solubility and hydrophobicity are other issues of concern with plant based biosurfactants (Xu et al. 2011).

3. Source of Biosurfactants

3.1a Bacterial

Bacteria have been studied well as the major source of biosurfactants amongst other microorganisms. Availability of nutrients such as carbon and nitrogen sources, varied environmental factors and fermentation conditions play an important role in biosurfactant production from microbes. The *Pseudomonas aeruginosa*, *Mycobacterium* sp., *Arthrobacter* sp., are some of the important bacteria producing biosurfactants and also utilized for hydrocarbon degradation quality widely (Fenibo et al. 2019).

3.1b Fungal

The fungal species have yet not been exploited as much as bacteria but active research is going on to explore their potential to produce eco-friendly biomolecules for sustainable agriculture. A few fungi and yeast species such as *Candida boimbicola, Pseudozyma rugulosa, Aspergillus*, *Corynebacterium* sp., *Dietzia* sp., *Gordonia* sp., *Williamsi* sp., *Tsukamurella* sp. have been exploited widely as the sources for biosurfactants (Kuyukina 2015, Morita et al. 2007).

4. Classification of Biosurfactants

The chemical structure and molecular weight can be the basis for the classification of Biosurfactants. Rosenberg and Ron (1999) classified biosurfactants in two categories; high molecular weight molecules (High-mass) and low molecular weight molecules (Low-mass). Glycolipids, lipopeptides and phospholipids are low-mass surfactants whereas high mass surfactants include polymeric and particulate surfactants (Fig. 1). Of these the most common and widely studied are glycolipids (Pacwa-Plociniczak et al. 2011, Cameotra et al. 2010).

5. Properties of Biosurfactants

Microbial surfactants are identified with their low poisonous quality, biodegradability, surface movement, resilience to pH, temperature, ionic quality, emulsifying and demulsifying capacity as well as for antimicrobial action (Cameotra and Makker 2004). The properties of biosurfactants (as reviewed by Roy 2017) are given in the table below:

Table 1. Important properties of biosurfactants.

Property	Description
Antiadhesive agents	The hydrophobicity of the surface can be changed by the biosurfactants which influence the bond of microorganisms over the surface (Kachholz and Schlingmann 1987).
Temperature and pH tolerance	Tolerant towards natural factors for, e.g., Lichenysin a product obtained from *Bacillus licheniformis* found tolerable to temperatures up to 50°C, pH range of 4.5 to 9.0, NaCl concentration maximum up to 50 g L^{-1} and Ca concentration 25 g L^{-1} (McInerney et al. 1990). Arthrobacter protophormiae produced biosurfactants have been observed to be thermostable at 30–100°C and they exhibited high stability in a variable pH (2 to 12) range (Das and Mukherjee 2007).
Eco-friendly Biodegradable	Sophorolipids, surfactins, and arthrofactins are rapidly degradable in the environment (Hirata et al. 2009) but rhamnolipids take 7–8 days to degrade (at a slow rate) Pei et al. (2009) without releasing any harmful byproducts.
Low toxicity	It is considered to be a low risk or harmless bio-active substance recommended as a corrective pharmaceutical. Poremba et al. (1991) observed 10 times higher toxicity displayed at LC50 by chemically derived surfactants against Photobacterium phosphoreum compared to rhamnolipids. The low toxicity profile of biosurfactants, sophorolipids from Candida bombicola made them helpful in nourishment ventures (Hatha et al. 2007).
Emulsifying and emulsion breaking	Emulsions are of two types: oil-in-water or water-in-oil emulsions. Addition of biosurfactants provide considerable stability to the molecules used for crop field applications and stored for a long time, even for years (Hu and Ju 2001).
Critical micelle concentration	Biosurfactants can form micelles from monomers at a very low concentration if mixed in water, this being the most critical parameter for any biosurfactant to solubilize hydrophobic pollutants in contaminated soil (Kumar and Lal 2014, Ehrhardt et al. 2015).
Surface and interface activity	Biosurfactants of low-molecular-mass bring down the surface and interfacial tensions to facilitate solubility of metals in water and soil, as well as reduce repulsive forces between two dissimilar phases. Reduction in repulsive forces between two immiscible compounds via higher emulsification activity is observed on application of high-molecular-mass biosurfactants, with higher emulsification support for easy mixing of elements (Calvo et al. 2009, Sober´on-Ch´avez and Maier 2011).

6. Applications of Biosurfactants in Agriculture

The use of biosurfactants in agriculture has been reviewed by a few workers (Deleu and Paquot 2004, Sachdev and Cameotra 2013) (Fig. 2). Their capacity to make micronutrients available to plants without exerting any side effect gives momentum to the overall growth and development of plants. Improving soil health by removing heavy metals and other harmful compounds is one of the major functions of microbial surfactants and used widely to improve agricultural soil. Pathogen elimination or suppression is another major quality of microbial surfactants without damaging the environment. Vast studies were conducted in each area in the last 20 years keeping in view the apocalyptic ecological conditions.

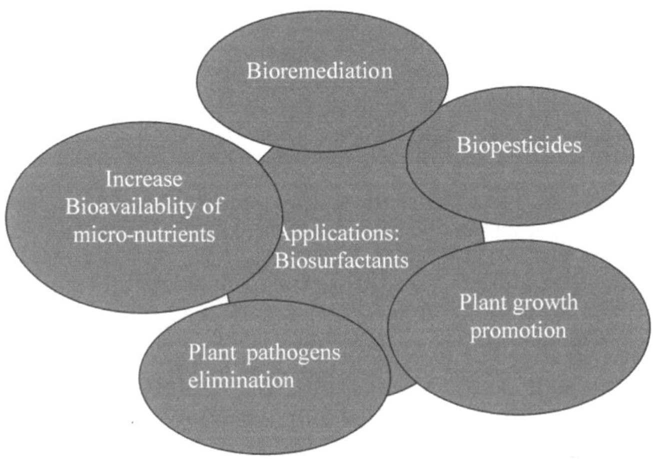

Figure 2. Applications of biosurfactants.

6.1 Plant Growth and Development

Seeds of different species of cultivars: corn, lettuce, soybean and sunflower were germinated *In vitro* to evaluate the effect of rhamnolipids, a biosurfactant obtained from *Pseudomonas aeruginosa* strain, LBI 2A1. The concentrations ranging from 0.25 to 1.00 $g \cdot L^{-1}$ were examined using distilled water as the control. The lettuce seeds germination increased at all tried concentrations from low concentrations of 0.75g/l/1.00g/l upto 75.50%, while for corn seeds 0.25 g/l of rhamnolipids was sufficient for stimulation. Presence of the biosurfactant did not influence germination of soybean seeds however a favourable effect on seedling development was recorded. The tensioactive property of biosurfactant increases with germination and development of sunflower seeds (Silva et al. 2015) and allow changes in the value of surface tension which leads to improved absorption of various substances important for further plant growth. The ability of biosurfactants (rhamnolipids) produced by *Pseudomonas* spp. has been appreciated to regulate the process of quorum sensing, i.e., cell to cell communication (Dusane et al. 2010). Biosurfactants induce increased motility of microorganisms which allows them to participate in signalling, differentiation and biofilm formation (Ron and Rosenberg 2011, Berti et al. 2007). A new *Pseudomonas rhizophila* S211 isolate is a plant growth-promoting *Rhizobacterium*, having potential in Pesticide-Bioremediation. About 215 genes have been isolated from the contaminated fields of Artichoke and analysed for their plant-growth promoting activities such as the synthesis of ACC deaminase, auxin, putative dioxygenases pyroverdin, exopolysaccharide (Hassen et al. 2018).

Many Endophytic microorganisms contribute by producing novel bioactive compounds to provide silent support for various activities throughout the growth period. Endophytic fungi *Xylaria regalis* (*X. regalis*) has been recognized from the cones of *Thuja plicata* and a green biosurfactant compound produced was evaluated *in vivo* for a wide range of growth parameters like enhancement in dry matter production of shoots and roots, chlorophyll, nitrogen, and phosphorus contents along with an increase in shoot and root length (Adnan et al. 2018). A synanthropic plant-*Chelidonium majus* L. possessing endophytic "isolate 2A", was identified as a *Bacillus pumilus*, produced biosurfactant with potential to degrade diesel oil and waste engine oil hydrocarbons very fast. It also promoted seed germination as well as growth of *Sinapis alba* in diesel and oil contaminated soil.

Lindane contamination is a threat to the agricultural soil. An indigenous bacterial strain *Paracoccus* sp. NITDBR1 isolated from agricultural field soil in Manipur, India, has been found capable of using lindane as a carbon source when grown with liquid mineral salt media containing 100 mg L^{-1} lindane. It took 8 days to degrade up to 90% of lindane incorporated in the media. *Paracoccus* sp. NITDBR is known to accelerate certain plant growth promoting traits viz. the solubilization of phosphorus, production of phytohormones (indole-3-acetic acid), ammonia, and nitrogen fixation. The biosurfactants produced by the strain NITDBR1 were of the nature of glycolipids and glycoproteins. The toxicity caused by lindane to plants also decreases which is indicated by the growth of roots (1.3 Fold) after soil treatment with NITDBR1 strain (Sahoo et al. 2019). *Bacillus* sp. strain J119 is known for its ability to boost plant growth in various plants such as sudan grass, canola, tomato and maize (Sheng et al. 2008). The biopreparation from *Pesudomonas* spp. containing biosurfactant was applied on seeds of five species *Zea mays, Lupinus luteus, Pisum sativum, Avena sativa,* and *Sinapsis alba* for 1 hour approximately inducing 4–6% improvement in seed germination, as noticed with good impact on further plant growth, manifested by an increase in the plant height and mass (Małgorzata et al. 2012). Biosurfactants isolated from *Rhodococcus erythropolis* CD 106 and *R. erythropolis* CD 111 showed enhanced ACC deaminase activity, IAA and siderophore production, phosphate solubilization plant growth promoting mechanisms (Krawczyńska et al. 2016). Biosurfactants produced by *Pseudomonas* sp. AJ15 were used for priming of *Withania somnifera* seeds grown under petroleum toxicity. The potential of biosurfactants to utilize petroleum as a carbon source for a high value for growth promoting parameters such as germination, shoot length, root length, fresh and dry weight and pigments (chlorophyll and carotenoid) as compared to non primed seeds (Das and Kumar 2016).

Thirty six *Rhizopseudomonas* Strains from date palm roots were isolated and evaluated for the abilities of biosurfactant production which affect various growth promoting activities, viz. mineral phosphate solubilization (63.8%), indole acetic acid

(IAA) production (88.8%), siderophores (83.3%) and ammonia (52.7%) released as well as for protease and cellulase activity (Ferjani et al. 2019).

Textile industry effluents contain a harmful dye azulene and heavy metal chromium, which are persistent toxic compounds badly influencing plant growth. The augmentation of biosurfactant lipopeptide with azulene and chromium promotes seed germination along with plant growth viz. specific leaf weight (SLW), plant biomass, protein content, chlorophyll content, soluble sugar and ascorbic acid concentration in chilli and wheat cultivars (Singh and Rathore 2019).

From the 23 isolates of microorganisms extracted from tea rhizosphere A6 was identified as *Brevibacterium sediminis* and screened for significant quality of plant growth promotion (PGP) and antifungal traits against three pathogenic fungi, viz. *Corticium rolfsii, Fomes lamaensis* and *Rhizoctonia solani.* The production of plant growth promoters like indole acetic acid (IAA) was quantified to 72.51 ± 2.18 µg/ml and phosphate solubilization was estimated at 60.644 ± 3.098 µg/ml from the isolate A6. Validation trials of PGP activity of A6 done *in vitro* on seedlings of rice had shown higher root/shoot mass and fresh weight without any significant difference in dry weight mass (Chopra et al. 2020). *Pseudomonas aeruginosa* RTE_4 produced di-rhamnolipid BS that induces PGPs like IAA with bio-control activity against a bacterial plant pathogen (*Xanthomonas campestris*) and two fungal pathogens of tea (*Corticium invisium* and *Fusarium solani*) directly attacking the foliar region (Chopra et al. 2020).

6.2 Biological Tools to Increase Availability of Micronutrients in Soil

All over the world, millions of hectares of arable land have inadequate levels of trace elements such as Cu, Fe, Mn, and Zn (Li et al. 2013). Specific deficiency of Zn and F announced as one of the major risk factors for disease in low-income countries when they consume food grown in soil having poor levels of Zn and Fe by WHO (2002). It is necessary to augment agricultural soil with essential nutrients, in adequate amounts to get an enriched plant produce which may otherwise lead to widespread malnutrition to human beings (Miller and Welch 2013) and also affecting crop yield (PAIS 2012). The four sources viz. natural, agricultural, industrial and domestic effluents contribute to the addition of many macro and micro-nutrients in the soil (Singh et al. 2018).

The reason for soil being deficient of micronutrients is not only the absence of adequate amounts, but also attributed to poor solubility, which may be affected by the soil organic matter content, adsorptive surface area, nutrient interactions, pH, and texture (Ayele et al. 2013). Biosurfactants interact with complex soil nutrients to solubilize form and mobilize metal-biosurfactant complexes to make them available for various plant activities. This is an active research area as far as the mechanisms behind many biosurfactant-induced effects are concerned. The elucidation on the differences in nutrient specificity (e.g., Cu, Fe, Mn, and Zn) and toxins (e.g., Cd and Pb) needs exploration (Singh et al. 2018). The biosurfactants of root-associated bacteria increase nutrient availability and uptake to support efficient distribution of metals and micronutrients in the soil, to promote plant growth, protect against toxic substances, and serve as a carbon source for bacteria survival (Singh et al. 2018). The chelation of metal and complexes with available trace metals, is followed by their adsorption, desorption, and removal from the soil surface interfaces for incorporation into the micelle to increase micronutrient concentrations and their bioavailability in soil (Rufino et al. 2012). Venkatesh and Vedaraman (2012) had observed that Rhamnolipids of *Pseudomonas aeruginosa* origin increase the availability of copper to plants. Rhamnolipids form different *Pseudomonas* species which also help plants utilize Zn & Cd from Maize and Sunflower (Wen et al. 2010, 2016).

Sometimes micronutrient deficiency in soil occurs due to the alkaline nature of the soil. Application of biosurfactants in this case breaks the metal-soil complex in order to form new bioavailable metal biosurfactant-complexes (Rufino et al. 2011). Stacey et al. (2007) investigated monorhamnosyl and dirhamnosyl rhamnolipids from lipophilic complexes with metals viz. Copper, Manganese, and Zinc, to facilitate their uptake by roots of *Brassica napus* and *Triticum durum*. *Bacillus* sp. strain J119 biosurfactants were identified for their trace element uptake capability (Sheng et al. 2008). Application of some of the microbial biosurfactants serve to help the uptake of biogenic substances like phosphorus by plants (Mukherjee et al. 2007).

6.3 Soil Bio-remediation

Biosurfactants from various microbial isolates have been considered for their potential role in plant growth promotion and other applications related to improving the effectiveness of agricultural soils. It is necessary to keep a check on the presence of organic and inorganic pollutants in agricultural land/soil to avoid abiotic stress, which gravely affects crop cultivation. To improve the quality of soil, contaminated with hydrocarbons and heavy metals, the bioremediation process is being utilized for soil quality improvement utilizing microbial biosurfactants for effective removal of hydrocarbons and heavy metals (Sun et al. 2006).

Presence of hydrocarbons and heavy metals is an all time problem for agricultural soil while biosurfactants play an important role in their effective and efficient removal from it (Sun et al. 2006). Different technology based methods like soil washing and clean up combined technology utilize biosurfactants for effective removal of hydrocarbon and metal, respectively (Aşçi et al. 2008, Coppotelli et al. 2010, Gottfried et al. 2010, Kanget et al. 2010, Liu et al. 2010, Partovinia et al. 2010, Pacwa-Plociniczak et al. 2011, Pei et al. 2009, Camilios et al. 2009, Zhao and Wong 2009, Robles González 2008, Santos et al. 2008). According to Sachdev et al. (2013) nutrient uptake has been reported to be aided by biosurfactants, on inducing substantial changes in root cell differentiation.

The biosurfactants adopt two mechanisms to biodegrade oil-derived hydrocarbons. The first involves reduction in surface tension of the medium present around the bacterium as well as a decrease in the interfacial tension between hydrocarbon molecules and the cell wall. As a consequence it increases bioavailability of the hydrophobic substrate to microorganisms. In the second mechanism, the interaction between the cell surface and biosurfactant, induces changes in the membrane which facilitate hydrophobicity in hydrocarbons and reduces the lipopolysaccharide index without causing any damage to the cell wall membrane. Thus, it allows hydrophobic-hydrophilic interactions due to blockage in the hydrogen bridges formation caused by biosurfactants, and all this leads to molecular rearrangements for promoting bioavailability and ensuant biodegradability due to reduced surface tension of the liquid by an increase in surface area (Aparna et al. 2011, Franzetti et al. 2009).

The biosurfactants increase the imperative physical process of desorption and accelerate the binding of hydrophobic pollutants tightly to soil particles which is a significant part of the bioremediation. The chemical insecticides keep accumulating in the agricultural soil as part of regular care of important crops; as a result the biosurfactants can be applied with insecticides to enhance insecticide degradation (Zhang et al. 2011). The *Lactobacillus pentosus* produced biosurfactant accelerates the property of biodegradation significantly to reduce octane hydrocarbon content in the soil by 58.6–62.8% (Moldes et al. 2011). The *Burkholderia* isolated from oil-contaminated soil produces a biosurfactant which has potential to remove varieties of pesticide contaminations (Wattanaphon et al. 2008). A maximum recovery level of 91.5% of endosulphan was reported on applying surfactin produced by *Bacillus subtilis* in the soil (Jayashree et al. 2006). The endosulphan comprises of two stereo isomers, α and β-endosulfan in a ratio of 7:3, and is less soluble in water but both are extremely toxic to aquatic organisms and the human central nervous system.

About forty-two hydrocarbon-degrading bacterial strains were isolated from the soil contaminated with petroleum hydrocarbon. All of them belong to the genera *Rhodococcus, Rahnella, Serratia* and *Proteu*. Of these twelve were considered effective for production of biosurfactants affecting the plant growth promoting activities (Krawczyńska et al. 2016). Surfactant-enhanced phytoremediation is counted as a green technology for the treatment of contaminated soil. In a pot experiment, two biosurfactants (rhamnolipid and soybean lecithin) were used to increase phytoremediation as green technology to degrade hydrocarbons using plants of *Zea mays* L. Both enhanced the microbial population in soil for an increased removal of petroleum hydrocarbons (Liao et al. 2016).

A property of the biosurfactant, such as a low critical micelle concentration helps to increase the contact angle of the soil–oil system but brings down the capillary force responsible for binding the soil and oil together (Usman et al. 2016). Rhamnolipid obtained from *Pseudomonas rhizophila* S211, due to its low-cost of production, is popular and a promising choice as an emulsifying and solubilization agent for chemical pesticides to bioremediate pesticide-contaminated agricultural soils (Haeesn et al. 2018). The interfacial surface reduction by biosurfactant solublilizes the hydrocarbons to make them available to cells. The non-aqueous phase liquid contaminants undergo emulsification and facilitate transport of the pollutants in the solid phase. Bioavailability greatly enhances microbial degradation and phytoremediation of hydrocarbons in the soil matrix, given that all other factors viz. soil organic matter content, pH, nutrient interaction, soil texture, adsorption surface and nutrients are optimal (Fenbino et al. 2019).

The process of copper ore enrichment generates lots of toxic sludge which gets deposited in the specific areas called landfills. A combination of biosurfactants and microorganisms termed as biopreparations is applied in the landfill area to assess their effect on germination and growth of plants to revitalize contaminated soil after post-flotation of sludge. The two species of *Azospirillum brasilense* and *Azospirillum lipoferum* were observed for their biotensoactive biosurfactant producing character which favours micelle formation and allows microbial cells to metabolise hydrocarbons effectively on coming in contact (Ojeda-Morale et al. 2015). The biosurfactant from *Streptomyces* sp. V2 was obtained from polyaromatic hydrocarbons contaminated soils using molasses as a carbon source. Purified biosurfactants were examined for properties like surface activeness, antimicrobial potency and potential to alleviate stress of phenanthrene and anthracene on plantlets. *Trigonellafoenum-graecum* and *Triticum* spp. viability and growth was enhanced significantly in soil amended with phenanthrene and anthracene, which directly points towards V2's bioremediation property (Bhuyan-Pawar et al. 2015). A synanthropic plant-*Chelidonium majus* L. having endophytic isolate 2A, identified as *Bacillus pumilus*, produces a biosurfactant with potential to degrade diesel oil and waste engine oil hydrocarbons very fast (Marchut 2018).

Soil washing is an *ex situ* remediation method applied to separate contaminating hazardous compounds from the excavated soil by washing it with a liquid often incorporated with chemicals. The contaminants found bonded to fine-grained soils like clay, sand, silt and gravels (Abdel-Moghny 2012) can easily be removed by washing. The soil contaminated with metals, fuels, semi-volatile organic compounds, chlorinated hydrocarbons and pesticides may be washed with fluid composed of water, water/chelating agents, water/surfactants, acids or bases or organic solvents (Hazrina et al. 2018) decided on the basis of target contaminant type present in it.

6.4 Plant Pathogen Elimination

As a result of their antifungal activity towards phytopathogenic fungi and larvicidal and mosquitocidal potential, glycolipid biosurfactants do not allow pest invasion in crop plants and animals. Emulsifying and antibacterial activities of glycolipids give them great potential to be used as food additives and preservatives. Furthermore, the valorization of food byproducts via the production of glycolipid biosurfactants has received much attention, as the bioconversion of byproducts into valuable compounds decreases the cost of production too. Generally, the use of glycolipids in many fields requires their retention in fermentation media therefore, different strategies have been developed to extract and purify glycolipids (Mnif and Ghribi 2016).

The role of biosurfactants towards plant microbes as an antibiotic helps them to be efficient in self-defence, hinder pathogenesis, increase intraspecific competition, protect them from disease and behave as a toxin during a plant–pathogen interaction. *Pseudomonas* strains derived biosurfactants, primarily the cyclic lipopeptide and rhamnolipid type, are used widely for biocontrol in plants and even these biosurfactants can be effectively applied as antibacterial molecules against an increasing number of pathogenic *Pseudomonas strains* (Jolien et al. 2001, Raaijmakers et al. 2006). *Pseudomonas* CMR12a and *P. aeruginosa* PNA1 produced cyclic lipopeptides and rhamnolipids were noticed to be effective in controlling *Pythium* and *Phytophtora* spp. causing rotting of fruits and flowers respectively (Perneel 2006). As biological controls biosurfactants create channels in the cell wall and disquiet the cell surface of the pathogen (Raaijmakers et al. 2006).

The biosurfactants extracted from *Pseudomonas* sp. were considered good for Inhibition of the viability of *Verticillium* sp. *in vitro* (Debode et al. 2007) by terminating the growth of *Rhizoctonia solani* a noxious pathogenic fungus of several plants and *Phythium ultimum*. This becomes possible by the production of dual functioning compounds such as viscosin, viscosinamid and tensin (Andersen et al. 2003).

Soil bacterial species *Bacillus* are known to produce diverse groups of secondary metabolites with high biological activity (bioactive compounds) possessing antagonistic activities against several pathogenic fungi (Sarwar et al. 2018). *Bacillus subtilis* and *B. amyloliquefaciens* are some of the strains known to produce antifungal peptides, viz. bacilysin and rhizocticin; surfactins, iturins, and fengycins; and antimicrobial polypeptides such as subtilin (Sharma et al. 2018). Anthracnose causing *Colletotrichum gloeosporioides* on papaya leaves was controlled by one of the soil isolates of *Bacillus subtilis* produced biosurfactants (Kim et al. 2010).

About 143 species of zoosporic plant pathogens have been identified, taxonomically belonging to diverse families viz. *Albuginaceae, Pythiaceae, Peronosporaceae, Saprolegniaceae, Synchytriaceae* (Stanghellini and Miller 1997). They were given the name zoosporic, due to a unicellular, motile, uni- or biflagellate common asexual stage in their life cycle. The production of Zoospores either occurs in a vesicle or a sporangium, from where upon release in the presence of free water in the form of irrigation water, rain, or dew, they get an opportunity to swim for varying time periods, i.e., few minutes to hours and ultimately encyst on a suitable host using a primarily chemotactic mechanism. The downy mildews causing zoosporic pathogens were found most devastating which primarily attack leaves of many crops. Besides, several species in the genera *Phytophthora* and *Pythium* were also considered most damaging pathogens of roots, foliage, and fruits. The notorious zoosporic pathogen *Phytophthora infestans*, is the cause of late potato blight, and *Plasmopara viticola*, causes downy grape mildew.

Zoospores of *Phytophthora capsici* were lysed by biosurfactants of *Pseudomonas putida* origin to get rid of damping off disease of cucumber. Biosurfactants produced by strains of *P. fluorescens* were used for effective management of *Pythium ultimum*, *Fusarium oxysporum*, and *Phytophthora cryptogea*, recognized as notorious pathogens to many important crop plants (Kruijt et al. 2009, Nielsen and Sorensen 2003).

The antagonistic activity of biosurfactants produced by *Xylaria regalis* was evaluated against *Fusarium oxysporum* and *Aspergillus niger* like phytopathogens for improving crop performance (Adnan et al. 2018). Surfactin isoform, a lipopeptide biosurfactant of *Brevibacillus brevis* strain HOB1 origin had powerful antibacterial and antifungal properties exploited for the control of phytopathogens (Haddad 2008).

Inhibitory roles of biosurfactants on aflatoxin producing *Aspergillus* spp., were also noticed which infect commercially important crops, such as peanuts, cottonseed, and corn, during storage (Sachdev and Cameotra 2013). A plant pathogen *Pseudomona aeruginosa* was reported to be inhibited by biosurfactant produced by *Staphylococcus* sp., isolated from crude oil-contaminated soil (Eddouaouda et al. 2012). Biosurfactant-producing rhizospheric isolates of *Pseudomonas* and *Bacillus* have exhibited biocontrol of soft rot causing *Pectobacterium* and *Dickeya* spp. (Krzyzanowska et al. 2012).

6.5 *Biosurfactants as Insecticides*

Biosurfactants have received much practical attention as biopesticides for controlling plant diseases and protecting stored products from insects. The large number of insect populations are known for their devastating nature to economically important crops. Due to environmental concerns microbial bioinsecticides have been used to control agricultural pests for decades (Ghribi et al. 2011). The biosurfactants obtained from different bacterial species have been reviewed for insecticidal activities (Table 2) (Edosa et al. 2019). The low toxicity and eco-friendly nature is the major concern to use them as a safe option for insect pest management. An insecticidal potential of di-rhamnolipids isolated from a strain of *Pseudomonas* was reported insecticidal against the green peach aphid (*Myzus persicae*) (Kim et al. 2010, 2011). The most famed *Bacillus* species are notable biosurfactant-producing bacteria. Their broad-spectrum lipopeptides, including bacillomycin, fengycin, iturin, lichenysin and surfactin (Manonmani et al. 2011, Mukherjee and Das 2005) are bioactive metabolites (which) ensue hemolysis with potent larvicidal activity (Chung et al. 2008, Geetha et al. 2011). Mnif and Ghribi (2015) have intensively reviewed the role of *Bacillus* and *Pseudomonas* bacterial species-derived biopesticides in pest management. Khedher et al. (2015, 2020) have reported the potential of *B. amyloliquefaciens* AG1 and *B. subtilis* V26 derived biosurfactants to control the *Tuta absoluta* larvae by causing histological damage in the larval midgut. The potential of biosurfactants was used to clean up *cypermethrin* insecticidal residue present on Lettuce, a leafy vegetable crop.

Pseudomonas rhizophila S211, was isolated from a pesticide contaminated artichoke field that showed biofertilization, biocontrol and bioremediation potentialities. The S211 genome was sequenced, annotated and key genomic elements related to plant growth promotion and biosurfactant (BS) synthesis were elucidated. The S211 genome comprises of 5,948,515 bp with 60.4% G+C content, 5306 coding genes and 215 RNA genes. The genome sequence analysis confirmed the presence of genes involved in plant-growth promoting and remediation activities such as the synthesis of ACC deaminase, putative dioxygenases, auxin, pyroverdin, exopolysaccharide l and rhamnolipid (Hassen et al. 2018).

7. Future Prospects

Biosurfactants are being considered as the future molecules having attributes to contribute as soil remediation agents by removing heavy metal and chemical pesticides, fungicides, and bactericides. Their antagonistic ability against plant disease causing microorganisms and devastating insect pests without affecting the environment badly counted as meritorious over the use of various harmful chemicals. The apocalyptic vision of ecologists has raised an alarm, “All over the world, to put on hold the practices, which are not eco-friendly”. The agrochemical industries should adopt this important by product of microbial metabolism to formulate eco friendly pesticides with the aid of biosurfactants for wide application on agricultural fields. Combinations (mixture) of different biosurfactants judiciously with complex polymers can formulate the effective product to achieve the target of minimum damage to the environment. The change in production scheme may provide highly target specific green surfactant/s by altering the chemical composition of produced bio-surfactants for use as a potent biocontrol against a wide range of pest populations.

There is a dire need to divert efforts to increase the production and reduce the cost of green surfactants manufacturing on a large scale to achieve a net economic gain from application of biosurfactants not only in agriculture but also in some other spheres. Utilizing the agri-waste, harmless industrial effluents and domestic waste as a source of microbial growth for overproduction of biosurfactants needs to be a matter of concern. Rhizospheres with a high prevalence of biosurfactant producing bacteria is a good sign for sustainable agriculture. Extensive and exclusive study on bacterial and fungal populations other than *Pseudomonas* and *Bacillus* need to be exploited to full potential for reshaping the futuristic agricultural practices. Categorically, exploring the functional metagenomics by applying state of art technology is the need of time to speed up research on novel green eco-friendly surfactants to curb the adverse effects of synthetic surfactants which are largely unavoidable in several commercial sectors including agrochemical industries for field crop protection.

References

Abdel-Mawgoud, A.M., F. Lepine and E. Deziel. 2010. Rhamnolipids: Diversity of structures, microbial origins and roles. Appl. Microbiol. Biotechnol. 86: 1323–1336.

Adnan, M., E. Alshammari, S.A. Ashraf, K. Patel, K. Lad et al. 2018. Physiological and molecular characterization of biosurfactant producing endophytic fungi *Xylaria regalis* from the cones of Thuja plicata as a potent plant growth promoter with its potential application. Biomed. Res. Int. 2018: 7362148. Published 2018 May 13. doi:10.1155/2018/7362148.

Ankita Chopra, Uday Kumar Vandanaa, Praveen Rahib, Surekha Satputec, Pranab Behari Mazumder et al. 2020. Plant growth promoting potential of *Brevibacterium sediminis* A6 isolated from the tea rhizosphere of Assam, India. Biocatalysis and Agricultural Biotechnology Volume 27, August 2020, 101610.

Aparna, A., G. Srinikethan and S. Hedge. 2011. Effect of addition of biosurfactant produced by *Pseudomonas* ssp. on biodegradation of crude oil. In International Proceedings of Chemical, Biological & Environmental Engineering 6: 71–75.

Aşçi, Y., M. Nurbaş and Y.S. Açikel. 2008. A comparative study for the sorption of Cd(II) by K-feldspar and sepiolite as soil components and the recovery of Cd(II) using rhamnolipid biosurfactant. J. Environ. Manage 88: 383–392.

Ayele, T., M. Ayana, T. Tanto and D. Asefa. 2013. Evaluating the status of micronutrients under irrigated and rainfed agricultural soils in Abaya Chamo Lake Basin, South-west Ethiopia. J. Sci. Res. Rev. 3: 18–27.

Banat, I.M., R.S. Makkar and S.S. Cameotra. 2000. Potential commercial applications of microbial surfactants. Appl. Microbiol. Biotechnol. 53: 495–508.

Berti, A.D., N.J. Greve, Q.H. Christensen and M.G. Thomas. 2007. Identification of a biosynthetic gene cluster and the six associated lipopeptides involved in swarming motility of *Pseudomonas syringae* pv. tomato DC3000. J. Bacteriol. 189: 6312–6323.

Calvo, C., M. Manzanera, G.A. Silva-Castro, I. Uad, J. Gonz´alezL´opez et al. 2009. Application of bioemulsifiers in soil oil bioremediation processes. Future prospects. Sci. Total Environ. 407: 3634–3640.

Cameotra, S.S. and R.S. Makkar. 2004. Recent applications of biosurfactants as biological and immunological molecules. Curr. Opin. Microbiol. 7: 262–266.

Cameotra, S.S., R.S. Makkar, J. Kaur and S.K. Mehta. 2010. Synthesis of biosurfactants and their advantages to microorganisms and mankind. Adv. Exp. Med. Biol. 672: 261–280.

Camilios Neto, D., J.A. Meira, E. Tiburtius, P.P. Zamora, C. Bugay et al. 2009. Production of rhamnolipids in solid-state cultivation: characterization, downstream processing and application in the cleaning of contaminated soils. Biotechnol. J. 4: 748–755.

Chopra, A., S. Bobate, P. Rahi, A. Banpurkar, P.B. Mazumder et al. 2020. *Pseudomonas aeruginosa* RTE4: A tea rhizobacterium with potential for plant growth promotion and biosurfactant production frontiers in bioengineering and biotechnology. Vol. 8, article 861 https://doi.org/10.3389/fbioe.2020.00861.

Chung, S., H. Kong, J.S. Buyer, D.K. Lakshman, J. Lydon et al. 2008. Isolation and partial characterization of *Bacillus subtilis* me488 for suppression of soilborne pathogens of cucumber and pepper. Applied Microbiology and Biotechnology 80(1): 115–23.

Das Amar Jyoti and Kumar Rajesh. 2016. Bioremediation of petroleum contaminated soil to combat toxicity on Withania somnifera through seed priming with biosurfactant producing plant growth promoting rhizobacteria. Journal of Environmental Management 174: 79–86.

Das, K. and A.K. Mukherjee. 2007. Crude petroleum-oil biodegradation efficiency of *Bacillus subtilis* and *Pseudomonas aeruginosa* strains isolated from a petroleum-oil contaminated soil from North-East India. Bioresour. Technol. 98: 1339–1345.

Deleu, M. and M. Paquot. 2004. From renewable vegetables resources to microorganisms: New trends in surfactants. Computers Rendus Chimie 7: 641–646.

Desai, J.D. and I.M. Banat. 1997. Microbial production of surfactants and their commercial potential. Microbiol. Mol. Biol. Rev. 61: 47–64.

Dusane, D., P. Rahman, S. Zinjarde, V. Venugopalan, R. McLean et al. 2010. Quorum sensing; implication on rhamnolipid biosurfactant production. Biotech. Genetic Eng. Rev. 27: 159–184.

Eddouaouda, K., S. Mnif, A. Badis, S.B. Younes, S. Cherif et al. 2012. Characterization of a novel biosurfactant produced by *Staphylococcus* sp. strain 1E with potential application on hydrocarbon bioremediation. J. Basic Microbiol. 52: 408–418.

Edwards, K.R., J.E. Lepo and M.A. Lewis. 2003. Toxicity comparison of biosurfactants and synthetic surfactants used in oil spill remediation to two estuarine species. Mar. Pollut. Bull. 46: 1309–1316.

Ehrhardt, D.D., J.F.F. Secato and E.B. Tambourgi. 2015. Production of biosurfactant by *Bacillus subtilis* using the residue from processing of pineapple, enriched with glycerol, as substrate. Chem. Eng. Tran. 43: 277–282.

Fenibo, E.O., G.N. Ijoma, R. Selvarajan and C.B. Chikere. 2019. Mcrobial surfactants: the next generation multifunctional biomolecules for applications in the petroleum industry and its associated environmental remediation. Microorganisms 7: 581.

Ferjani, R., H. Gharsa, V. Estepa-Pérez et al. 2019. Plant growth promoting *Rhizopseudomonas*: expanded biotechnological purposes and antimicrobial resistance concern. Ann Microbiol 69: 51–59. https://doi.org/10.1007/s13213-018-1389-0.

Franzetti, A., P. Caredda, C. Ruggeri, P. La Colla, E. Tamburini et al. 2009. Potential applications of surface active compounds by *Gordonia* sp. strain BS29 in soil remediation technologies. Chemosphere 75: 810–807.

Geetha, S.J., I.M. Banat and S.J. Joshi. 2018. Biosurfactants: Production and potential applications in microbial enhanced oil recovery (MEOR). Biocatal. Agric. Biotechnol. 14: 23–32.

Ghribi, D., L. Abdelkefi-Mesrati, H. Boukedi, M. Elleuch, S. Chaabouni Ellouz Chaabouni and S. Tounsi. 2011. The impact of the *Bacillus subtilis* SPB1 biosurfactant on the midgut histology of *Spodoptera littoralis* (Lepidoptera: Noctuidae) and determination of its putative receptor. J. Invertebr. Pathol., 109: 183–186.

Glick, B.R., B. Todorovic, J. Czarny, Z. Cheng, J. Duan et al. 2007. Promotion of plant growth by bacterial ACC deaminase. Crit. Rev. Plant Sci. 26: 227–242.

Gottfried, A., N. Singhal, R. Elliot and S. Swift. 2010. The role of salicylate and biosurfactant in inducing phenanthrene degradation in batch soil slurries. Appl. Microbiol. Biotechnol. 86: 1563–157.

Haddad, N.I. 2008. Isolation and characterization of a biosurfactant producing strain, *Brevibacilis brevis* HOB1. J. Ind. Microbiol. Biotechnol. 35: 1597–1604.

Hassen, W., M. Neifar, H. Cherif, A. Najjari, H. Chouchane et al. 2018. *Pseudomonas rhizophila* S211, a new plant growth-promoting rhizobacterium with potential in Pesticide-Bioremediation. Front. Microbiol. 9: 34. doi: 10.3389/fmicb.2018.00034.

Hatha, A.A.M., G. Edward and K.S.M.P. Rahman. 2007. Microbial biosurfactants. Review. J Mar. Atmos. Res. 3: 1–17.

Hazrina, H.Z., M.S. Noorashikin, S.Y. Beh, S.H. Loh, N.N. Zain et al. 2018. Formulation of chelating agent with surfactant in cloud point extraction of methylphenol in water. Royal Society Open Sci. 5: 180070. doi: 10.1098/rsos.180070.

Hirata, Y., M. Ryu, Y. Oda, K. Igarashi, A. Nagatsuka et al. 2009. Novel characteristics of sophorolipids, yeast glycolipid biosurfactants, as biodegradable low-foaming surfactants. J. Biosci. Bioeng. 108(2): 142–6. doi: 10.1016/j.jbiosc.2009.03.012. PMID: 19619862.

Hu, Y. and L.K. Ju. 2001. Purification of lactonic sophorolipids by crystallization. J. Biotechnol. 87: 263–272.

Jayashree, R., N. Vasudevan and S. Chandrasekaran. 2006. Surfactants enhanced recovery of endosulfan from contaminated soils. Int. J. Environ. Sci. Tech. 3(3): 251–259.

Jolien D'aes, Katrien De Maeyer, Ellen Pauwelyn and Monica Höfte. 2010. Biosurfactants in plant–Pseudomonas interactions and their importance to biocontrol. Environmental Microbiology Reports In Special Issue: Pseudomonas. Editors: Professors Burkhard Tummler, Victor de Lorenzo, Alain Filloux and Joyce Loper. Pages 359–372 June 2010.

Kachholz, T. and M. Schlingmann. 1987. Possible food and agriculture application of microbial surfactants: an assessment. *In*: Kosaric, N., W.L. Cairns and N.C.C. Gray (eds.). Surfactant Sciences Series: Biosurfactants and Biotechnology. Dekker, Basel 25: 183–210.

Kang, S.W., Y.B. Kim, J.D. Shin and E.K. Kim. 2010. Enhanced biodegradation of hydrocarbons in soil by microbial biosurfactant, sophorolipid. Appl. Biochem. Biotechnol. 160: 780–790.

Kim, S.K., Y.C. Kim, S. Lee, J.C. Kim, M.Y. Yun et al. 2010. Insecticidal activity of rhamnolipid isolated from *Pseudomonas* sp. EP-3 against green peach aphid (*Myzus persicae*) J. Agric. Food Chem. 59: 934–938. doi: 10.1021/jf104027x.

Kim, S.K., Y.C. Kim, S. Lee, J.C. Kim, M.Y. Yun et al. 2011. Insecticidal activity of rhamnolipid isolated from *Pseudomonas* sp. EP-3 against green peach aphid (*Myzus persicae*). J. Agric. Food Chem. 59: 934–938.

Kosaric, N. 2001. Biosurfactants and their application for soil bioremediation. Food Technol. Biotechnol. 39: 295–304.

Krawczyńska Małgorzata, Barbara Kołwzan1, Justyna Rybak1, Krzysztof Gediga, Natalia S. Shcheglova et al. 2012. The influence of biopreparation on seed germination and growth. Pol. J. Environ. Stud. 21(6): 1697–1702.

Krishnaswamy, M., G. Subbuchettiar, T.K. Ravi and S. Panchaksharam. 2008. Biosurfactants properties, commercial production and application. Current Science 94: 736–747.

Kruijt, M., H. Tran and J.M. Raaijmakers. 2009. Functional, genetic and chemical characterization of biosurfactants produced by plant growth-promoting *Pseudomonas putida* 267. J. Appl. Microbiol. 107: 546–556. doi: 10.1111/j.1365-2672.2009.04244.x.

Krzyzanowska, D.M., M. Potrykus, M. Golanowska, K. Polonis et al. 2012. Rhizosphere bacteria as potential biocontrol agents against soft rot caused by various *Pectobacterium* and *Dickeya* spp. strains. J. Plant Pathol. 94(2): 367–378. doi: 10.4454/JPP.FA.2012.042.

Kumar, R. and S. Lal. 2014. Synthesis of organic nanoparticles and their applications in drug delivery and food nanotechnology: A review. J. Nanomater. Mol. Nanotechnol. 3: 11–2.

Kuyukina, M.S., I.B. Ivshina, T.A. Baeva, S.V. Gein, V.A. Chereshnev et al. 2015. Trehalolipid biosurfactants from nonpathogenic *Rhodococcus* actinobacteria with diverse immunomodulatory activities. New Biotechnol. 32: 559–568. doi: 10.1016/j.nbt.2015.03.006.

Li, Z.B., Q.L. Zhang and P. Li. 2013. Distribution characteristics of available trace elements in soil from a reclaimed land in a mining area of north Shaanxi, China. Int. Soil Water Conserv. Res. 1: 65–75.

Liao Chang Jun, Xu WenDing, Lu GuiNing, Deng FuCai, Liang XuJun et al. 2016. Biosurfactant-enhanced phytoremediation of soils contaminated bycrude oil using maize (*Zea mays* L). Ecological Engineering 92: 10–17.

Manonmani, A., I. Geetha and S. Bhuvaneswari. 2011. Indian Journal of Medical Research 134: 476–82.

Marchut-Mikolajczyk, O., P. Drożdżyński, D. Pietrzyk et al. 2018. Biosurfactant production and hydrocarbon degradation activity of endophytic bacteria isolated from *Chelidonium majus* L. Microb. Cell Fact. 17: 171 https://doi.org/10.1186/s12934-018-1017-5.

McInerney, M.J., M. Javaheri and D.P. Nagle. 1990. Properties of the biosurfactant produced by *Bacillus licheniformis* strain JF-2. Journal of Industrial Microbiology 5: 95–101 https://doi.org/10.1007/BF01573858.

Miller, D.D. and R.M. Welch. 2013. Food system strategies for preventing micronutrient malnutrition. Food Policy 42: 115–128.

Mnif, I. and D. Ghribi. 2015. Lipopeptides biosurfactants: Mean classes and new insights for industrial, biomedical, and environmental applications. Biopolymers 104: 129–147.

Mnif, I. and D. Ghribi. 2016. Glycolipid biosurfactants: Main properties and potential applications in agriculture and food industry. J. Sci. Food Agric. 96: 4310–4320.

Moldes, A.B., R. Paradelo, D. Rubinos, R. Devesa-Rey, J.M. Cruz et al. 2011. *Ex situ* treatment of hydrocarbon-contaminated soil using biosurfactants from *Lactobacillus pentosus*. J. Agric. Food Chem. 59: 9443–9447.

Morita, T., M. Konishi, T. Fukuoka, T. Imura, D. Kitamoto et al. 2007. Physiological differences in the formation of the glycolipid biosurfactants, mannosylerythritol lipids, between *Pseudozyma antarctica* and *Pseudozyma aphidis*. Appl. Microbiol. Biotechnol. 74: 307–315. doi: 10.1007/s00253-006-0672-3.

Mukherjee, A.K. and K. Das. 2005. Correlation between diverse cyclic lipopeptides production and regulation of growth and substrate utilization by *Bacillus subtilis* strains in a particular habitat. FEMS Microbiology Ecology 54: 479–89.

Mukherjee, A.K. and K. Das. 2010. Microbial surfactants and their potential applications: an overview. *In*: Sen, R. (eds.). Biosurfactants. Advances in Experimental Medicine and Biology, vol. 672. Springer, New York, NY. https://doi.org/10.1007/978-1-4419-5979-9_4.

Mukherjee, S., P. Das and R. Sen. 2007. Towards commercial production of microbial surfactants. Trends Biotechnol. 24: 1199.

Mulligan, C.N. 2005. Environmental applications for biosurfactants. Environ. Pollut. 133: 183–198.

Mungray, A.K. and P. Kumar. 2009. Fate of linear alkylbenzene sulfonates in the environment: A review. International Biodeterioration and Biodegradation 63(8): 981–987.

Nielsen, T.H. and J. Sorensen. 2003. Production of cyclic lipopeptides by *Pseudomonas fluorescens* strains in bulk soil and in the sugar beet rhizosphere. Appl. Environ. Microbial. 69: 861–868. doi: 10.1128/AEM.69.2.861-868.2003.

Nitschke, M., C. Ferraz and G.M. Pastore. 2004. Selection of microorganisms for biosurfactant production using agroindustrial wastes. Braz. J. Microbiol. 435: 81–85.

Ojeda-Morales, M.E., M. Domínguez-Domínguez, M.A. Hernández-Rivera et al. 2015. Biosurfactant production by strains of Azospirillum isolated from petroleum-contaminated sites. Water Air Soil Pollut. 226: 401. https://doi.org/10.1007/s11270-015-2659-0.

Pacw-Plociniczak, M., G.A. Plaza, Z. Piotrowska-Seget and S.S. Cameotra. 2011. Environmental applications of biosurfactants: Recent advances. Int. J. Mol. Sci. 13: 633–654.

Partovinia, A., F. Naeimpoor and P. Hejazi. 2010. Carbon content reduction in a model reluctant clayey soil: slurry phase n-hexadecane bioremediation. J. Hazard Mater 181: 133–139.

Pei, X., Xa Zhan and L. Zhou. 2009. Effect of biosurfactant on the sorption of phenanthrene onto original and H2O2 treated soils. J. Environ. Sci. (China) 21: 1378–1385.

Perneel, M., J. Heyrman, A. Adiobo, K. De Maeyer, J.M. Raaijmakers et al. 2007. Characterization of CMR5c and CMR12a, novel fluorescent *Pseudomonas* strains from the cocoyam rhizosphere with the biocontrol activity. J. Appl. Microbiol. 103: 1007–1020.

Priyam Vandana and Dinesh Singh. 2018. Review on biosurfactant production and its application. Int. J. Curr. Microbiol. App. Sci. 7(8): 4228–4241. doi: https://doi.org/10.20546/ijcmas.708.443.

Ron, E. and E. Rosenberg. 2011. Natural roles in biosurfactants. Environ. Microbiol. 3: 229–236.

Rosenberg, E. and E.Z. Ron. 1999. High- and low-molecular-mass microbial surfactants. Appl. Microbiol. Biotechnol. 52(2): 154–62. doi: 10.1007/s002530051502.

Roy, A. 2017. Review on the biosurfactants: properties, types and its applications. J. Fundam. Renewable Energy Appl. 8: 248. doi:10.4172/20904541.1000248.

Rufifino, R.D., G.I.B. Rodrigues, G.M. Campos-Takaki, L.A. Sarubbo, S.R.M. Ferreira et al. 2011. Application of a yeast biosurfactant in the removal of heavy metals and hydrophobic contaminant in a soil used as slurry barrier. Appl. Environ. Soil Sci. 939648.

Rufino, R.D., J.M. Luna, G.M. Campos-Takaki, S.R. Ferreira, L.A. Sarubbo et al. 2012. Application of the biosurfactant produced by *Candida lipolytica* in the remediation of heavy metals. Chem. Eng. 27: 61–66.

Rufifino, R.D., J.M. de Luna, G.M. de Campos Takaki and L.A. Sarubbo. 2014. Characterization and properties of the biosurfactant produced by *Candida lipolytica* UCP 0988. Electron. J. Biotechnol. 17: 34–38.

Sachdev, D.P. and S.S. Cameotra. 2013. Biosurfactants in agriculture. Appl. Microbiol. Biotechnol. 97:1005–1016. doi: 10.1007/s00253-012-4641-8.

Sahoo, B., R. Ningthoujam and S. Chaudhuri. 2019. Isolation and characterization of a lindane degrading bacteria *Paracoccus* sp. NITDBR1 and evaluation of its plant growth promoting traits. Int. Microbiol. 22(1): 155–167. doi:10.1007/s10123-018-00037-1.

Santos, E.C., R.J. Jacques, F.M. Bento, C. Peralba Mdo, P.A. Selbach et al. 2008. Anthracene biodegradation and surface activity by an iron-stimulated *Pseudomonas* sp. Bioresour. Technol. 99: 2644–2649.

Sarwar, A., G. Brader, E. Corretto, G. Aleti, M. Abaidullah et al. 2018. Qualitative analysis of biosurfactants from Bacillus species exhibiting antifungal activity. PLoS ONE 13, e0198107.

Sharma, R., J. Singhb and N. Verma. 2018. Production, characterization and environmental applications of biosurfactants from *Bacillus amyloliquefaciens* and *Bacillus subtilis*. Biocatal. Agric. Biotechnol. 16: 132–139.

Sheng, X.F., L.Y. He, Q.Y. Wang, H.S. Ye, C.Y. Jiang et al. 2008. Effects of inoculation of biosurfactant-producing *Bacillus* sp. J119 on plant growth and cadmium uptake in a cadmium-amended soil. J. Hazard Mater 155: 17–22.

SilvaVinícius Luiz da, Roberta Barros Lovaglio, Henrique Hespanhol Tozzi, Massanori Takaki, Jonas Contiero et al. 2015. Rhamnolipids: a new application in seeds development. Journal of Medical and Biological Science Research 1(8): 100–106.

Singh, R., B.R. Glick and D. Rathore. 2018. Biosurfactants as a biological tool to increase micronutrient availability in soil: A review. Pedosphere 28: 170–189. doi: 10.1016/S1002-0160(18)60018-9.

Singh Ratan and Rathore Dheeraj. 2019. Impact assessment of azulene and chromium on growth and metabolites of wheat and chilli cultivars under biosurfactant augmentation. Ecotoxicology and Environmental Safety 186.

Smita Bhuyan-Pawar, Reshma P. Yeole, Vynkatesh M. Sanam, Shradha P. Bashettil and Shilpa S. Mujumdar. 2015. Biosurfactant Mediated Plant growth promotion in soils amended with polyaromatic hydrocarbons. Int. J. Curr. Microbiol. App. Sci. Special Issue 2: 343–356.

Sober´on-Ch´avez, G. and R.M. Maier. 2011. Biosurfactants: A general overview. pp. 1–11. *In*: Sober´on-Ch´avez, G. (ed.). Biosurfactants: From Genes to Applications. Springer, Berlin, Heidelberg.

Stacey, S., M.J. McLaughlin, I. Cakmak, E. Lombi, C. Johnston et al. 2007. Novel chelating agent for trace element fertilizers. Available online at https://www.ionainteractive.com (verified on Oct. 28, 2016).

Stanghellini, M.E. and M. MillerRaina. 1997. Biosurfactants: their identity and potential efficacy in the biological control of Zoosporic plant pathogens. Plant Dis. 81(1): 4–12.

Sun, X., L. Wu and Y. Luo. 2006. Application of organic agents in remediation of heavy metals-contaminated soil. Ying Yong Sheng Tai Xue Bao 17: 1123–1128.

Usman, M.M., A. Dadrasnia, K.T. Lim, A.F. Mahmud, S. Ismail et al. 2016. Application of biosurfactants in environmental biotechnology; remediation of oil and heavy metal. AIMS Bioeng. 3: 289–304. doi: 10.3934/bioeng.

Venkatesh, N.M. and N. Vedaraman. 2012. Remediation of soil contaminated with copper using rhamnolipids produced from *Pseudomonas aeruginosa* MTCC 2297 using waste frying rice bran oil. Ann. Microbiol. 62: 85–91.

Volkering, F., A.M. Breure and W.H. Rulkens. 1998. Microbiological aspects of surfactant use for biological soil remediation. Biodegradation 8: 401–417.

Wattanaphon, H.T., A. Kerdsin, C. Thammacharoen, P. Sangvanich, A.S. Vangnai et al. 2008. A biosurfactant from *Burkholderia cenocepacia* BSP3 and its enhancement of pesticide solubilization. J. Appl. Microbiol. 105: 416–423.

Wen, J., M.J. McLaughlin, S.P. Stacey and J.K. Kirby. 2010. Is rhamnolipid biosurfactant useful in cadmium phytoextraction? J. Soils Sediment. 10: 1289–1299.

Wen, J., M.J. McLaughlin, S.P. Stacey and J.K. Kirby. 2016. Aseptic hydroponics to assess rhamnolipid-Cd and rhamnolipidZn bioavailability for sunflower (*Helianthus annuus*): A phytoextraction mechanism study. Environ. Sci. Pollut. Res. 23: 21327–21335.

World Health Organization (WHO). 2002. The World Health Report 2002: Reducing Risks, Promoting Healthy Life. WHO, Geneva.

Xu, Q., M. Nakajima, Z. Liu and T. Shiina. 2011. Biosurfactant from microbubble preparation and application. International J. of Molecul. Sci. 12: 462–475.

Ying, G.G. 2006. Fate, behavior and effects of surfactants and their degradation products in the environment. Environ. Int. 32: 417–431.

Zhang, C., S. Wang and Y. Yan. 2011. Isomerization and biodegradation of beta-cypermethrin by *Pseudomonas aeruginosa* CH7 with biosurfactant production. Bioresour. Technol. 102: 7139–7146.

11

Rhamnolipid Biosurfactants

Production and Application in Agriculture

Siti Syazwani Mahamad,[1] *Shobanah Menon Baskaran,*[2] *Izzah Nurfarahiyah Md Isa,*[1] *Mohd Rafein Zakaria*[1,2,*] and *Mohd Ali Hassan*[1,2]

1. Introduction

Surfactants have a significant role as cleaning, wetting, dispersing, emulsifying, foaming and antifoaming agents in numerous functional applications and products. Most surfactants present today are manufactured from petrochemical sources where, due to their low biodegradability, they contribute to environmental problems (Kłosowska-Chomiczewska et al. 2017). Rhamnolipids, a class of glycolipids among biosurfactants known primarily from Pseudomonas genus bacteria, is a promising biosurfactant. Rhamnolipids' chemical properties resemble petroleum-derived surfactants suitably used as emulsifiers, spreaders and dispersing agents in pesticide formulations. Rhamnolipids also act as biopesticide against agricultural pests, showing insecticidal activity against green peach aphids, and biocontrol agents against several phytopathogenic fungi (Oluwaseun et al. 2017). The addition of rhamnolipids or cell-free broth-containing rhamnolipids is effective in suppressing the development of phytopathogens, e.g., *Fusarium oxysprum, Botyris cinerea, Mucor* sp. and many more (Crouzet et al. 2020, Borah et al. 2016).

Studies conducted over the last two decades reported that rhamnolipids (RLs) are produced dominantly from *P. aeruginosa*. Rhamnolipid yields as well as

[1] Department of Bioprocess Technology, Faculty of Biotechnology and Biomolecular Sciences, Universiti Putra Malaysia, 43400, UPM Serdang, Selangor, Malaysia.

[2] Laboratory of Processing and Product Development, Institute of Plantation Studies, Universiti Putra Malaysia, 43400, UPM Serdang, Selangor, Malaysia.

* Corresponding author: mohdrafein@upm.edu.my

congeners produced are varied by species level, substrates consumed, environmental fermentation conditions (Chong and Li 2017, Tan and Li 2018, Irorere et al. 2017). The growth and production of RLs was optimized in shake flasks and pilot-scale bioreactors for specific fermentation periods (Bazsefidpar et al. 2019, Pathaka and Pranav 2015, Gong et al. 2020). The RLs produced are further extracted and purified using a wide variety of chemical solvents and the purified samples are characterised using advanced analysing tools to determine their physicochemical and biological properties (Irorere et al. 2017, Smyth et al. 2014). Rhamnolipids showed both antimicrobial and antifungal properties when tested on wide ranges of microbial pathogens and their activity efficiency was dependent on the concentration of RLs, types of congeners present in the extracted rhamnolipids as well as the types of microorganisms tested (Díaz De Rienzo et al. 2016, Fracchia et al. 2015, Sajid et al. 2020).

The present antimicrobial and antifungal agents which have been used are highly toxic and nonbiodegradable thus prone to environmental nuisance (Malviya et al. 2020). Many researchers have proved that Rhamnolipids have diverse and unique properties including sequestering, detergency, emulsifying, wetting, vesicle forming and phase dispersion that could be an alternative to synthetic chemicals used in controlling plant pathogens—hence they have been recognized as safe biocontrol agents (Oluwaseun et al. 2017). The utilization of RLs in the management of bacterial and fungal crop pathogens has been recently driven by interdisciplinary studies leading to many discoveries of plant protection and improvement. For example, the ability of RLs derived from *P. aeruginosa* strains has been able to induce nonspecific immunity in plants (Sinumvayo et al. 2015). Leite et al. (2016) described the efficacy of RLs against *Ralstonia solanacearum*, a pathogen that is responsible for bacterial wilt in tomato, tobacco, pepper, and potato plants in their *in vitro* studies.

The studies also have been extended to investigate the antifungal properties of RLs by conducting *in vivo* plant systems, and the effectiveness of RLs in protecting the pepper plants from *Pytopthora* blight disease, and their success in preventing the development of *Colletotrichum orbiculare* on the cucumber leaves. In particular, in the plant bioassay test, RLs were found to be capable of controlling the sugarcane plants red rot disease caused by *Colletotrichum falcatum*. The *C. falcatum* fungal spore treated with Rhamnolipid—DS9 showed a reduced number of disease scores as the concentration of RL-DS9 increased (Goswami et al. 2015). Though several biological control agents have been successfully identified, further research on their application towards plant pathogens—controlling is required to help in the discovery of various strains for industrial scale production with better attributes.

In this review, a specific overview is given regarding rhamnolipid biosurfactants and their potential use in agricultural applications. The mechanism and factors affecting efficient lysis of the wide range of plant pathogens are described in this chapter.

2. Rhamnolipids

Rhamnolipids (RLs) are the best-known biosurfactants and initially isolated from an opportunistic pathogen, *Pseudomonas aeruginosa*, a gram-negative bacterium, and this is still a significant producer (Sałek and Euston 2019). The bacteria synthesize RLs as secondary metabolites in the stationary stage of microbial growth, which either stay bound to the surface of the microbial cells or are excreted outside the cells (Varjani and Upasani 2017). The type of RLs produced depends on the cultivation conditions, the source of carbon used and the producing strain (Silva et al. 2010). Rhamnolipids consist of units of rhamnose linked to units of 3-hydroxyl fatty acid via O-glycosidic bond. α-1,2--glycosidic bonds link the rhamnose units to each other, while the 3-hydroxyl fatty acids are connected by an ester bond with one another. Based on the number of rhamnose units within the molecule, RLs exist as mono-RLs and di-RLs and further differentiated into various congeners depending on the fatty acid units composition (Abdel-Mawgoud et al. 2010, Irfan-Maqsood and Seddiq-Shams 2014).

Rhamnolipids have found use in a broad range of applications, in which they demonstrate marked advantages over other surfactants. With advantageous properties like biodegradability, active under a wide range of conditions, and no additional costs needed for disposal made RLs a suitable substitute to chemical-based surfactants (Kłosowska-Chomiczewska et al. 2017). The additional advantage of producing RLs is that cheap or waste materials (cheese whey, used canola oil and corn steep liquor) can be used as carbon substrates for producing them (de Silva et al. 2014, George and Jayachandran 2013, Marchant and Banat 2012).

2.1 *Physicochemical Properties of Rhamnolipids*

The hydrophile-lipophile equilibrium in the compound molecule determines some of the basic physicochemical properties which characterize the rhamnolipid solutions (Benincasa et al. 2010). Critical micellar concentration (CMC), surface tension, wetting properties, foaming behaviour, and emulsification activity are important physicochemical features of biosurfactants (Tiso et al. 2017a). The properties of RLs depend on the form and ratio of the constituent homologs of RLs (mono- and di-rhamnose), which differ according to the bacterial strain, cultivation conditions, fermentation process strategy and the medium composition, and in particular the source of carbon (Nicolò et al. 2017). The CMC of RLs has been reported to vary from 10 to 200 mg/L and is based on the chemical composition of the different organisms and their chemical environment (Gudiña et al. 2015). Studies by Kłosowska-Chomiczewska et al. (2017) have shown that the most significant effect on the CMC was the hydrophobicity of the carbon substrate used for biosynthesis of the RLs. Rhamnolipids have been shown to reduce the water surface tension from 72 to 28 mN/m, and the water-oil system interfacial tension from 43 to $<$ 1 mN/m (Kaskatepe and Yildiz 2016). Nevertheless, the values are in the range of an effective and often used chemical-based surfactant, sodium dodecyl sulfate (SDS) (Tiso et al. 2017a).

A variety of analytical methods can be used to fractionate and characterize the forms of RLs in the mixture. In the past, the most frequently used techniques were high performance liquid chromatography (HPLC) equipped with a photodiode array detector or UV detector and gas chromatography-mass spectrometry (GC-MS), but they are time-consuming and do not provide accurate quantification analysis

(Pornsunthorntawee et al. 2010). Recently, as effective methods for studying species of RLs, HPLC equipped with an evaporative light scattering detector (ELSD) and liquid chromatography-mass spectrometry (LC/MS) has been developed. Fourier transform infrared (FTIR) spectroscopy and nuclear magnetic resonance (NMR) analysis were also conducted to characterize the chemical structures of RLs (Irorere et al. 2017, Smyth et al. 2014).

2.2 *Biological Properties of Rhamnolipids*

Rhamnolipids have interesting biological activities in addition to their good physicochemical properties. Rhamnolipids are heat tolerant glycolipids which have antimicrobial, haemolytic, zoosporicidal, cytotoxic, algicidal, antiviral, antibiofilm, antiamoebic, and antiadhesive properties (Sajid et al. 2020). The antimicrobial property of RLs is due to their permeabilizing effect, resulting in bacterial cell plasma membrane disruption (Díaz De Rienzo et al. 2016, Fracchia et al. 2015), their ability to compromise the charge of the cell surface (Kaczorek 2012) and to alter the hydrophobicity of bacterial cells (Elshikh et al. 2017). They are also able to prevent and inhibit the formation of biofilms, causing the constituent bacteria to be more susceptible to antimicrobial agents (Naughton et al. 2019). Rhamnolipids are well known for their antifungal activity against phytopathogenic fungi, which enables them to be used in agriculture for plant protection. Rhamnolipids derived from *P. aeruginosa* have been reported to suppress growth of plant pathogens such as *Oomycetes*, *Ascomycota* and *Mucor* spp. fungi (Sha et al. 2012). Recent research has also demonstrated that several RLs have defensive impacts on plants against phytopathogenic fungi and bacterial infections by activation of the plant immune response. *P. aeruginosa* RLs have been identified for enhancing protection mechanisms in tobacco, wheat, and *Arabidopsis thaliana* (Vatsa et al. 2010).

3. Production of Rhamnolipids

Rhamnolipids were found in 1949 by Jarvis and Johnson and are the most studied biosurfactants, produced by *Pseudomonas aeruginosa* (Abdel-Mawgoud et al. 2010). In the past few years, the interest in RLs production is gaining attention due to their high potential in commercial applications including pharmaceutical and agro-industries. Antimicrobial properties of RLs against both phytopathogenic fungi and bacteria have been studied as they seem to meet the criteria for environmental remediation and biological control. Besides, their surface-active compounds can be derived from a variety of cost-effective substrates such as organic wastes, which provide dual benefits of natural-based surfactant production and controlling environmental pollution (Gudiña et al. 2015). Many strategies are designed to estimate optimum processes for RLs production and to improve the yield and productivity together with product quality. Rhamnolipids are mainly produced by *Pseudomonas* species for example *P. aeruginosa* and reported as promising glycolipids containing rhamnose and 3-hydroxy fatty acids (Irfan-Maqsood and Seddiq-Shams 2014). Despite its pathogenicity, *P. aeruginosa* has been mentioned by studies as the most advanced production strain to produce two groups of RLs: mono-RLs and di-RLs with a high surface activity performance, using various strategies and different carbon

Table 1. Summary of RLs production from different strains of *P. aeruginosa* and their corresponding carbon sources.

Carbon sources	**Microorganism**	**Maximum concentration yield (g/L)**	**References**
Sunflower seed oil, soybean oil and glycerol	*P. aeruginosa* DS10-129	4.31, 2.98 and 1.77	(Rahman et al. 2002)
Waste cooking oils (olive and sunflower seed oil)	*P. aeruginosa* 47T2 NCIB 40044	2.7	(Haba et al. 2000)
Brazil nut and maracuja oils	*P. aeruginosa* LBI	9.9 and 9.2	(Costa et al. 2006)
Maize steep liquor (10% (v/v)) and molasses syrup (10% (w/v))	*P. aeruginosa strain* #112	3.2	(Gudiña et al. 2015)
Glycerol and NH_4NO_3	*P. aeruginosa* DAUPE 614	3.9	(Monteiro et al. 2007)
Glucose and glycerol	*P. aeruginosa* EM1	7.5 and 4.9	(Wu et al. 2008)
Soap stock waste (soybean)	*P. aeruginosa LBI*	*11.72*	(Nitschke et al. 2005)
Canola oil	*P. species* DSM 2874	45	(Trummler et al. 2003)

sources such as sugar, glycerol, and vegetable oils for RLs production (Henkel et al. 2012). This microbial species produces RLs as their secondary metabolite and the production coincides with the stationary growth phase. Rhamnolipids are produced via the fermentation process, where the bacteria is added into the fermentation tank that is provided with nutrient sources and proper environmental conditions.

During production, the downstream processes involving recovery, concentration and purification are considered as critical steps. Many downstream processes rely on the type of substrates used and their properties, and also their fermentation mode such as batch or fed-batch processes (Sarachat et al. 2010). A summary of RLs production derived from *P. aeruginosa* using various carbon sources as their substrates is shown in **Table 1**.

3.1 Growth and Production Conditions

To be able to redirect the metabolism of microbial species towards the production of RLs in liquid culture, their growth and production processes depend on several specific parameters and environmental conditions. The growth factors are mainly focusing on nutritional requirements. Meanwhile, the operating conditions such as pH, temperature, oxygen availability and agitation speed must be administered to ensure the production at an optimum level.

3.1.1 Nutritional Factors

Nutrients are the most important factors to ensure the microbial growth is in a good condition. Optimum aerobic and anaerobic processes require sufficient macro-nutrients including carbon (C), nitrogen (N), and phosphorus (P), and micronutrients like iron (Fe) (Martínez-Toledo et al. 2018). Carbon sources are needed by microorganisms to be utilized as energy for their growth. Glucose, ethanol, glycerol, and mannitol are examples of the water-soluble carbon sources mainly used for RLs production by *P. aeruginosa* (Silva et al. 2010). Meanwhile, it has been reported that the production levels fall when using immiscible water carbon sources such as olive oil and alkanes (Martínez-Toledo et al. 2018). Nitrogen sources like nitrates are important to support production. It has been observed that the culture of *P. aeruginosa* showed overproduction of biosurfactant in the stationary growth phase because of nitrogen limitation, and reflecting nitrates are the best sources of nitrogen (Maqsood and Jamal 2011). Phosphate (P) limitation in the culture media was reported more effective than nitrogen limitation for biosurfactant production by *P. aeruginosa*, only when palmitic acid or hexadecane were used as carbon sources in an anaerobic state (Mulligan 2009). Furthermore, Fe plays an important role for biosurfactant production in particular microorganisms. Iron will act as an enzyme activator, specifically isocitrate lyase, which is an enzyme entailed in the growth of cells on hydrophobic substrates. This type of enzyme is necessary for a cell to deal with acetyl coenzyme A (Co-A) and converting it into a C4 units during the process of biosurfactant synthesis (Hommel and Ratledge 2010).

3.1.2 Environmental Conditions

Environmental factors for instance pH, temperature, oxygen availability and agitation speed affect the biosurfactants production on both cell and growth activities. It was learned that the production of RLs by *P. aeruginosa* maximizes in a pH range of 6.0 to 6.5 (Joy et al. 2020). In another study, the production of RLs was increased within a pH range of 6.2 to 6.4 and lowered drastically for a pH more than 7.0 (Kaskatepe et al. 2016). The metabolism of microbial species is pH sensitive where it is the key factor that affects the chemical reactions of the living cells (Maqsood and Jamal 2011). Surrounding temperature influences the production of RLs, and pH dependence. However, the optimum temperature is reported differently with the variety of microbial species. A study reported that the RLs production from *P. aeruginosa* increased within a temperature range of 25 to 37°C (Wu et al. 2008, Gudiña et al. 2015). In addition, the agitation rate which affected the RLs production has been described in most studies. Agitation rate will affect the efficiency of mass transfer from both the medium culture and oxygen, specifically during the cell growth of microorganisms and RLs formation. Wei and his colleagues (2005) reported that the RLs production increased up to 80% and the rate of cell growth showed improvement from 0.22–0.72 per hour when the agitation rate was raised to between 50–200 rpm. The study also stated that dissolved oxygen (DO) levels increased from 0.12–0.55 mg/L when a similar agitation rate was applied, showing that high concentration of DO appeared to have a significant influence on cell growth and RLs production.

There are differences in structural composition resulting from aerobic and anaerobic cultivation of RLs production. The physicochemical properties of RLs for example surface activity are determined by their structural composition. Any changes in the composition have potential to affect their emulsifying activity (de Santana-Filho et al. 2014). Nevertheless, most of the RLs are produced by aerobic microorganisms which are grown in the liquid media, which means that oxygen availability is essential to the RLs yield. Oxygen limitation can impact the RLs-producer strains genes expression, hence modifying the RL yield, structural formation and both their physical and chemical properties (Zhao et al. 2018).

3.2 Production Scale and Yields of Rhamnolipids

3.2.1 Cultivation Process Strategies

To make them competitive with chemical surfactants, researchers have focused on enhancing strain improvement and process optimization to increase the RLs titres and yields (Li et al. 2019). Many works need to be conducted to acquire sufficient yield of RLs, from laboratory to industrial scale. Certain factors and interdisciplinary research approaches between the technologies and genetic engineering are needed to be considered to transfer and commercialize the production to pilot scale; the potency of RLs-producing strains to enhance the production capacity, cost of effective raw material and downstream processes (Ahmad Khan et al. 2014). Since RLs are recovered from the fermentation broth, many processes are required to separate mixtures from unused substrates, and other metabolic compounds and products which are causing purification to be a complex step especially for low RLs titres (Beuker et al. 2016). According to Kaskatepe et al. (2015), the RLs yield depends heavily on the selection of the type of RLs producing strain, the culture components, and the fermentation controls, as the cultivation strategies reported affect the titres of RLs in the range from 3.9 to 78.56 g/L. Meanwhile, Chong and Li (2017) have informed the production of RLs from *P. aeruginosa* using soybean oil as a substrate reached the highest RLs titre of 112 g/L. Furthermore, the RLs yield and productivity are varied among different batches. Randhawa (2014) stated the fermentation of biosurfactants in batch cultures is not commercially viable for industrial-scale production. It was reported the yield of RLs produced in the continuous process was several folds larger than the batch culture (Pinzon et al. 2013). In contrast, the fed-batch process by feeding and adding more nutrients to the fermenter during the fermentation has been claimed as the most efficient mode to enhance RLs yield and productivity (Eslami et al. 2020).

A study reported about 22.7 g/L of RLs concentration was obtained through fed-batch fermentation mode using *P. aeruginosa* BYK-2 KCTC 18012P and fish oil as carbon sources (Lee et al. 2004). A fermentation using *P. aeruginosa* without genetic manipulation is reported by Bazsefidpar et al. (2019), showing maximum RLs concentrations which obtained a yield 4.8-fold higher than the batch cultivation, with overall RLs yield and productivity of 240 g/L and 0.9 $g/L^{-1}h^{1}$, respectively by using sunflower oil as a substrate. Nevertheless, both conversion yield and productivity are needed to be focused on as important parameters in determining RLs production expenditure. For instance, high RLs titre (45 g/L), is considered insufficient for commercial production when the volumetric productivity and conversion yields are only 0.14 $g/L^{-1}h^{1}$ and 22.3%, respectively (Trummler et al. 2003). Low yield conversion caused by a huge amount of unused raw material, and limited products resulting from low productivity, are leading to the high production cost (Jiang et al. 2019).

3.2.2 *Carbon sources for Rhamnolipids Production*

Rhamnolipids are synthesized by isolates from the environment, for instance, hydrocarbon polluted soil, or produced using agricultural wastes that are rich in carbon sources waste (Sachdev and Cameotra 2013). Studies have shown carbon sources as a major factor of RLs production categorized into hydrocarbons, carbohydrates, and hydroxyl compounds. Isolating the same microbial producing species growing on different carbon sources may cause them to behave differently and therefore, dissimilar RLs may be produced. Rhamnolipids are described as the best emulsification agents when using olive oil as their carbon sources, with emulsification index E_{24} and yield concentrations of 65% and 0.8 g/L, respectively (Abouseoud et al. 2008). However, studies conducted by (Parthasarathi and Sivakumaar 2009) are reported contrarily, where the rhamnolipids produced by *Pseudomonas* strain using cashew apple juice showed maximum rhamnolipids yields of 9.35 g/L, high emulsifying index E_{24} of 63%, and the best surface tension reduction at 40.0 mN/m, compared to other carbon sources glucose; fructose, mannitol, glycerol, and olive oil.

Glycerol is one of the hydroxyl compounds that can be obtained from renewable sources or side products of biodiesel manufacturing which is usually generated from transesterification of vegetable oils like rapeseed oil or soybean oil using methanol (Rodrigues et al. 2017). Generally, the degree of impurities of generated glycerol as a by-product may be different. A study identified RLs production using crude glycerol from biodiesel production as a carbon source without undergoing a pre-treatment process, successfully producing final RLs concentrations ranging from 1.0 to 8.5 g/L. However the rate of a substrate to product conversion ($Y_{P/S}$) reported was low between 0.1 to 0.15 (Avili et al. 2012). Differently, oils, fats and fatty acids are reported as the best substrates for RLs production by *P. aeruginosa* with high yields and final concentrations. It was informed that the final RLs concentration obtained from *P. aeruginosa* was 13.93 g/L using waste cooking oil as a sole carbon source (Lan et al. 2015).

Nitschke and co-workers (2005), reported that the final concentration of RLs production using soap stock from diverse vegetable oil industries reached 11.7 g/L. However, strong emulsion compounds of RLs extensively require organic solvent extraction to separate and purify products from oily substrates that may increase the production cost (Sarachat et al. 2010). Apart from that, sugar-containing wastes utilization has been targeted by many researchers in RLs production compared to the oil or glycerol containing wastes concerning their cost. Rhamnolipids production yields from different types of *Pseudomonas* strains and culture conditions are summarized in Table 2.

Sugar and starch processing industries are reported to generate huge quantities of carbohydrate-rich wastewater that can be used as a substrate for fermentation processes (Preethi et al. 2019). The average of substrate to product rate conversion is much lower than glycerol with the $Y_{P/S}$ observed in studies at 0.24 and 0.61 g/g, respectively (Hintermayer and Weuster-Botz 2017). Santos and co-workers (2016) in their studies explained low yields of RLs production from sugar type substrates. It has been informed that cell metabolism suppresses both lipogenic and glycolytic pathways when glucose is used as a carbon source (Yin et al. 2012). The glucose will be degraded to form intermediate glucose-6-phosphate. Before converting

Table 2. Summary of RLs production from different types of *Pseudomonas* sp., fermentation conditions and product yields.

Carbon source	Strain	Growth conditions	Fermentation mode	Working volume	$Y_{P/S}$ (g/g)	Max. yield (g/L)	References
Soybean soap stock waste	*P. aeruginosa* LBI	pH 6.8 30°C 200 rpm	Batch	125 mL	0.75	11.72	(Nitschke et al. 2005)
Rapeseed oil	*Pseudomonas* sp. DSM 2874	30°C 120 rpm	Fed - batch	150 mL	0.22	45	(Trummler et al. 2003)
Soybean oil and glycerol	*P. aeruginosa* DS10-129	pH 7.5 30°C 200 rpm	Batch	250 mL	0.72 0.29	4.31 1.77	(Rahman et al. 2002)
Waste cooking oil	*Pseudomonas* SWP-4	pH 7.0 35°C 150 rpm	Batch	250 mL	0.88	13.93	(Lan et al. 2015)
Corn steep liquor (10% (v/v)) and molasses (10% (w/v))	*P. aeruginosa* strain #112	pH 7.0 37°C 180 rpm	Batch	500 mL	NA	3.2	(Gudiña et al. 2015)
Glucose and glycerol	*P. aeruginosa* EM1	pH 6.8 37°C 200 rpm	Batch	500 mL	NA	7.5 4.9	(Wu et al. 2008)
Cashew apple juice	*P. fluorescens*	30°C 200 rpm	Batch	500 mL	0.47	9.35	(Parthasarathi and Sivakumaar 2009)
Sunflower seed oil	*P. aeruginosa* HAK02	pH 8.5 25°C 180 rpm	Fed - batch	5 L	0.64	223	(Bazsefidpar et al. 2019)
Fish oil	*P. aeruginosa* BYK-2 KCTC 18012P	pH 7.0 25°C 200 rpm	Fed - batch	7 L	0.75	22.7	(Lee et al. 2004)

The non-reported $Y_{P/S}$ is labeled as NA

to acetyl-CoA and transforming to fatty acid, glucose needs to be oxidized to pyruvate to form lipids. In this case, if oils are used, fatty acids will be oxidized via β-oxidation to acetyl-CoA to produce sugars by an activity called gluconeogenesis. Polysaccharide precursor glucose-6-phosphate will be synthesized accordingly towards glucose formation (Pelley 2012). This hypothesis clarifies that oil produced higher biosurfactant yields than sugars-containing carbon sources, wherein both hydrophilic and hydrophobic substrates create different metabolic pathways and form different precursors for biosurfactant production. Despite the challenges to obtaining pure mono-RLs and di-RLs, sugar-type carbon sources are still promising for easier downstream processes compared to other carbon sources.

4. Properties of Rhamnolipid-producing Strains

4.1 Pseudomonads

The *Pseudomonas* genus is one of the most diverse bacterial genera and is a Gram-negative bacterium with the highest number of species. The list of species in the genus has increased every year (Gomila et al. 2015). *Pseudomonas* species are abundant in soil, although some have long been identified as plant pathogens (Peix et al. 2018), others emerge as plant-associated growth promoters with possible biocontrol functions (Agaras et al. 2015). They are small, rod-shaped, non-spore-forming Gram-negative bacilli with 57–68 mol% guanine/cytosine content, motile using one or more polar flagella. *Pseudomonas* species are strictly aerobic, although anaerobic growth is possible in some cases if a source of nitrate can be used (Wisplinghoff 2017). The genus *Pseudomonas* with its metabolic versatility played a significant role in biotechnology and pharmaceutical industries (Mozejko-Ciesielska et al. 2019).

The first recorded producer of RLs was *P. aeruginosa* (previously known as *P. pyocyanea*), which was discovered by Bergstorm et al. (1946) as an 'oily glycolipid' named piolipic acid, consisting of β-hydroxydecanoic acid and L-rhamnose (Irorere et al. 2017). The other *Pseudomonas* species known to produce RLs are *P. nitroreducens* (Onwosi and Odibo 2012), *P. fluorescens* (Prabakaran et al. 2015), *P. putida* (Beuker et al. 2016), *P. chlororaphis* (Lan et al. 2015) and *P. alcaligenes* (Oliveira et al. 2009). *P. aeruginosa* remains the most efficient producer of RLs amongst the available strains (Chong and Li 2017).

4.2 Production Cost of Biosurfactants

Application of RLs in farming activities have been suggested to be a good alternative for synthetic or chemical pesticides, especially when dealing with environmental and food safety issues. In 1885, chemical pesticides began to control downy mildew by using copper on grapes. By 1980, copper reached about 15% of the fungicides market for the zoosporic pathogens-controlling purpose (Zwieten et al. 2007). The continuity and high reliance on chemicals in agro industries, led to the growth of numerous chemical pesticides in the market, which is helping in the pathogens-resistant strains development and environmental pollution (Kaur and Garg 2014). Though biosurfactants are less toxic, highly biodegradable and have diverse chemical structures, many researchers are working hard toward improving their production to make them cost effective. Biosurfactants are generally produced by microbial species growing in hydrocarbons as their carbon sources, which are expensive hence contributing to a high production cost. The cost of production of biosurfactants emulsan using 5 L fermenter with fed-batch mode has been learned to be three times bigger than the cost of existing commercialized emulsifiers (Banat et al. 2014). Meanwhile, de Gusmão et al. (2010) reported that the production costs of high-volume low value biosurfactants were estimated at 20–30% higher than those of chemical surfactants. However, the production of RLs by *P. aeruginosa* utilizing soybean oil as substrates showed a different perspective in another analysis, where the cost was decreased as the production size increased. (Nitschke et al. 2010). Next, selection of raw materials from industrial or agro-based wastes, and cheap renewable substrates from their study can be inferred as critical in reducing the operating cost since the raw materials represent 30 to 80 percent of the total production expenditure (Mukherjee et al. 2006). Comparing the choice and targeted purification steps also become key factors to establish a cost-effective downstream process. Downstream costs accounted for approximately 60% of the total production cost, meaning that a high purity grade of biosurfactants is not necessary for irrelevant industrial applications and thus reducing purification costs. High-volume low value biosurfactants, for example, can be targeted to suitable areas such as bioremediation and agriculture, whereas high purity molecules should be used in biomedical applications (Randhawa 2014). Concerning the cost of biosurfactants manufacturing, it has been suggested that the biosurfactant price must be equal to or lower than that of chemical surfactants to render them competitive and economically acceptable in the surfactant market.

4.3 Genes and Enzymes Involved in Rhamnolipids Production

In order to obtain the genes which are essential for the biosynthesis of RLs, *P. aeruginosa* was used as a model strain. Production of RLs in *P. aeruginosa* relies on central metabolic pathways, such as fatty acid synthesis and dTDP-activated sugars (Pérez-Armendáriz et al. 2019). A very complex genetic regulatory system (QSR) regulates the biosynthesis of RLs, which also regulates various *P. aeruginosa* virulence-associated traits (Williams and Cámara 2009).

P. aeruginosa rhamnolipid pathway contains three main enzymes and is based on the two precursors β-hydroxy-fatty acid and rhamnose (Chong and Li 2017). The de novo synthesis of fatty acids produces the activated β-hydroxy-fatty acid hydroxyacyl-ACP (Fig. 1). 3-hydroxyacyl-ACP:3hydroxyacyl-ACP O-3-hydroxy-acyl-transferase (RhlA), the first specific enzyme subsequently forms hydroxyalkanoyloxy (HAA), a dimer by binding two hydroxyacyl-ACP molecules. This molecule contains no rhamnose structure and is therefore not a rhamnolipid. Due to the presence of carboxyl, ester, and hydroxyl groups and eventually an amphiphilic structure, it is still a biosurfactant. The six reactions of glucose produce the second precursor. The activated dTDP-L-rhamnose is then fused via rhamnosyltransferase I (RhlB) into the HAA molecule to generate a mono-rhamnolipid. The second rhamnosyltransferase (RhlC) adds to the mono-rhamnolipid a second sugar, which

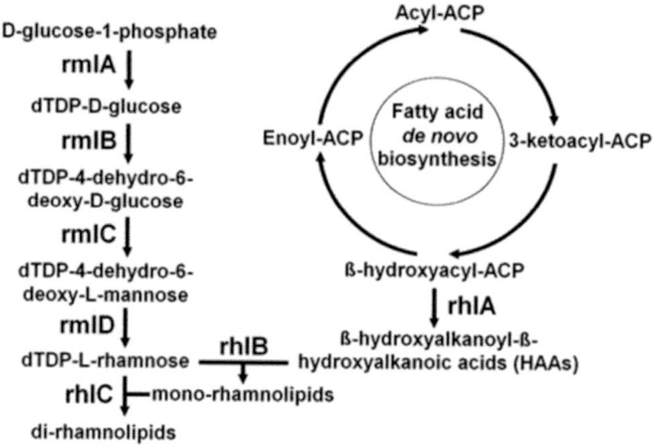

Figure 1. Biosynthesis pathways of rhamnolipids (Bahia et al. 2018).

eventually results in the di-rhamnolipid biosurfactant (Bahia et al. 2018). Most bacteria have the enzymes required to synthesise the precursors in the biosynthesis of RLs, but the enzymes responsible for the synthesis of HAA, mono- and di-RLs are present almost specifically in species of *Pseudomonas* and *Burkholderia* (Fig. 1) (Chong and Li 2017).

4.4 *Quorum Sensing*

Bacterial quorum sensing (QS) is a mechanism for bacterial communication that secretes signal molecules known as auto-inducers to create synchronised behaviours within a population of bacteria (Chong and Li 2017). In many gram-negative bacteria, the QS system consists of a signal synthase, a signal receptor protein and molecule.

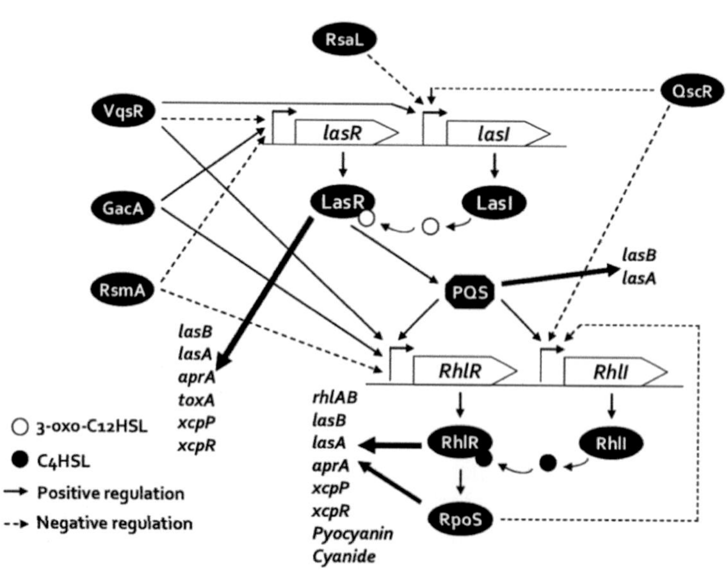

Figure 2. Schematic diagram in *P. aeruginosa* of the *las* and *rhl* genes and the quorum sensing molecules for RLs production (Dusane et al. 2010).

The major QS system in *P. aeruginosa* is *las* and *rhl* (Fig. 2) (Dobler et al. 2016). Las operon consists of two transcriptional activator proteins in *P. aeruginosa*, LasR and Lasl, which guide the autoinducer synthesis of N-3-oxododecanoyl homoserine lactone (PAI-1 or 3-oxo-C12-HSL). Induction of the lasB gene encoding the elastase enzyme and other virulence genes includes the expression of LasR and PAT-1 autoinducer (Dusane et al. 2010). The production of rhamnolipids is controlled by the *rhl* system. The *rhl* system consists of RhlR transcriptional activator proteins coding the receptor for N-butyrylhomoserine lactone (PAI-2 or C4-HSL) and Rhll coding the signal receptor for C4-HSL (Colmer-Hamood et al. 2016). Two enzymes synthesize rhamnolipids: Rhamnosyltransferase I and Rhamnosyltransferase II. The transcriptional activator RhlR activates rhlAB operon gene transcription, coding Rhamnosyltransferase I, which is known for generating mono-RLs, and another gene, rhlC coding Rhamnosyltransferase II, which transforms mono-RLs to di-RLs (Alhede et al. 2014).

5. Antimicrobial and Antifungal Properties

Biosurfactants have been used in various applications, of which the most notable are in biomedical fields. The diverse structures of biosurfactants make them have strong antifungal, antimicrobial, and antiviral activities (Rodrigues et al. 2006, Mnif and Ghribi 2015). Besides lipopeptides, RLs are cited more often in studies as new antimicrobial and antifungal agents that have potential to contribute significantly in the therapy against microbial infections and virulence spoilage factors on crops (Vatsa et al. 2010, Crouzet et al. 2020). Several medical studies have shown the positive effect of RLs on microbial behaviour and tested its application for biomedical use.

5.1 *Antimicrobial Properties*

In earlier studies, the antibacterial activities of RLs did show excellent properties of *in vitro* antimicrobial actions towards Gram-positive and Gram-negative bacterium when they were evaluated by the minimum inhibitory concentration (MIC) (Haba et al. 2003). As a result of their surface tension activity and membrane permeability, RLs derived from *P. aeruginosa* were discussed as potent antimicrobial agents in inhibiting growth of *Staphylococcus aureus*, *Proteus vulgaris* and *Streptococcus faecalis* with a MIC of approximately 8 mg/mL, 8 mg/mL and 4 mg/mL respectively (Benincasan et al. 2004). Rhamnolipids also have been identified as very effective bacteriostatic agents against *Salmonella Enteritidis* (Zezzi et al. 2012), *Listeria monocytogenes* and *Klebsiella pneumoniae* (Bharali et al. 2013) which are pathogens associated with foodborne diseases. Another study showed synergistic interaction of RLs and nisin, which is a broad-spectrum bacteriocin (Magalhães and Nitschke 2013). Bharali et al. (2013) described RLs, as compounds that kill *Staphylococcus aureus* (MTCC3160) by destroying its cell wall structure and components that lead to metabolites leakage from the intracellular region and to the lysis of the cell as a whole, indicating that RLs have a strong membrane permeability rate. Another study also reported that RLs produced by *Pseudomonas* sp. BUP6 as an antimicrobial agent was tested against *E. coli* and *Staphylococcus aureus* and showed 43% and 42% inhibitory effects (Priji et al. 2016).

5.2 Antifungal Properties

Rhamnolipids are good antifungal agents for fighting phytopathogenic fungi, allowing them to be used in the agro industry specifically for crops protection, and its application has been attributed to zoospore lysis (Miao et al. 2015). Sharma et al. (2007) are suggesting the potential use of RLs derived from *Pseudomonas* sp. GRP3 is a biofungicide agent against the soil-borne fungal disease, caused by several pathogens that weaken or kill chili and tomato seeds and new seedlings, before and after they germinate. Rhamnolipids from the *Pseudomonas* species are reported to be involved in the lysis of the zoospores cell membrane of both *Pythium* and *Phytophthora* fungi (Crouzet et al. 2020). Rhamnolipids have been identified as biological control agents that are responsible for plant defense by suppressing the viability of *Botrytis cinerea*, the pathogenic fungus (Yoo et al. 2005). The biosurfactant—based biofungicide was successfully launched, namely Zonix, a rhamnolipid—based fungicide from Jeneil Biotech Inc, USA. Zonix's targeted protected plants include cane crops, leafy vegetables, citrus and tropical fruits from any zoosporic plant—pathogenic microorganism including *Phytophthora, Plasmodiophora, Pythium, Rhizophydium, Sclerophthora, Spongospora, and Trachysphaera* (Thavasi et al. 2014).

5.3 Mechanisms of Reaction of Rhamnolipids Towards Pathogens and Pests

The biological control activities of RLs against pathogens have been evaluated through their surface and interfacial tension, particle size critical micelle concentration (CMC), and emulsification ability (Mendes et al. 2015). Rhamnolipids possess very low surface tension (28–30mN/m), minimal CMC (10–200 mg/L), but elevated emulsifying activity indexes range from 60 to 70% (Gudiña et al. 2015). Understanding the mode of action of RLs toxicity towards microorganisms has been linked to their ability to break down the intracellular contents of targeted organisms by promoting cell lyses and disruption (Banat et al. 2010).

Rhamnolipids cause the disruption of the lipid content and protein of cell membranes of targeted pathogens, consequently hindering the configuration of the cell wall and cell osmolarity (Abdel-Mawgoud et al. 2010). According to Liu et al. (2012), RLs are capable of altering cell surface hydrophobicity (CSH), an essential factor in bioprocesses of microorganisms such as cell to cell contact and nutrient uptake, and the surface charge of bacteria by changing their cell surface's functional groups and element concentrations through the process of adsorption of RLs on the cellular envelope. Figure 3 illustrates the mechanism of action of RLs towards cell surface properties. The cell surface charges of microorganisms are neutralized as the RLs have been attached on the cell surface. During the adsorption activities, the RLs interact with functional groups on cell surfaces such as hydroxyl, carboxyl, and amines groups (Liu et al. 2012). The RLs then increase the permeability of the biomembrane, allowing the vital intracellular components to be released outside and to stay on the surface of the cell (Liu et al. 2011). Some chemical components, such as proteins, are then released again from the cell membrane, therefore changing the charge of the cell surface (Liu et al. 2011).

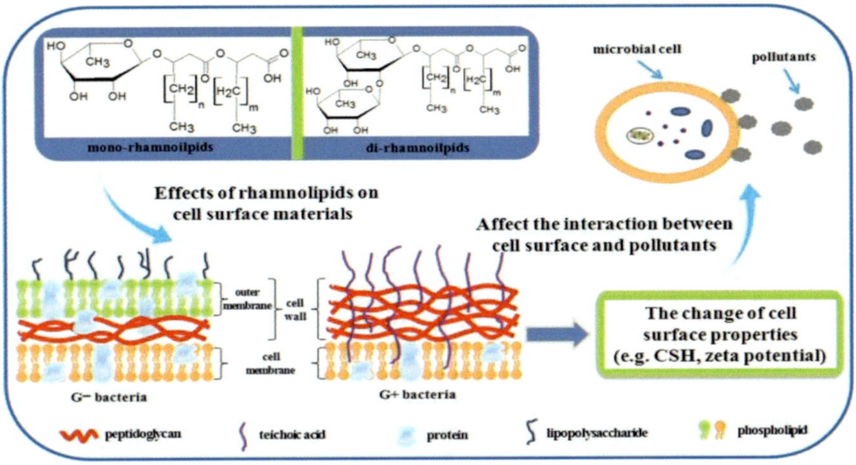

Figure 3. Illustration mode of action of RLs affects the cell surface properties (Shao et al. 2017).

The properties of cell surfaces including CSH, cell surface charge and compounds, and morphological structure are dependent on each other to interact with external environments, which may affect their material transfer, metabolism rate and growth (Liu et al. 2012). However, the efficacy of RLs to serve against targeted pathogens depends on the specific conditions applied including concentration and types of RLs, also the types of pathogen strains (Yuan et al. 2007). In addition, the application of RLs in controlling pests and insects is still new and developing. AgSciTech from US has filed patent (US 2005/0266036 A1) towards the using of RLs on insecticidal activity by proposing direct application on pests such as ants, whiteflies, lice, termites, ticks, nematodes, houseflies, spiders, cockroaches and aphids. The experimental results showed that the significance of insecticidal activity of RLs against these pests was effective at concentrations of 0.0075 to 5% (Thavasi et al. 2014). In another study by Kim et al. (2011), more than 80% of green peach aphids showed a mortality rate within 24 hours of treatment using RLs that were isolated from *Pseudomonas* EP-3. The study has identified the insecticidal mechanism of RLs on pests through cuticle membrane damage.

6. Conclusion

This chapter provided a comprehensive explanation of rhamnolipid biosurfactants and their potential uses in agricultural applications. An earlier research and development at laboratory scale and the attempt to produce in a larger scale process was also highlighted with the purpose to see trends and progress of rhamnolipids research. The mechanism and factors affecting efficient lysis of the wide range of microbial and plant pathogens tested in laboratories are successfully reviewed. Field trial applications of rhamnolipids at certain dosages (100–1000 μg/ml) on selected cash crops such as chilli, tomatoes, cucumbers, lettuce and others have successfully eliminated the widespread of the plant disease caused by plant pathogens. It is expected that RLs can be used widely in agricultural sectors and will gradually replace chemical-based pesticides in the future if the total production cost could be reduced. The use of renewable and cheap carbon sources, development of high yield producer strains and low-cost and solventless extraction processes are under development to make rhamnolipids more cost-effective, competitive, and widely used in the agricultural sectors.

Acknowledgements

This work is partially supported by the Ministry of Science, Technology and Innovation (MOSTI) Malaysia under International Collaborative Funds (IF1019E1159). Siti Syazwani Mahamad is the recipient of a Special Graduate Research Assistant from Universiti Putra Malaysia.

References

Abdel-Mawgoud, A.M., F. Lépine and E. Déziel. 2010. Rhamnolipids: Diversity of structures, microbial origins and roles. Applied Microbiology and Biotechnology 86(5): 1323–1336.

Abouseoud, M., R. Maachi, A. Amrane, S. Boudergua, A. Nabi et al. 2008. Evaluation of different Bacarbon and nitrogen sources in production of biosurfactant by *Pseudomonas fluorescens*. Desalination 223(1): 143–151.

Agaras, B.C., M. Scandiani, A. Luque, L. Fernández, F. Farina et al. 2015. Quantification of the potential biocontrol and direct plant growth promotion abilities based on multiple biological traits distinguish different groups of *Pseudomonas* spp. isolates. Biological Control 90: 173–186.

Ahmad Khan, M.S., B. Singh and S.S. Cameotra. 2014. Biological applications of biosurfactants and strategies to potentiate commercial production. pp. 279–288. *In*: Kosaric, N. and F.V. Sukan. (eds.). Biosurfactants: Production and Utilization—Processes, Technologies, and Economics (Vol. 159). CRC Press.

Alhede, M., T. Bjarnsholt, M. Givskov and M. Alhede. 2014. *Pseudomonas aeruginosa* biofilms: Mechanisms of immune evasion. pp. 1–40. *In*: Advances in Applied Microbiology (Vol. 86).

Avili, M.G., M.H. Fazaelipoor, S.A. Jafari and S.A. Ataei. 2012. Comparison between batch and fed-batch production of rhamnolipid by *Pseudomonas aeruginosa*. Iranian Journal of Biotechnology 10(4): 263–269.

Bahia, F.M., G.C. de Almeida, L.P. de Andrade et al. 2018. Rhamnolipids production from sucrose by engineered *Saccharomyces cerevisiae*. Scientific Report 8: 2905.

Banat, I.M., A. Franzetti, I. Gandolfi, G. Bestetti, M.G. Martinotti et al. 2010. Microbial biosurfactants production, applications and future potential. Applied Microbiology and Biotechnology 87: 427–444.

Banat, I.M., S.K. Satpute, S.S. Cameotra, R. Patil, N.V. Nyayanit et al. 2014. Cost effective technologies and renewable substrates for biosurfactants' production. Frontiers in Microbiology 5: 697.

Bazsefidpar, S., B. Mokhtarani, R. Panahi and H. Hajfarajollah. 2019. Overproduction of rhamnolipid by fed-batch cultivation of *Pseudomonas aeruginosa* in a lab-scale fermenter under tight DO control. Biodegradation 30(1): 59–69.

Benincasa, M., A. Abalos, I. Oliveira and A. Manresa. 2004. Chemical structure, surface properties and biological activities of the biosurfactant produced by *Pseudomonas aeruginosa* LBI from soapstock. Antonie van Leeuwenhoek. International Journal of General and Molecular Microbiology 85(1): 1–8.

Benincasa, M., A. Marqués, A. Pinazo and A. Manresa. 2010. Rhamnolipid surfactants: alternative substrates, new strategies. pp. 170–184. *In*: Biosurfactants. Springer, New York.

Bergstorm, S., H. Toerell and H. Davide. 1946. Pyolipic acid, a metabolic product of *Pseudomonas pyocyanea*, active against *Mycobacterium tuberculosis*. Archives of Biochemistry 10(1): 165–166.

Beuker, J., A. Steier, A. Wittgens, F. Rosenau, M. Henkel et al. 2016. Integrated foam fractionation for heterologous rhamnolipid production with recombinant *Pseudomonas putida* in a bioreactor. AMB Express 6(1): 11.

Bharali, P., J.P. Saikia, A. Ray and B.K. Konwar. 2013. Rhamnolipid (RL) from *Pseudomonas aeruginosa* OBP1: A novel chemotaxis and antibacterial agent. Colloids and Surfaces B: Biointerfaces 103: 502–509.

Borah, S.N., D. Goswami, H.K. Sarma, S.S. Cameotra, S. Deka et al. 2016. Rhamnolipid biosurfactant against *Fusarium verticillioides* to control stalk and ear rot disease of maize. Frontiers in Microbiology 7: 1505.

Chong, H. and Q. Li. 2017. Microbial production of rhamnolipids: Opportunities, challenges and strategies. Microbial Cell Factories 16(137): 1–12.

Colmer-Hamood, J.A., N. Dzvova, C. Kruczek and A.N. Hamood. 2016. *In vitro* analysis of *Pseudomonas aeruginosa* virulence using conditions that mimic the environment at specific infection sites. In Progress in Molecular Biology and Translational Science (Vol. 142, pp. 151–191).

Costa, S., M. Nitschke, R. Haddad, M.N. Eberlin, J. Contiero et al. 2006. Production of *Pseudomonas aeruginosa* LBI rhamnolipids following growth on Brazilian native oils. Process Biochemistry 41(2): 483–488.

Crouzet, J., A. Arguelles-Arias, S. Dhondt-Cordelier, S. Cordelier, J. Pršić et al. 2020. Biosurfactants in plant protection against diseases: rhamnolipids and lipopeptides case study. Frontiers in Bioengineering and Biotechnology, 8.

de Gusmão, C.A.B., R.D. Rufino and L.A. Sarubbo. 2010. Laboratory production and characterization of a new biosurfactant from *Candida glabrata* UCP1002 cultivated in vegetable fat waste applied to the removal of hydrophobic contaminants. World Journal of Microbiology and Biotechnology 26(9): 1683–1692.

de Santana-Filho, A.P., D. Camilios-Neto, L.M. de Souza, G.L. Sassaki, D.A. Mitchell et al. 2014. Evaluation of the structural composition and surface properties of rhamnolipid mixtures produced by *Pseudomonas aeruginosa* UFPEDA 614 in different cultivation periods. Applied Biochemistry and Biotechnology 175(2): 988–995.

de Silva, N.M.P.R., R.D. Rufino, J.M. Luna, V.A. Santos, L.A. Sarubbo et al. 2014. Screening of *Pseudomonas* species for biosurfactant production using low-cost substrates. Biocatalysis and Agricultural Biotechnology 3(2): 132–139.

Díaz De Rienzo, M.A., P. Stevenson, R. Marchant and I.M. Banat. 2016. Antibacterial properties of biosurfactants against selected Gram-positive and -negative bacteria. FEMS Microbiology Letters 363(2).

Dobler, L., L.F. Vilela, R.V. Almeida and B.C. Neves. 2016. Rhamnolipids in perspective: gene regulatory pathways, metabolic engineering, production and technological forecasting. New Biotechnology 33(1): 123–135.

Dusane, D.H., S.S. Zinjarde, V.P. Venugopalan, R.J.C. McLean, M.M. Weber et al. 2010. Quorum sensing: Implications on rhamnolipid biosurfactant production. Biotechnology and Genetic Engineering Reviews 27(1): 159–184.

Elshikh, M., S. Funston, A. Chebbi, S. Ahmed, R. Marchant et al. 2017. Rhamnolipids from non-pathogenic *Burkholderia thailandensis* E264: Physicochemical characterization, antimicrobial and antibiofilm efficacy against oral hygiene related pathogens. New Biotechnology 36: 26–36.

Eslami, P., H. Hajfarajollah and Shayesteh Bazsefidpar. 2020. Recent advancements in the production of rhamnolipid biosurfactants by *Pseudomonas aeruginosa*, RSC Advance 56: 34014–34032.

Fracchia, L., J.J. Banat, M. Cavallo and I.M. Banat. 2015. Potential therapeutic applications of microbial surface-active compounds. AIMS Bioengineering 2(3): 144–162.

George, S. and K. Jayachandran. 2013. Production and characterization of rhamnolipid biosurfactant from waste frying coconut oil using a novel *Pseudomonas aeruginosa* D. Journal of Applied Microbiology 114(2): 373–383.

Gomila, M., A. Pena, M. Mulet, J. Lalucat, E. Garcia-Valdes et al. 2015. Phylogenomics and systematics in *Pseudomonas*. Frontiers in Microbiology 6: 214.

Gong, Z., Q. He, C. Che, J. Liu, G. Yang et al. 2020. Optimization and scale-up of the production of rhamnolipid by *Pseudomonas aeruginosa* in solid-state fermentation using high-density polyurethane foam as an inert support. Bioprocess and Biosystems Engineering 43(3): 385–392.

Goswami, D., S.N. Borah, J. Lahkar and P.J. Handique. 2015. Antifungal properties of rhamnolipid produced by *Pseudomonas aeruginosa* DS9 against *Colletotrichum falcatum*. Journal of Basic Microbiology 55: 1265–1274.

Gudiña, E.J., A.I. Rodrigues, E. Alves, M.R. Domingues, J.A. Teixeira et al. 2015. Bioconversion of agro-industrial by-products in rhamnolipids toward applications in enhanced oil recovery and bioremediation. Bioresource Technology 177: 87–93.

Haba, E., M.J. Espuny, M. Busquets and A. Manresa. 2000. Screening and production of rhamnolipids by *Pseudomonas aeruginosa* 47T2 NCIB 40044 from waste frying oils. Journal of Applied Microbiology 88(3): 379–387.

Haba, E., A. Pinazo, O. Jauregui, M.J. Espuny, M.R. Infante et al. 2003. Physicochemical characterization and antimicrobial properties of rhamnolipids produced by *Pseudomonas aeruginosa* 47T2 NCBIM 40044. Biotechnology and Bioengineering 81(3): 316–322.

Henkel, M., M.M. Müller, J.H. Kügler, R.B. Lovaglio, J. Contiero et al. 2012. Rhamnolipids as biosurfactants from renewable resources : Concepts for next-generation rhamnolipid production 47: 1207–1219.

Hintermayer, S.B. and D. Weuster-Botz. 2017. Experimental validation of *in silico* estimated biomass yields of *Pseudomonas putida* KT2440. Biotechnology Journal 12(6).

Höfte, M. and P. De Vos. 2006. Plant pathogenic *Pseudomonas* species. pp. 507–533. *In*: Plant-Associated Bacteria. Springer, Dordrecht.

Hommel, R.K. and C. Ratledge. 2010. Biosynthetic mechanisms of low molecular weight surfactants and their precursor molecules. pp. 19–80. *In*: Biosurfactants : Production, properties and applications. CRC Press.

Irfan-Maqsood, M. and M. Seddiq-Shams. 2014. Rhamnolipids: Well-characterized glycolipids with potential broad applicability as biosurfactants. Industrial Biotechnology 10(4): 285–291.

Irorere, V.U., L. Tripathi, R. Marchant, S. McClean, I.M. Banat et al. 2017. Microbial rhamnolipid production: A critical re-evaluation of published data and suggested future publication criteria. Applied Microbiology and Biotechnology 101(10): 3941–3951.

Joy, S., S.K. Khare and S. Sharma. 2020. Synergistic extraction using sweep-floc coagulation and acidification of rhamnolipid produced from industrial lignocellulosic hydrolysate in a bioreactor using sequential (fill-and-draw) approach. Process Biochemistry 90: 233–240.

Jiang, J., Y. Zu, X. Li, Q. Meng, X. Long et al. 2019. Recent progress towards industrial rhamnolipids fermentation : process optimization and foam control. Bioresource Technology, pp. 122–394.

Kaczorek, E. 2012. Effect of external addition of rhamnolipids biosurfactant on the modification of Gram-positive and Gram-negative bacteria cell surfaces during biodegradation of hydrocarbon fuel contamination. Polish Journal of Environmental Studies 21(4): 901–909.

Kaskatepe, B., S. Yildiz, M. Gumustas and S.A. Ozkan. 2015. Biosurfactant production by *Pseudomonas aeruginosa* in kefir and fish meal. Brazilian Journal of Microbiology 46(3): 855–859.

Kaskatepe, B. and S. Yildiz. 2016. Rhamnolipid biosurfactants produced by *Pseudomonas* species. Brazilian Archives of Biology and Technology 59: 1–16.

Kaur, H. and H. Garg. 2014. Pesticides: Environmental impacts and management strategies. *In*: Marcelo L. Larramendy and Sonia Soloneski (eds.). Pesticides - Toxic Aspects. InTech.

Kim, B.S., J.Y. Lee and B.K. Hwang. 2000. *In vivo* control and in vitro antifungal activity of rhamnolipid B, a glycolipid antibiotic, against *Phytophthora capsici* and *Colletotrichum orbiculare*. Pest Management Science 56(12): 1029–1035.

Kim, S.K., Y.C. Kim, S. Lee, J.C. Kim, M.Y. Yun et al. 2011. Insecticidal activity of rhamnolipid isolated from *Pseudomonas* sp. EP-3 against green peach aphid (*Myzus persicae*). Journal of Agricultural and Food Chemistry 59(3): 934–938.

Kłosowska-Chomiczewska, I.E., K. Mędrzycka, E. Hallmann, E. Karpenko, T. Pokynbroda et al. 2017. Rhamnolipid CMC prediction. Journal of Colloid and Interface Science 488: 10–19.

Lan, G., Q. Fan, Y. Liu, C. Chen, G. Li et al. 2015. Rhamnolipid production from waste cooking oil using *Pseudomonas* SWP-4. Biochemical Engineering Journal 101: 44–54.

Lee, K.M., S.H. Hwang, S.D. Ha, J.H. Jang, D.J. Lim et al. 2004. Rhamnolipid production in batch and fed-batch fermentation using *Pseudomonas aeruginosa* BYK-2 KCTC 18012P. Biotechnology and Bioprocess Engineering 9(4): 267–273.

Leite, G.G., J.V. Figueirôa, T.C. Almeida, J.L. Valões, W.F. Marques et al. 2016. Production of rhamnolipids and diesel oil degradation by bacteria isolated from soil contaminated by petroleum. Biotechnology Progress 32(2): 262–270.

Li, Z., Y. Zhang, J. Lin, W. Wang, S. Li et al. 2019. High-yield di-rhamnolipid production by *Pseudomonas aeruginosa* YM4 and its potential application in MEOR. Molecules 24(7).

Liu, Z.-F., G.-M. Zeng, H. Zhong, X.-Z. Yuan, L. Jiang et al. 2011. Effect of saponins on cell surface properties of *Penicillium simplicissimum*: Performance on adsorption of cadmium(II). Colloids and Surfaces B: Biointerfaces 86(2): 364–369.

Liu, Z., Z. Zeng, G. Zeng, J. Li, H. Zhong et al. 2012. Influence of rhamnolipids and Triton X-100 on adsorption of phenol by *Penicillium simplicissimum*. Bioresource Technology 110: 468–473.

Magalhães, L. and M. Nitschke. 2013. Antimicrobial activity of rhamnolipids against *Listeria monocytogenes* and their synergistic interaction with nisin. Food Control 29(1): 138–142.

Malviya, D., P.K. Sahu, U.B. Singh, S. Paul, A. Gupta et al. 2020. Lesson from ecotoxicity: Revisiting the microbial lipopeptides for the management of emerging diseases for crop protection. International Journal of Environmental Research and Public Health 17(4): 1434.

Maqsood, M.I. and A. Jamal. 2011. Factors affecting the rhamnolipid biosurfactant production. Journal of Biotechnology 8(1): 1–5.

Marchant, R. and I.M. Banat. 2012. Microbial biosurfactants: challenges and opportunities for future exploitation. Trends in Biotechnology 30(11): 558–565.

Martínez-Toledo, A., R. Rodríguez-Vázquez and I.C. Hernández. 2018. Culture media formulation and growth conditions for biosurfactants production by bacteria. International Journal of Environmental Science and Natural Resources 10(3): 117–125.

Mendes, A.N., L.A. Filgueiras, J.C. Pinto and M. Nele. 2015. Physicochemical properties of rhamnolipid biosurfactant from *Pseudomonas aeruginosa* PA1 to applications in microemulsions. Journal of Biomaterials and Nanobiotechnology 6(1): 64–79.

Mia, S., S.S. Dashtbozorg, N.V. Callow and L.K. Ju. 2015. Rhamnolipids as platform molecules for production of potential anti-zoospore agrochemicals. Journal of Agricultural and Food Chemistry 63(13): 3367–3376.

Mnif, I. and D. Ghribi. 2015. Review lipopeptides biosurfactants: mean classes and new insights for industrial, biomedical, and environmental applications. Peptide Science 104(3): 129–147.

Monnier, N., A. Furlan, C. Botcazon, A. Dahi, G. Mongelard et al. 2018. Rhamnolipids from *Pseudomonas aeruginosa* are elicitors triggering *Brassica napus* protection against *Botrytis cinerea* without physiological disorders. Frontiers in Plant Science 9: 1170.

Monteiro, S., G. Sassaki, L. de Souza, J. Meira, J. de Araújo, D. Mitchell, L. Ramos and N. Krieger. 2007. Molecular and structural characterization of the biosurfactant produced by *Pseudomonas aeruginosa* DAUPE 614. Chemistry and Physics of Lipids, 147(1): 1–13.

Mozejko-Ciesielska, J., K. Szacherska and P. Marciniak. 2019. Pseudomonas species as producers of eco-friendly polyhydroxyalkanoates. Journal of Polymers and the Environment 27(6): 1151–1166.

Mukherjee, S., P. Das and R. Sen. 2006. Towards commercial production of microbial surfactants. Trends in Biotechnology 24 (11): 509–515.

Mulligan, C.N. 2009. Recent advances in the environmental applications of biosurfactants. Current Opinion in Colloid & Interface Science 14: 372–378.

Naughton, P.J., R. Marchant, V. Naughton and I.M. Banat. 2019. Microbial biosurfactants: Current trends and applications in agricultural and biomedical industries. Journal of Applied Microbiology 127(1): 12–28.

Nicolò, M.S., M.G. Cambria, G. Impallomeni, M.G. Rizzo, C. Pellicorio et al. 2017. Carbon source effects on the mono/dirhamnolipid ratio produced by *Pseudomonas aeruginosa* L05, a new human respiratory isolate. New Biotechnology 39: 36–41.

Nitschke, M., S.G. Costa and J. Contiero. 2005. Rhamnolipid surfactants: an update on the general aspects of these remarkable biomolecules. Biotechnology Progress 21(6): 1593–1600.

Nitschke, M., S.G. Costa and J. Contiero. 2010. Structure and applications of a rhamnolipid surfactant produced in soybean oil waste. Applied Biochemistry and Biotechnology 160: 2066–2074.

Oliveira, F.J.S., L. Vazquez, N.P. de Campos and F.P. de França. 2009. Production of rhamnolipids by a *Pseudomonas alcaligenes* strain. Process Biochemistry 44(4): 383–389.

Oluwaseun, A., P. Phazang and N. Sarin. 2017. Significance of rhamnolipids as a biological control agent in the management of crops/plant pathogens. Current Trends in Biomedical Engineering & Biosciences 10(3).

Onwosi, C.O. and F.J.C. Odibo. 2012. Effects of carbon and nitrogen sources on rhamnolipid biosurfactant production by *Pseudomonas nitroreducens* isolated from soil. World Journal of Microbiology and Biotechnology 28(3): 937–942.

Parthasarathi, R. and P.K. Sivakumaar. 2009. Effect of different carbon sources on the production of biosurfactant by *Pseudomonas fluorescens* isolated from mangrove forests (Pichavaram), Tamil Nadu, India. Global Journal of Environmental Research (Vol. 3).

Pathaka, A.N. and H. Pranav. 2015. Optimization of rhamnolipid: A new age biosurfactant from *Pseudomonas aeruginosa* MTCC 1688 and its application in oil recovery, heavy and toxic metals recovery. Journal of Bioprocessing and Biotechniques 5: 1–29.

Peix, A., M.H. Ramírez-Bahena and E. Velázquez. 2018. The current status on the taxonomy of *Pseudomonas* revisited: an update. Infection, Genetics and Evolution 57: 106–116.

Pelley, J.W. 2012. Fatty acid and triglyceride metabolism. In Elsevier's Integrated Review Biochemistry (Second, pp. 49–55, 81–88). Elsevier.

Pérez-Armendáriz, B., C. Cal-y-Mayor-Luna, E.G. El-Kassis and L.D. Ortega-Martínez. 2019. Use of waste canola oil as a low-cost substrate for rhamnolipid production using *Pseudomonas aeruginosa.* AMB Express 9(1): 61.

Pinzon, N.M., A.G. Cook and L.-K. Ju. 2013. Continuous rhamnolipid production using denitrifying *Pseudomonas aeruginosa* cells in hollow-fiber bioreactor. Biotechnology Progress 29(2): 352–358.

Pornsunthorntawee, O., P. Wongpanit and R. Rujiravanit. 2010. Rhamnolipid biosurfactants: Production and their potential in environmental biotechnology. In Biosurfactants (pp. 211–221). Springer, New York, NY.

Prabakaran, G., S.L. Hoti, H.S.P. Rao and S. Vijjapu. 2015. Di-rhamnolipid is a mosquito pupicidal metabolite from *Pseudomonas fluorescens* (VCRC B426). Acta tropica 148: 24–31.

Preethi, Usman, T.M.M., J. Rajesh Banu, M. Gunasekaran and G. Kumar. 2019. Biohydrogen production from industrial wastewater: An overview. Bioresource Technology Reports, 7.

Priji, P., S. Sajith, K.N. Unni, R.C. Anderson, S. Benjamin et al. 2016. *Pseudomonas* sp. BUP6, a Novel Isolate from Malabari Goat Produces an Efficient Rhamnolipid Type Biosurfactant, pp. 1–13.

Rahman, K.S.M., T.J. Rahman, S. McClean, R. Marchant, I.M. Banat et al. 2002. Rhamnolipid biosurfactant production by strains of *Pseudomonas aeruginosa* using low-cost raw materials. Biotechnology Progress 18(6): 1277–1281.

Randhawa, K.K.S. 2014. Biosurfactants produced by genetically manipulated microorganisms: Challenges and opportunities. pp. 49–72. *In*: Kosaric, N. and F.V. Sukan (eds.). Biosurfactants: Production and Utilization—Processes, Technologies, and Economics (Vol. 159). CRC Press.

Rikalović, M.G., M.M. Vrvić and I.M. Karadžić. 2015. Rhamnolipid biosurfactant from *Pseudomonas aeruginosa*—From discovery to application in contemporary technology. Journal of the Serbian Chemical Society 80(3): 279–304.

Rodrigues, L., I.M. Banat, J. Teixeira and R. Oliveira. 2006. Biosurfactants: potential applications in medicine. Journal of Antimicrobial Chemotherapy 57(4): 609–618.

Rodrigues, A., J.C. Bordado and R.G. Santos. 2017. Upgrading the glycerol from biodiesel production as a source of energy carriers and chemicals. Energies 10(11): 1817.

Sachdev, D.P. and S.S. Cameotra. 2013. Biosurfactants in agriculture. Applied Microbiology and Biotechnology 97: 1005–1016.

Sajid, M., M.S. Ahmad Khan, S. Singh Cameotra and A. Safar Al-Thubiani. 2020. Biosurfactants: Potential applications as immunomodulator drugs. Immunology Letters 223: 71–77.

Sałek, K. and S.R. Euston. 2019. Sustainable microbial biosurfactants and bioemulsifiers for commercial exploitation. Process Biochemistry 85: 143–155.

Sanchez, L., B. Courteaux, J. Hubert, S. Kauffmann, J.H. Renault et al. 2012. Rhamnolipids elicit defense responses and induce disease resistance against biotrophic, hemibiotrophic, and necrotrophic pathogens that require different signaling pathways in *Arabidopsis* and highlight a central role for salicylic acid. Plant Physiology 160(3): 1630–1641.

Santos, D.K.F., R.D. Rufino, J.M. Luna, V.A. Santos, L.A. Sarubbo et al. 2016. Biosurfactants: Multifunctional biomolecules of the 21st century. International Journal of Molecular Sciences 17(3).

Santoyo, G., M. Orozco-Mosqueda, C. del and M. Govindappa. 2012. Mechanisms of biocontrol and plant growth-promoting activity in soil bacterial species of *Bacillus* and *Pseudomonas*: A review. Biocontrol Science and Technology 22(8): 855–872.

Sarachat, T., O. Pornsunthorntawee, S. Chavadej and R. Rujiravanit. 2010. Purification and concentration of a rhamnolipid biosurfactant produced by *Pseudomonas aeruginosa* SP4 using foam fractionation. Bioresource Technology 101(1): 324–330.

Sha, R., L. Jiang, Q. Meng, G. Zhang, Z. Song et al. 2012. Producing cell-free culture broth of rhamnolipids as a cost-effective fungicide against plant pathogens. Journal of Basic Microbiology 52(4): 458–466.

Sha, R. and Q. Meng. 2016. Antifungal activity of rhamnolipids against dimorphic fungi. Journal of General and Applied Microbiology 62(5): 233–239.

Shao, B., Z. Liu, H. Zhong, G. Zeng, G. Liu et al. 2017. Effects of rhamnolipids on microorganism characteristics and applications in composting: A review. Microbiological Research 200: 33–44.

Sharma, A., R. Jansen, M. Nimtz, B.N. Johri, V. Wray et al. 2007. Rhamnolipids from the rhizosphere bacterium *Pseudomonas* sp. GRP3 that reduces damping-off disease in chilli and tomato nurseries. Journal of Natural Products 70(6): 941–947.

Silva, S.N.R.L., C.B.B. Farias, R.D. Rufino, J.M. Luna, L.A. Sarubbo et al. 2010. Glycerol as substrate for the production of biosurfactant by *Pseudomonas aeruginosa* UCP0992. Colloids and Surfaces B: Biointerfaces 79(1): 174–183.

Silva, V., R.B. Lovaglio, V. Luiz Da Silva, R.B. Lovaglio, H.H. Tozzi et al. 2015. Rhamnolipids: A new application in seeds development. Journal of Medical and Biological Sciences 1(8): 100–106.

Sinumvayo, J.P. and N. Ishimwe. 2015. Agriculture and Food Applications of Rhamnolipids and its Production by *Pseudomonas Aeruginosa* 6(223).

Smyth, T.J.P., M. Rudden, K. Tsaousi and I.M. Banat. 2014. Protocols for the detection and chemical characterisation of microbial glycolipids. pp. 29–60. *In*: Hydrocarbon and Lipid Microbiology Protocols. Springer, Berlin, Heidelberg.

Tan, Y.N. and Q. Li. 2018. Microbial production of rhamnolipids using sugars as carbon sources. Microbial Cell Factories 17(1): 89.

Thavasi, R., R. Marchant and I.M. Banat. 2014. Biosurfactant applications in agriculture. pp. 313–326. *In*: Naim Kosaric and Fazilet Vardar Sukan (eds.). Biosurfactants: Production and Utilization—Processes, Technologies, and Economics. CRC Press.

Tiso, T., S. Thies, M. Müller, L. Tsvetanova, L. Carraresi et al. 2017a. Rhamnolipids: Production, performance, and application. Consequences of Microbial Interactions with Hydrocarbons, Oils, and Lipids: Production of Fuels and Chemicals. Handbook of Hydrocarbon and Lipid Microbiology (pp. 587–622). Springer International Publishing.

Tiso, T., R. Zauter, H. Tulke, B. Leuchtle, W.J. Li et al. 2017b. Designer rhamnolipids by reduction of congener diversity: Production and characterization. Microbial Cell Factories 16(1): 1–14.

Trummler, K., F. Effenberger and C. Syldatk. 2003. An integrated microbial/enzymatic process for production of rhamnolipids and L-(+)-rhamnose from rapeseed oil with *Pseudomonas* sp. DSM 2874. European Journal of Lipid Science and Technology 105(10): 563–571.

Varjani, S.J. and V.N. Upasani. 2017. Critical review on biosurfactant analysis, purification and characterization using rhamnolipid as a model biosurfactant. Bioresource Technology 232: 389–397.

Vatsa, P., L. Sanchez, C. Clement, F. Baillieul, S. Dorey et al. 2010. Rhamnolipid biosurfactants as new players in animal and plant defense against microbes. International Journal of Molecular Sciences 11(12): 5095–5108.

Wei, Y.H., C.L. Chou and J.S. Chang. 2005. Rhamnolipid production by indigenous *Pseudomonas aeruginosa* J4 originating from petrochemical wastewater. Biochemical Engineering Journal 27(2): 146–154.

Williams, P. and M. Cámara. 2009. Quorum sensing and environmental adaptation in *Pseudomonas aeruginosa*: A tale of regulatory networks and multifunctional signal molecules. In Current Opinion in Microbiology, 12(2): 182–191.

Wisplinghoff, H. 2017. *Pseudomonas* sp., *Acinetobacter* sp. and miscellaneous gram-negative bacilli. In Infectious diseases (pp. 1579–1599). Elsevier.

Wu, J.Y., K.L. Yeh, Lu, W. Bin, C.L. Lin, J.S. Chang et al. 2008. Rhamnolipid production with indigenous *Pseudomonas aeruginosa* EM1 isolated from oil-contaminated site. Bioresource Technology 99(5): 1157–1164.

Yan, F., S. Xu, J. Guo, Q. Chen, Q. Meng et al. 2015. Biocontrol of post-harvest *Alternaria alternata* decay of cherry tomatoes with rhamnolipids and possible mechanisms of action. Journal of the Science of Food and Agriculture 95(7): 1469–1474.

Yin, C., S. Qie and N. Sang. 2012. Carbon source metabolism and its regulation in cancer cells. Critical Reviews in Eukaryotic Gene Expression. Begell House Inc.

Yoo, D.-S., B.-S. Lee and E.-K. Kim. 2005. Characteristics of microbial biosurfactant as an antifungal agent against plant pathogenic fungus. Journal of Microbiology Biotechnology 15(6): 1164–1169.

Yuan, X.Z., F.Y. Ren, G.M. Zeng, H. Zhong, H.Y. Fu et al. 2007. Adsorption of surfactants on a *Pseudomonas aeruginosa* strain and the effect on cell surface lypohydrophilic property. Applied Microbiology and Biotechnology 76(5): 1189–1198.

Zezzi do Valle Gomes, Milene and Nitschke, Marcia. 2012. Evaluation of rhamnolipid and surfactin to reduce the adhesion and remove biofilms of individual and mixed cultures of food pathogenic bacteria. Food Control 25: 441–447.

Zhao, F., R. Shi, F. Ma, S. Han, Y. Zhang et al. 2018. Oxygen effects on rhamnolipids production by *Pseudomonas aeruginosa*. Microbial Cell Factories 17(1).

Zwieten, M. Van, G. Stovold and L. Van Zwieten. 2007. Alternatives to copper for disease control in the Australian organic industry. A report for the Rural Industries Research and Development Corporation. Australia.

12

Microbial Biosurfactants

A New Frontier for Safe and Sustainable Agriculture

Shabeena Farooq,[1] *Sabah Fatima,*[1] *Muzafar Zaman,*[1] *Basharat Hamid,*[1,*] *Shah Ishfaq,*[1] *Zahoor Ahmad Baba,*[2] *R Z Sayyed*[3] and *Rahul Datta*[4]

1. Introduction

The current necessity of countries world wide is to improve and sustain agricultural production in order to satisfy the ever rising food demands of the human population. In recent years, natural surfactants developed by living cells have gained much more attention compared to synthetic chemical surfactants, because of growing environmental consciousness and a focus on a healthy society in harmony with the global climate. Biosurfactants are surface active biomolecules, produced by a variety of microorganisms and play a crucial role in different fields. Biosurfactants (BS) are mostly produced by bacteria, yeasts, and fungi that are known to act as green surfactants. They have been swept up as potential alternatives to conventional surfactants and their use in various industrial fields, including oil regeneration, pharmaceuticals, food, cosmetics, agriculture and livestock (Martinez-Trujillo et al. 2015, Bustos-Vázquez et al. 2018). There are also many benefits of using biosurfactants (BS) over conventional surfactants such as are bio-degradable, cost

[1] Department of Environmental Science, University of Kashmir, Hazratbal, Srinagar-190006, Jammu and Kashmir, India.
[2] Division of Basic Science and Humanities, FOA, Wadura, Sher-e-Kashmir University of Agricultural Sciences and Technology-193201, Wadura, Jammu and Kashmir, India.
[3] Department of Microbiology, PSGVP Mandal's Arts, Science and Commerce College, Shahada 425409, Maharashtra, India.
[4] Department of Geology and Pedology, Mendel University in Brno, Czech Republic.
* Correspondence: basharat384@gmail.com

efficient and less toxic and can be generated at low costs from renewable sources. In the stationary process of microbial development, BSs are secondary metabolites synthesized by a range of microorganisms like bacteria, filamentous fungi and yeasts in different substrates, primarily those immiscible in water, such as oils, alkanes, sugars and waste, extracellularly secreted or attached to cellular components (Becerra and Horna 2016, Varjani and Upasani 2017). At present, the growing interest in the use of BSs in agriculture is due to their possible use as an alternative for conventional surfactants in field conditions.

Biosurfactants are structurally diverse compounds classified by: (i) Chemical composition of glycolipids, fatty acids or lipids, polymeric surfactants and surfactant particles (Santos et al. 2016, Varjani and Upasani 2017) and (ii) On the basis of the molecular weight: (a) low molecular compounds (for example: lipopeptides, glycolipides, shorter chain-level acids and phospholipids) and (b) high-level molecular compounds (for example: polymer surfaces, lipopolysaccharide surfactants or bio emulsifiers) (Varjani and Upasani 2017, Kubicki et al. 2019, Fenibo et al. 2019). Glycolipids are the largest class of biosurfactants commonly recognized and studied and these compounds include, in their design, carbohydrates (hydrophilic fraction) connected to long chains of aliphatic or hydroxy aliphatics (hydrophobical fraction) by ester groups. Within the group of glycolipids there are rhamnolipids, trehalolipids and sophorolipids (Vijayakumar and Saravanan 2015, Santos et al. 2016). More awareness is being established by advancement in cell and molecular biology of the underlying functions of biosurfactants in microbial physiology and in bio-degradation. Rhamnolipids (*Pseudomonas*), sophorolipids (*Torulopsis*), lipid mannosylerythritol (mainly *Candida*), surfactin (*Bacillus*), and emulsans (from *Acinetobacter*) are the most frequently used biosurfactants. Their usage in numerous biotechnological applications can reduce the existing environmental emissions caused by conventional surfactants and thereby contribute to sustainability.

The interesting factors of biosurfactants, due to their functions of biodegradability, low toxicity, ecological acceptability and the ability to be naturally occurring in a more biocompatible and biodegradable manner compared with chemical surfactants (Luna et al. 2015); low critical micellar concentration (CMC), biological activity, detergent capacity, emulsifying ability, biocompatibility and digestibility, besides being synthesized from comparatively cheap biomass or industrial waste sources (Luna et al. 2015, Martinez-Trujillo et al. 2015, Santos et al. 2016, Varjani and Upasani 2017, Bustos-Vázquez et al. 2018, Das et al. 2018). Biosurfactants are also not cost-competitive as opposed to their synthetic counterparts, considering their characteristics and possible use in a wide variety of environmental and industrial processes. The physiological role of BSs is not fully understandable in the generating cell yet, however, the goal is to synthesize and release these compounds by means of solubilization and emulsification procedures to promote the transport, absorption and metabolism of hydrophobic compounds, and serve as biocidal agents in survival and competitiveness in ecological niches (Santos et al. 2016). This book chapter therefore, focuses to explore the possible role of biosurfactants from environmental isolates in the promotion of plant growth and for other agricultural applications.

2. Biosurfactants in Agriculture

Improved farming methods, plant propagation, and agrochemicals including pesticides and fertilizers have dramatically improved yields thus causing widespread ecological and environmental harm. Soil contamination from agricultural fertilizers and activities in pest control have increasingly become a major obstacle to the development of a healthy and pollution free environment. The future role and implementation of biosurfactants in agriculture is encouraging when we look through their various aspects. Firstly, for sustainable and profitable agriculture, the soil fertility status is significant. The application of biosurfactants, a multifunctional microbial metabolite, is a strategy that can enhance the supply of nutrients and thereby promote agricultural performance provided the toxic elements such as Cd, Pb and Ni are also taken into account. They are employed to solubilize and mobilize soil nutrients (e.g., Cu, Fe, Mn, and Zn) through micelle formation and make them accessible to plants from nutrient deficient soils through the creation of complexes of metal-biosurfactants. Secondly, biosurfactants from bacteria and fungi can be easily implemented for soil remediation to increase the consistency of agricultural soil. These biomolecules can replace the synthetic surfactants already used by the pesticides industry saving millions of dollars.

Biosurfactants (such as rhamnolipids, sophorolipids and lipopeptides) enhance soil quality and serve as nutrients (Araújo et al. 2019). In addition, root-linked bacteria generate biosurfactants that enhance nutrient supply and absorption and promote the fair distribution of micronutrients and metals that support the plant growth (Singh et al. 2018), and defend against toxic materials and act as a carbon source. Thirdly, the compounds from microbes can exert microbial influence on the pathogens, or through activating the immune system of the plant that protects it from infection (Keswani et al. 2019). Furthermore, plant-associated bacteria, actinomycetes, yeast and fungi produce biosurfactants that are less toxic and environment-friendly and have prominent applications in agriculture. Many of the rhizosphere micro-organisms (i.e., soil microbes controlled by roots of plant) are partners of reciprocal interaction with plants that have a profound impact on them. Various pathways known are used by rhizospheric bacteria that help to promote growth of plants (Bhattacharyya et al. 2020, Rai et al. 2020). Some important applications of biosurfactants are represented in Fig. 1.

2.1 *Soil Quality Improvement*

The existence of organic and inorganic contaminants that inflict abiotic stress to a growing crop plant affects the productivity of agricultural land. Bioremediation processes are needed to improve the consistency of the soil that is saturated with hydrocarbons and heavy metals. Biosurfactants and/or biosurfactant producing microorganisms can potentially be used for hydrocarbon and heavy metal removal (Lal et al. 2018). Since the bioavailability of biosurfactants is proven to boost and biodegrade hydrophobic substances, various technologies such as soil-washing and cleanup technology use biosurfactants to efficiently eliminate hydrocarbons and metals, respectively (Das et al. 2017, Liu et al. 2018, Silva et al. 2018). Biosurfactants intensify an interesting event of hydrophobic contaminant desorption closely bound

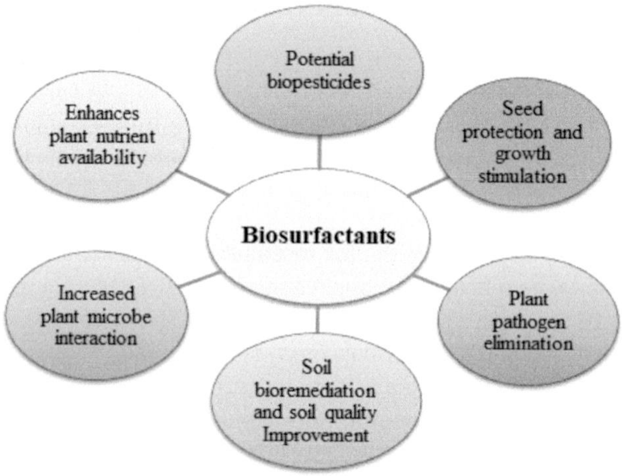

Figure 1. Potential of agricultural applications of biosurfactants.

to soil particles and this is a very important method for the bioremediation process. The biohazardous chemicals like polycyclic aromatic hydrocarbons (PAHs); anthracene (Salamat et al. 2018, Kim et al. 2019), phenanthrene (Golshan et al. 2016, Mehetre et al. 2019), and heavy metals (Lal et al. 2018, Ravindran et al. 2020) are reported for being successfully removed from soil by biosurfactants. In the soil the biosurfactants enhance the bioavailability of different contaminants from clay particles and clay fragments to the cells of microbes (Araújo et al. 2019). These biomolecules then decrease surface-interface stress and allow greater dispersion, which leads to higher emulsifying of soil-eliminated contaminants (Sarubbo et al. 2015) and the soil appears contaminant-free, sustainable and ideal for planting crops and plants (Fenibo et al. 2019, Jimoh and Lin 2019). A number of bio-fungicides have been used as bio-control agents for the control of plant diseases like Fengycins synthesized by *Bacillus subtilis* (Fan et al. 2017), surfactin and fengycin produced by *B. subtilis* (Li et al. 2019a,b) and iturin which is a lipopeptide has also been reported as an antifungal biosurfactant produced by *B. subtilis* (Dunlap et al. 2019). The bio-friendly farming practice of biosurfactants preserves and encourages the growth and development of plants and thus benefits the nature (Köhl et al. 2019). The effectiveness of biosurfactants as agents of plant growth-promotion in soils polluted with PAHs was tested by Bhuyan-Pawar et al. (2015). Some representative examples of various biosurfactants involved in overall improvement of soil status through soil clean-up functions are provided in Table 1.

2.2 *Bioavailability of Plant Nutrients*

In order to fulfill the food demands of the rising world population, modern food production technology is needed. Globally, a vast area (millions of hectares) of agricultural land is deficient in plant-available trace elements such as Cu, Fe, Mn, and Zn (Ali et al. 2020). In nutrient deficient soils the biosurfactants perform a vital role in supplying micronutrients; with trace metals they form chelate and complexes, followed by adsorption, desorption, and removal of the metals from interfaces

Table 1. Improvement of soil status by biosurfactants through clean-up of contaminated soil.

Biosurfactant	**Source microbe**	**Soil biodegradation**	**References**
Rhamnolipids	*Rhodococcus* sp. D-1	Carbendazim	Bai et al. 2017
Anionic Glycolipid	*Pseudomonas* sp. B0406	Endosulfan and methyl parathion	García-Reyes et al. 2017
Rhamnolipid	*Pseudomonas rhizophila* strain S211	Pesticides	Hassen et al. 2018
Rhamnolipid	Lysinibacillus sphaericus-IITR51	γ-hexachlorocyclohexane and endosulfan	Gaur et al. 2019
Surfactin	*Bacillus velezensis-MHNK1*	Atrazine	Jakinala et al. 2019
Rhamnolipids	*Arthrobacter globiformis*	PAH and DDT	Wang et al. 2018
Rhamnolipids	*Pseudomonas aeruginosa*-R25, *Pseudomonas aeruginosa*-R21 *and Pseudomonas aeruginosa*-R7	Petroleum hydrocarbons	Tahseen et al. 2016
Rhamnolipid	Pseudomonas aeruginosa-SR17	Petroleum hydrocarbons	Patowary et al. 2018
Lipopeptide	*Bacillus subtilis*-PB1	Diesel	Mnif et al. 2017
Lipopeptide	*Bacillus cereus* SPL-4	Polycyclic aromatic hydrocarbons (PAHs)	Bezza and Chirwa 2017
Rhamnolipid	*Pseudomonas aeroginosa*	Mercury	Babapoor et al. 2020
Glycolipid	*Burkholderia* sp. Z-90	Heavy metals	Yang et al. 2018
Rhamnolipids	*Acidithiobacillus ferrooxidans and Acidithiobacillus thiooxidans*	Zinc	Diaz et al. 2015
Rhamnolipid	*Pseudomonas aeruginosa*-J4	Copper	Haryanto and Chang 2015

of the soil surface, thus leading to elevated concentrations of micronutrients and bioavailability in soil (Bhandari et al. 2020, Etesami et al. 2020) by lowering the interfacial stress between metals and the soil and forming micelles (Mnif et al. 2018, Mańko-Jurkowska et al. 2019). Biosurfactants can bind directly to absorbed micronutrients under conditions of low interfacial stress and move them to the interface between root zones (Lal et al. 2018). The application of bacterial strains capable of developing biosurfactants is a strategy that increases micronutrient supply in plants (Etesami et al. 2020). Biosurfactants and inorganic ligands are effective and are involved in solubilizing soil micronutrients and trace metals simultaneously (Lal et al. 2018, Etesami et al. 2020). This can also be used to describe desorption, from solid soil surfaces of weakly bound trace metals/micronutrients (Kumar et al. 2018). Biosurfactants can increase the supply of metals such as Cu, Fe, Zn and Mn, to plants in different ways in the soil. Biosurfactants form chelates and complexes with ionic and non ionic metals and free them from the soil according to Le Chatelier's

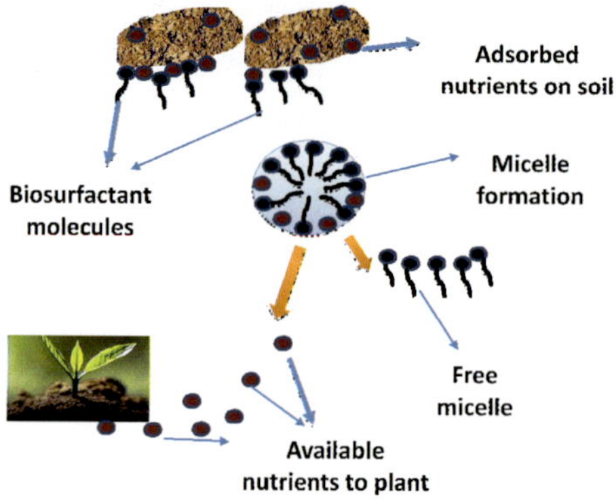

Figure 2. Role of biosurfactants in plant nutrient availability through micellation via adsorption/desorption.

principle (Hogan et al. 2016). Biosurfactants can create complexes with positively charged metals using anionic properties. This allows biosurfactants to be used as metallurgical sequesters, allowing the metal to desorb from soil and polluted water. Soil alkalinity is a significant cause of soil depletion in micronutrients. In this situation, biosurfactants aid to untie the metals from the clay complexes and develop new bio-available complexes for metal biosurfactants (Luna et al. 2016, Hogan et al. 2017) as shown in Fig. 2. A significant way of improving sustainable agriculture would be the method of increasing the absorption of micronutrients through a facility utilizing biosurfactants in the soil around the plant core. Efficiency and abundance of micronutrients in soil (to plants) may be boosted either by bioaugmentation of bacteria that produce biosurfactants or a modification in biosurfactants in soils that are deficient in micronutrients. Several studies indicate that possible bacteria in soil have been bio-enhanced to improve plant growth and soil consistency (Shaikh et al. 2016). For example, Verma et al. (2017) reported *Bacillus* sp. strain J119 as a biosurfactant producing bacteria that was able to promote plant growth and uptake of trace metals in maize, tomato, canola and sudan grass. Moreover, lipophilic meta-rhamnolipid complex formation and its plant absorption were observed and in this case, lipophilic complexes with Cu, Mn, and Zn were formed by mono-rhamnosyl and dirhamnosyl rhamnolipids, encouraging the absorption of such metals by *Brassica napus* and *Triticum durum* roots (Chen et al. 2017). The mechanism involved in bioavailability of nutrients by biosurfactants is depicted in Fig. 2.

2.3 *Biosurfactants as Biopesticides*

In the age of awareness regarding safe foods free of pesticides and the rise in pesticide resistant pathogens, the rhizospheric microorganisms or their metabolites act significantly for managing the plant diseases. The plant pathogens are responsible for significant losses (10 to 40%, depending on crop type) in agricultural production resulting in high economic losses (Savary et al. 2019). Their eco-friendly qualities, high biodegradability and output from renewable energy supplies are of primary importance in the application of bio-surfactants as biopesticides in disease control. Bulk surfactants are used for agriculture to boost microbial and microbial products' antagonistic behaviours (Caulier et al. 2018, Dukare et al. 2019). In the sense of crop defense, rhamnolipids and lipopeptides have been widely investigated. Other biosurfactants, including sophorolipids, MEL and lipids are marginalized for bearing antibiotic features in the direction of phytopathogens (Yoshida et al. 2015, Mnif and Ghribi 2016, Sen et al. 2017, Chen et al. 2020, Penha et al. 2020). Due to the strong balance amongst industrial development, effectiveness and environmental protection, rhamnolipids and lipopeptides are particularly interesting performers in bio-control strategies. Rhizobacteria of the genera *Bacillus* and *Pseudomonas* are recognized as plant growth promoters and fungicidal agents for developing versatile metabolites and their application in agriculture (Shameer et al. 2018). Bio-control is a promising technique focused on the use of organisms to mitigate stress by competition with space pathogens and nutrients, through the production of antimicrobial materials and induction of a natural protective mechanism for the plant (Berg et al. 2017, Bonanomi et al. 2018, Ab Rahman et al. 2018). The pesticidal activities involved in protection of various crop plants are represented in Table 2.

2.3.1 *Antimicrobial Activity*

Many biosurfactants have bio-control importance on the grounds that they have antimicrobial activities against plant pathogens in sustainable agriculture (Fira et al. 2018). The capacity of biosurfactants to identify and inhibit certain pathogens makes them antimicrobial. Biosurfactants, in addition to producing biofilms in their source bacteria, also have the ability to destroy biofilms of other bacteria. The biopesticide value of biosurfactants lies in the fact that the molecular signal leads to defense genes and the accumulation of antimicrobial metabolites (Chakraborty et al. 2020) and these molecular signals are known by microbe-associated molecular patterns (MAMPs). Biopesticide activity of lipopeptides like, fengycin, surfactin and iturin, and from *Bacillus* sp. is caused by interference with membrane structures (Penha et al. 2020). They induce antimicrobial activities against bacteria, oomycetes, fungi and viruses. Surfactins (in which heptapeptides are linked to β-hydroxy fatty acid) show marked antagonistic activity against bacteria and viruses, and limited antifungal activity however, iturins (in which heptapeptides are linked to a β-amino fatty acid), show prominent antagonism against various fungi but to a lesser extent towards viruses and bacteria (Asari et al. 2017). Fengycins in which decapeptides are linked to β-hydroxy fatty acid mostly show efficient activities against yeasts and filamentous fungi (Zihalirwa Kulimushi et al. 2017) by directly binding to cell membrane of fungi resulting in leakage and consequent lysis (Sur et al. 2018).

2.3.2 *Insecticidal Activity*

Biosurfactants can be used to substitute chemical surfactants in conventional insecticides as well as act like insecticides themselves (Jimoh and Lin 2019, de Vasconcelos et al. 2020). Various *in vitro* and *in situ* experiments showed the importance of surfactants in the enhancement of insecticide activities in other processes (Chaudhari et al. 2020). For agricultural applications, much research has been based on glycolipids, in particular rhamnolipids that are biosurfactants acting as insect and mosquito invasion monitor biopesticides. Among lipopeptides, surfactins, bacillomycins fengycins, and iturins are commonly studied for this purpose. Many

Table 2. Biopesticidal potential of microbial biosurfactants on plant pests.

Biosurfactants	**Source organisms**	**Phytopathogens**	**Effects**	**References**
Rhamnolipids	*Pseudomonas* sp. DSM2874	*Glomerella cingulata*	Conidial germination inhibition, Growth inhibition (MIC)	Crouzet et al. 2020
Fengycin A and Iturin A	*Bacillus megaterium* WL-3	*Bacillus megaterium WL-3*	Potato	Wang et al. 2020
Orfamides and Sessilins	*Pseudomonas* sp. CMR12a	*Pythium myriotylum*	cocoyam	Oni et al. 2019
Rhamnolipids	*Pseudomonas aeruginosa*-B5	*Cercospora kikuchii, Cladosporium cucumerinum, Colletotrichum orbiculare, Cylindrocarpon destructans, Magnaporthe grisea, Phytophthora capsici*	Zoospore lysis, spore germination and hyphal growth inhibition	Soltani-Dashtbozorg 2015
Biosurfactant PRO1 (formulation of 25% RLs)	*Pseudomonas aeruginosa*	*Phytophthora cryptogea*	Zoospore lysis, reduction of sporangia formation	Crouzet et al. 2020
Rhamnolipids	*Pseudomonas aeruginosa*	*Botrytis cinerea*	Spore germination and mycelial growth inhibition	Monnier et al. 2018
Semipurified rhamnolipid mixture	*Pseudomonas aeruginosa*	*Leptosphaeria maculans*	Mycelial growth inhibition	Monnier et al. 2020
Liipopeptides fengycins, iturins surfactins and bacillomycins	*Bacillus atrophaeus*-L193	*Rhopalosiphum padi*	affecting cuticle membranes	Rodríguez et al. 2018
Cyclic lipopeptide	*Bacillus tequilensis*-CH	*Anopheles culicifacies*	Larvicidal activity	Pradhan et al. 2018

organisms from the superfamily of aphids and lepidopterna have been documented to be sensitive to such molecules. Biosurfactants are known to prevent various pests by direct growth inhibition (Kissoyan et al. 2019); cuticle damages (Marcelino et al. 2017, Rodríguez et al. 2018) and/or histological damages to larvae (Khedher et al. 2020).

2.4 *Stimulation of Plant Immunity*

Plants have developed complex defense mechanisms to enhance resistance to phytopathogens. Activation of the plant immune system involves invasion patterns (IPs) molecules also known as elicitors which can originate from or be produced by

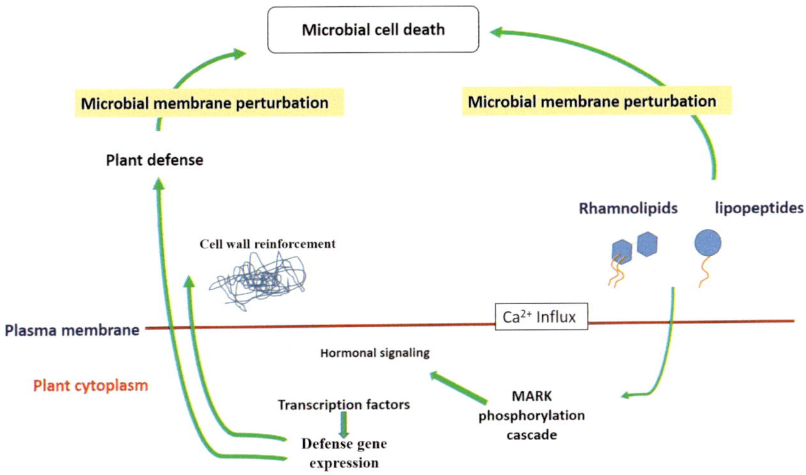

Figure 3. Biosurfactants as inducers of Plant Immunity Responses.

the microbe and the plants (Cook et al. 2015, Kanyuka and Rudd 2019, Schellenberger et al. 2019) that signal the plant about the pathogen invasion and trigger an immune response (Kanyuka and Rudd 2019). Recently, the efficiency of biosurfactants in the defense of the seed and triggering growth has been studied, showing that lipopeptides (Toral et al. 2018) have an efficacy against phytopathogens like *Botrytis cinerea* and that of rhamnolipids (Borah et al. 2016). Biosurfactants are known to be involved in both local as well as systemic immunity responses (Fig. 3). The role of biosurfactants in stimulation of plant immunity in various crop plants against phytopathogens is presented in Table 3.

2.4.1 *Inducers of Local Immunity*

The manner in which plant cells experience biosurfactant induced immune response still remains uncertain. It is hypothesised that they may communicate with plant membrane lipids, considering their amphiphilic existence (Schellenberger et al. 2019). After microbial perception, early signaling events are set up including ion fluxes, reactive oxygen species (ROS) accumulation and phosphorylation cascades (Bigeard et al. 2015). Biosurfactants such as rhamnolipids and lipopeptides activate plant immunity via lipid receptors and/or possible disruptions in the membrane lipid-raft structure (Farace et al. 2015). Insertion of rhamnolipids into a fat bilayer does not alter lipid dynamics strongly, but phytosterols may have a detrimental effect on the effect of glycolipids on the destabilization of plant plasma membrane and these slight shifts in lipid dynamics may be related to the induction of plant defense (Monnier et al. 2019). It has been reported that when used alone, rhamnolipids result in antioxidant reactions in cherry tomato fruit that lead to a substantial reduction of fungal disease (Yan et al. 2015) and this safety is attributed to a higher activation of defense-related enzymes. Although most research on glycolipid biosurfactants has focused on their antimicrobial and antifouling behaviors, rhamnolipids have recently been found to promote plant-inherent immunity as well (Crouzet et al. 2020). After plant sensing, early signals like ROS accumulation in grapevine and *Brassica napus*

Table 3. Biosurfactants in stimulation of plant immunity against various phytopathogens.

Biosurfactant	**Source**	**Immunity against sp.**	**Crop**	**References**
Rhamnolipid	*Pseudomonas aeruginosa*	*Zymoseptoria tritici*	Wheat	Platel et al. 2021
Fengycin	*B. amyloliquefaciens* FZB42	*S. sclerotiorum*	Tomato	Farzand et al. 2019
Orfamide	*Pseudomonas* sp. CMR12a	*R. solani* AG2-2	Bean	Ma et al. 2016
Orfamide	*Pseudomonas* sp. CMR12a	*Cochliobolus miyabeanus*	Rice	Ma et al. 2017
Surfactin and Iturin A	*B. amyloliquefaciens* S13-3	*Colletotrichum gloeosporioides*	Strawberry	Yamamoto et al. 2015
Surfactin	*B. amyloliquefaciens* S499	*Zymoseptoria tritici*	Wheat	Le Mire et al. 2018
Rhamnolipids	*P. aeruginosa*	*B. cinerea*	Brassica napus	Monnier et al. 2018
Fengycin and surfactin	*Bacillus subtilis* GLB191	*Erysiphe necator*	Grapevine	Li et al. 2019a
Cyclic lipopeptides	*Pseudomonas* sp.	*Magnaporthe oryzae*	Rice	Omoboye et al. 2019
Fengycin and surfactin	Bacillus subtilis BBG111	Magnaporthe oryzae	Rice	Chandler et al. 2015
Surfactin, plipastatin and mycosubtilin	*Bacillus subtilis*	*Bo. cinerea*	Grapevine	Farace et al. 2015

are activated by rhamnolipids (Monnier et al. 2018) along with a calcium influx and phosphorylation in grapevine. It has been experimented that in Arabidopsis, local tolerance towards *Botrytis filerea*, *Hyaloperonospora arabidopsidis* or *P. syringae* pv tomato mediated by rhamnolipid uses numerous pathways of signaling which rely on the pathogen type (Crouzet et al. 2020).

2.4.2 *Inducers of Plant Systemic Resistance*

Many biosurfactants are also able to stimulate systemic resistance (Crouzet et al. 2020). The system involving systemic resistance generally entails a series of actions starting from recognition at the cell surface of the plant; stimulating early cellular immune-reponses; systemic signaling of hormones; and consequent activation of the defense mechanisms (Pršic´ and Ongena 2020). Lipopeptides have been reported to protect the host plants through the development of induced systemic resistance (ISR) that in turn limits the growth of pathogens such as bacteria, fungi and viruses (Penha et al. 2020). ISR is mostly dependent on Jasmonic acid/ethylene dependent signaling mechanisms. CLPs play a vital role in ISR-eliciting activity (Rahman et al. 2015). The *B. amyloliquefaciens* produced LPs like surfactin were observed to trigger expression of plant defense gene (PDF2.1) which encodes for defensin (host defense peptide) against *R. solani* in lettuce (Chowdhury et al. 2015). Like rhamnolipids, lipoppeptides are able to induce perturbation in plant plasma membrane and thus can cause the triggering of protective mechanisms by a cascade of molecular events (Schellenberger et al. 2019). While plasma membrane receptors experience several elicitors, recent observations of amphiphilic biosurfactants like lipopeptides indicate that lipids in the plant plasma membrane bilayer have been sensed unusually to justify their singular behaviour (Crouzet et al. 2020). CLPs of the iturin community have often recorded induction of plant defenses.

2.5 *Benefits of Plant-microbe Interactions*

Plant associated (epiphytic as well as endophytic) microbes are well known for producing biosurfactants and play vital roles in enhancing plant-microbe interactions (Bee et al. 2019). It is highly essential for microbes to communicate with plant surfaces like the root to give the plants a beneficial impact by rhizobacteria (Backer et al. 2018, Bhat et al. 2018). Rhizobacterial biosurfactants have known antagonistic effects (Kour et al. 2019, Dahiya et al. 2020). The interaction of the plant is needed by microbial factors including motility, the ability to shape biofilms on the root surface and quorum sensing molecules. The biosurfactant (rhamnolipid) producing *Pseudomonas* spp. was recently reported by Wang et al. (2019) that controls the quorum sensing process (cell to cell communication). These natural surfactants therefore constitute essential variables for microorganisms so that they are integrated effectively with roots of the plants and enhance the plant health. In addition the bioavailability of hydrophobic molecules that can function as nutrients is improved by these biosurfactants formed by rhizobacteria. Soil microbial biosurfactants increase the moisture content of the soil and facilitate efficient application of synthetic fertilizers in soil, while promoting the growth of the plant. The investigations and the studies of biosurfactant functions suggest that these green compounds play a crucial role in sustainable agriculture.

3. Conclusions

Biosurfactants have many uses in the agriculture and agrochemical sectors. Biosurfactants are commonly regarded as multi-functional compounds due to their harmless properties as opposed to the synthetic surfactants. Biosurfactants, produced by bacteria, yeast, and fungi, are efficient biomolecules for diverse applications for their potential to be commercially produced at large scales, their low toxicity and high biodegradability and thus possess an appealing industrial application. In the agricultural industry, they have played an important role in improving soil fertility, and for sustainable and efficient agriculture, characterization of soil fertility is important. A multifunctional microbial metabolite, biosurfactant, is an approach which can improve nutrient abundance to facilitate farm success. Biosurfactants could be used for improved treatment of heavy metal contaminated soil and water. They are capable of being used as bio control agents and for removing different phytopathogens, which can increase crop production but can also replace the expensive and environmentally unlikely chemical fertilizers. Thus it can be concluded that application of biosurfactants in agriculture has a promising role in enhancing the agricultural production in a sustainable way. Thus, efforts must be made to make these biomolecules more realistic for diverse and agricultural applications for the health and environmental sustainability of human beings. Furthermore, it is also important to make it readily usable, cost-effective and economically viable, so that all farmers can easily adapt it for their agricultural applications.

References

Ab Rahman, S.F.S., E. Singh, C.M. Pieterse and P.M. Schenk. 2018. Emerging microbial biocontrol strategies for plant pathogens. Plant Science 267: 102–111.

Ali, W., K. Mao, H. Zhang, M. Junaid, N. Xu et al. 2020. Comprehensive review of the basic chemical behaviours, sources, processes, and endpoints of trace element contamination in paddy soil-rice systems in rice-growing countries. Journal of Hazardous Materials, 122720.

Araújo, H.W., R.F. Andrade, D. Montero-Rodríguez, D. Rubio-Ribeaux, C.A.A. da Silva et al. 2019. Sustainable biosurfactant produced by *Serratia marcescens* UCP 1549 and its suitability for agricultural and marine bioremediation applications. Microbial Cell Factories 18(1): 1–13.

Asari, S., M. Ongena, D. Debois, E. De Pauw, K. Chen et al. 2017. Insights into the molecular basis of biocontrol of *Brassica pathogens* by *Bacillus amyloliquefaciens* UCMB5113 lipopeptides. Annals of Botany 120(4): 551–562.

Babapoor, A., R. Hajimohammadi, S.M. Jokar and M. Paar. 2020. Biosensor design for detection of mercury in contaminated soil using rhamnolipid biosurfactant and luminescent bacteria. Journal of Chemistry, 2020.

Backer, R., J.S. Rokem, G. Ilangumaran, J. Lamont, D. Praslickova et al. 2018. Plant growth-promoting rhizobacteria: context, mechanisms of action, and roadmap to commercialization of biostimulants for sustainable agriculture. Frontiers in Pant Science 9: 1473.

Bai, N., S. Wang, R. Abuduaini, M. Zhang, X. Zhu et al. 2017. Rhamnolipid-aided biodegradation of carbendazim by *Rhodococcus* sp. D-1: Characteristics, products, and phytotoxicity. Science of the Total Environment 590: 343–351.

Becerra, L. and M. Horna. 2016. Isolation of biosurfactant producing microorganisms and lipases from wastewaters from slaughterhouses and soils contaminated with hydrocarbons. Scientia Agropecuaria 7(1): 23–31.

Bee, H., M.Y. Khan and R.Z. Sayyed. 2019. Microbial Surfactants and their significance in agriculture. In Plant Growth Promoting Rhizobacteria (PGPR): Prospects for Sustainable Agriculture (pp. 205–215). Springer, Singapore.

Berg, G., M. Köberl, D. Rybakova, H. Müller, R. Grosch et al. 2017. Plant microbial diversity is suggested as the key to future biocontrol and health trends. FEMS Microbiology Ecology 93(5).

Bezza, F.A. and E.M.N. Chirwa. 2017. The role of lipopeptide biosurfactant on microbial remediation of aged polycyclic aromatic hydrocarbons (PAHs)-contaminated soil. Chemical Engineering Journal 309: 563–576.

Bhandari, G. and P. Bhatt. 2021. Concepts and application of plant–microbe interaction in remediation of heavy metals. *In*: Microbes and Signaling Biomolecules Against Plant Stress (pp. 55–77). Springer, Singapore.

Bhat, M.A., V. Kumar, M.A. Bhat, I.A. Wani, F.L. Dar et al. 2020. Mechanistic insights of the interaction of plant growth-promoting rhizobacteria (PGPR) with plant roots toward enhancing plant productivity by alleviating salinity stress. Frontiers in Microbiology, 11.

Bhattacharyya, C., S. Banerjee, U. Acharya, A. Mitra, I. Mallick et al. 2020. Evaluation of plant growth promotion properties and induction of antioxidative defense mechanism by tea rhizobacteria of Darjeeling, India. Scientific Reports 10(1): 1–19.

Bhuyan-Pawar, S., R.P. Yeole, V.M. Sanam, S.P. Bashetti, S.S. Mujumdar et al. 2015. Biosurfactant mediated plant growth promotion in soils amended with polyaromatic hydrocarbons.

Bigeard J., J. Colcombet and H. Hirt. 2015. Signaling mechanisms in pattern-triggered immunity (PTI). Molecular Plant 8(4): 521–539.

Bonanomi, G., M. Lorito, F. Vinale and S.L. Woo. 2018. Organic amendments, beneficial microbes, and soil microbiota: toward a unified framework for disease suppression. Annual review of Phytopathology 56: 1–20.

Borah, S.N., D. Goswami, H.K. Sarma, S.S. Cameotra, S. Deka et al. 2016. Rhamnolipid biosurfactant against *Fusarium verticillioides* to control stalk and ear rot disease of maize. Frontiers in Microbiology 7: 1505.

Bustos-Vázquez, G., A. Vidal-Fontela, X. Vecino-Bello, J.M. Cruz-Freire, A.B. Moldes-Menduiña et al. 2018. Uso de biosurfactantes extraidos de los licores de lavado de maíz para la eliminación de aceite quemado de motor en suelo arenoso. Agrociencia 52(4): 581–591.

Caulier, S., A. Gillis, G. Colau, F. Licciardi, M. Liépin et al. 2018. Versatile antagonistic activities of soil-borne *Bacillus* spp. and *Pseudomonas* spp. against Phytophthora infestans and other potato pathogens. Frontiers in Microbiology 9: 143.

Chakraborty, B.N., U. Chakraborty and K. Sunar. 2020. Induced immunity developed by *Trichoderma* species in plants. In Trichoderma (pp. 125–147). Springer, Singapore.

Chandler, S., N. Van Hese, F. Coutte, P. Jacques, H. Monica et al. 2015. Role of cyclic lipopeptides produced by *Bacillus subtilis* in mounting induced immunity in rice (*Oryza sativa* L.). Physiological and Molecular Plant Pathology 91: 20–30.

Chaudhari, A.K., V.K. Singh, S. Das, J. Prasad, A.K. Dwivedy et al. 2020. Improvement of *in vitro* and *in situ* antifungal, AFB1 inhibitory and antioxidant activity of *Origanum majorana* L. essential oil through nanoemulsion and recommending as novel food preservative. Food and Chemical Toxicology 143: 111536.

Chen, J., X. Liu, S. Fu, Z. An, Y. Feng et al. 2020. Effects of sophorolipids on fungal and oomycete pathogens in relation to pH solubility. Journal of Applied Microbiology 128(6): 1754–1763.

Chen, Y.T., Y. Wang and K.C. Yeh. 2017. Role of root exudates in metal acquisition and tolerance. Current Opinion in Plant Biology 39: 66–72.

Chowdhury, S.P., J. Uhl and R. Grosch. 2015. Cyclic Lipopeptides of *Bacillus amyloliquefaciens* subsp. plantarum Colonizing the lettuce rhizosphere enhance plant defense responses toward the bottom rot pathogen *Rhizoctonia solani*. Molecular Plant Microbe Interactions 28: 984–995.

Cook, D.E., C.H. Mesarich and B.P.H.J. Thomma. 2015. Understanding plant immunity as a surveillance system to detect invasion. Annual Review of Phytopathology 53: 541–563.

Crouzet, J., A. Arguelles-Arias, S. Dhondt-Cordelier, S. Cordelier, J. Pršić et al. 2020. Biosurfactants in plant protection against diseases: rhamnolipids and lipopeptides case study. Frontiers in Bioengineering and Biotechnology, 8.

Dahiya, M.P., M.R. Kaushik and A. Sindhu. 2020. An Introduction to Plant growth promoting rhizobacteria, antifungal metabolites biosynthesis using PRPR with reference to *Pseudomonas* species and it's other characteristics like antagonistic and biocontrolling properties. International Research Journal on Advanced Science Hub 2: 95–100.

Das, A.J., S. Lal, R. Kumar and C. Verma. 2017. Bacterial biosurfactants can be an ecofriendly and advanced technology for remediation of heavy metals and co-contaminated soil. International Journal of Environmental Science and Technology 14(6): 1343–1354.

Dashtbozorg, S.S., S. Miao and L.K. Ju. 2016. Rhamnolipids as environmentally friendly biopesticide against plant pathogen Phytophthora sojae. Environmental Progress & Sustainable Energy 35(1): 169–173.

de Vasconcelos, G.M.D., J. Mulinari, V.K. de Oliveira Schmidt, R.D. Matosinhos, J.V. de Oliveira et al. 2020. Biosurfactants as Green Biostimulants for Seed Germination and Growth.

Diaz, M.A., I.U. De Ranson, B. Dorta, I.M. Banat, M.L. Blazquez et al. 2015. Metal removal from contaminated soils through bioleaching with oxidizing bacteria and rhamnolipid biosurfactants. Soil and Sediment Contamination: An International Journal 24(1): 16–29.

Dukare, A.S., S. Paul, V.E. Nambi, R.K. Gupta, R. Singh et al. 2019. Exploitation of microbial antagonists for the control of postharvest diseases of fruits: a review. Critical Reviews in Food Science and Nutrition 59(9): 1498–1513.

Dunlap, C., M. Bowman and A.P. Rooney. 2019. Iturinic lipopeptide diversity in the *Bacillus subtilis* species group–important antifungals for plant disease biocontrol applications. Frontiers in Microbiology 10: 1794.

Etesami, H. and S.M. Adl. 2020. Plant growth-promoting rhizobacteria (PGPR) and their action mechanisms in availability of nutrients to plants. In Phyto-Microbiome in Stress Regulation (pp. 147–203). Springer, Singapore.

Fan, H., J. Ru, Y. Zhang, Q. Wang, Y. Li et al. 2017. Fengycin produced by *Bacillus subtilis* 9407 plays a major role in the biocontrol of apple ring rot disease. Microbiological Research 199: 89–97.

Farace, G., O. Fernandez, L. Jacquens, F. Coutte, F. Krier et al. 2015. Cyclic lipopeptides from *Bacillus subtilis* activate distinct patterns of defence responses in grapevine. Molecular Plant Pathology 16(2): 177–187.

Farzand, A., A. Moosa, M. Zubair, A.R. Khan, V.C. Massawe et al. 2019. Suppression of sclerotinia sclerotiorum by the induction of systemic resistance and regulation of antioxidant pathways in tomato using fengycin produced by *Bacillus amyloliquefaciens* FZB42. Biomolecules 9(10): 613.

Fenibo, E.O., S.I. Douglas and H.O. Stanley. 2019. A review on microbial surfactants: production, classifications, properties and characterization. Journal of Advances in Microbiology pp. 1–22.

Fira, D., I. Dimkić, T. Berić, J. Lozo, S. Stanković et al. 2018. Biological control of plant pathogens by Bacillus species. Journal of Biotechnology 285: 44–55.

Franco-Marcelino, P.R., V.L. da Silva, R. Rodrigues Philippini, C.J. Von Zuben, J. Contiero et al. 2017. Biosurfactants produced by *Scheffersomyces stipitis* cultured in sugarcane bagasse hydrolysate as new green larvicides for the control of Aedes aegypti, a vector of neglected tropical diseases. Plos one 12(11): e0187125.

García-Reyes, S., G. Yáñez-Ocampo, A. Wong-Villarreal, R.K. Rajaretinam, C. Thavasimuthu et al. 2018. Partial characterization of a biosurfactant extracted from *Pseudomonas* sp. B0406 that enhances the solubility of pesticides. Environmental Technology 39(20): 2622–2631.

Gaur, V.K., A. Bajaj, R.K. Regar, M. Kamthan, R.R. Jha et al. 2019. Rhamnolipid from a *Lysinibacillus sphaericus* strain IITR51 and its potential application for dissolution of hydrophobic pesticides. Bioresource Technology 272: 19–25.

Golshan, M., R. Rezaei Kalantary, S. Nasseri, M. Farzadkia, A. Esrafili et al. 2016. Phenanthrene removal from liquid medium with emphasis on production of biosurfactant. Water Science and Technology 74(12): 2879–2888.

Haryanto, B. and C.H. Chang. 2015. Removing adsorbed heavy metal ions from sand surfaces via applying interfacial properties of rhamnolipid. Journal of Oleo Science 64(2): 161–168.

Hassen, W., M. Neifar, H. Cherif, A. Najjari, H. Chouchane et al. 2018. *Pseudomonas rhizophila* S211, a new plant growth-promoting rhizobacterium with potential in pesticide-bioremediation. Frontiers in Microbiology 9: 34.

Hogan, D.E. 2016. Biosurfactant (monorhamnolipid) complexation of metals and applications for aqueous metalliferous waste remediation.

Hogan, D.E., J.E. Curry, J.E. Pemberton and R.M. Maier. 2017. Rhamnolipid biosurfactant complexation of rare earth elements. Journal of Hazardous Materials 340: 171–178.

Jakinala, P., N. Lingampally, A. Kyama and B. Hameeda. 2019. Enhancement of atrazine biodegradation by marine isolate *Bacillus velezensis* MHNK1 in presence of surfactin lipopeptide. Ecotoxicology and Environmental Safety 182: 109372.

Jimoh, A.A. and J. Lin. 2019. Biosurfactant: A new frontier for greener technology and environmental sustainability. Ecotoxicology and Environmental Safety 184: 109607.

Kanyuka, K. and J.J. Rudd. 2019. Cell surface immune receptors: the guardians of the plant's extracellular spaces. Current Opinion in Plant Biology 50: 1–8.

Keswani, C., H.B. Singh, R. Hermosa, C. García-Estrada, J. Caradus et al. 2019. Antimicrobial secondary metabolites from agriculturally important fungi as next biocontrol agents. Applied Microbiology and Biotechnology 103(23-24): 9287–9303.

Khedher, S.B., H. Boukedi, A. Laarif and S. Tounsi. 2020. Biosurfactant produced by *Bacillus subtilis* V26: a potential biological control approach for sustainable agriculture development. Organic Agriculture, pp. 1–8.

Kim, C.H., D.W. Lee, Y.M. Heo, H. Lee, Y. Yoo et al. 2019. Desorption and solubilization of anthracene by a rhamnolipid biosurfactant from *Rhodococcus fascians*. Water Environment Research 91(8): 739–747.

Kissoyan, K.A., M. Drechsler, E.L. Stange, J. Zimmermann, C. Kaleta et al. 2019. Natural *C. elegans* microbiota protects against infection via production of a cyclic lipopeptide of the viscosin group. Current Biology 29(6): 1030–1037.

Köhl, J., R. Kolnaar and W.J. Ravensberg. 2019. Mode of action of microbial biological control agents against plant diseases: relevance beyond efficacy. Frontiers in Plant Science 10: 845.

Kour, D., K.L. Rana, N. Yadav, A.N. Yadav, A. Kumar et al. 2019. Rhizospheric microbiomes: biodiversity, mechanisms of plant growth promotion, and biotechnological applications for sustainable agriculture. In Plant Growth Promoting Rhizobacteria for Agricultural Sustainability (pp. 19–65). Springer, Singapore.

Kubicki, S., A. Bollinger, N. Katzke, K.E. Jaeger, A. Loeschcke et al. 2019. Marine biosurfactants: biosynthesis, structural diversity and biotechnological applications. Marine Drugs 17(7): 408.

Kumar, P. and P. Dwivedi. 2018. 26 Ameliorative effects. Soil Amendments for Sustainability: Challenges and Perspectives, 363.

Lal, S., S. Ratna, O.B. Said and R. Kumar. 2018. Biosurfactant and exopolysaccharide-assisted rhizobacterial technique for the remediation of heavy metal contaminated soil: An advancement in metal phytoremediation technology. Environmental Technology & Innovation 10: 243–263.

Le Mire, G., A. Siah, M.N. Brisset, M. Gaucher, M. Deleu et al. 2018. Surfactin protects wheat against *Zymoseptoria tritici* and activates both salicylic acid-and jasmonic acid-dependent defense responses. Agriculture 8(1): 11.

Li, Y., M.C. Héloir, X. Zhang, M. Geissler, S. Trouvelot et al. 2019a. Surfactin and fengycin contribute to the protection of a *Bacillus subtilis* strain against grape downy mildew by both direct effect and defence stimulation. Molecular Plant Pathology 20(8): 1037–1050.

Li, Z., Y. Zhang, J. Lin, W. Wang, S. Li et al. 2019b. High-yield di-rhamnolipid production by *Pseudomonas aeruginosa* YM4 and its potential application in MEOR. Molecules 24(7): 1433.

Liu, G., H. Zhong, X. Yang, Y. Liu, B. Shao et al. 2018. Advances in applications of rhamnolipids biosurfactant in environmental remediation: A review. Biotechnology and Bioengineering 115(4): 796–814.

Luna, J.M., R.D. Rufino, A.M.A. Jara, P.P. Brasileiro, L.A. Sarubbo et al. 2015. Environmental applications of the biosurfactant produced by *Candida sphaerica* cultivated in low-cost substrates. Colloids and Surfaces A: Physicochemical and Engineering Aspects 480: 413–418.

Luna, J.M., R.D. Rufino and L.A. Sarubbo. 2016. Biosurfactant from *Candida sphaerica* UCP0995 exhibiting heavy metal remediation properties. Process Safety and Environmental Protection 102: 558–566.

Ma, Z., G.K.H. Hua, M. Ongena and M. Hofte. 2016. Role of phenazines and cyclic lipopeptides produced by *pseudomonas* sp. CMR12a in induced systemic resistance on rice and bean. Environmental Microbiology Reports 8: 896–904.

Ma, Z., M. Ongena and M. Hofte. 2017. The cyclic lipopeptide orfamide induces systemic resistance in rice to *Cochliobolus miyabeanus* but not to *Magnaporthe oryzae*. Plant Cell Reports 36: 1731–1746.

Mańko-Jurkowska, D., E. Ostrowska-Ligęza, A. Górska and R. Głowacka. 2019. The role of biosurfactants in soil remediation.

Martinez-Trujillo, M.A., I.M. Venegas, S.E. Vigueras-Carmona, G. Zafra-Jiménez, M. García-Rivero et al. 2015. Optimization of a bacterial biosurfactant production. Revista Mexicana de Ingeniería Química 14(2): 355–362.

Mehetre, G.T., S.G. Dastager and M.S. Dharne. 2019. Biodegradation of mixed polycyclic aromatic hydrocarbons by pure and mixed cultures of biosurfactant producing thermophilic and thermo-tolerant bacteria. Science of the Total Environment 679: 52–60.

Mnif, I. and D. Ghribi. 2016. Glycolipid biosurfactants: main properties and potential applications in agriculture and food industry. Journal of the Science of Food and Agriculture 96(13): 4310–4320.

Mnif, I., R. Sahnoun, S. Ellouz-Chaabouni and D. Ghribi. 2017. Application of bacterial biosurfactants for enhanced removal and biodegradation of diesel oil in soil using a newly isolated consortium. Process Safety and Environmental Protection 109: 72–81.

Mnif, I., S. Ellouz-Chaabouni and D. Ghribi. 2018. Glycolipid biosurfactants, main classes, functional properties and related potential applications in environmental biotechnology. Journal of Polymers and the Environment 26(5): 2192–2206.

Monnier, N., A. Furlan, C. Botcazon, A. Dahi, G. Mongelard et al. 2018. Rhamnolipids from *Pseudomonas aeruginosa* are elicitors triggering *Brassica napus* protection against *Botrytis cinerea* without physiological disorders. Frontiers in Plant Science 9: 1170.

Monnier, N., M. Cordier, A. Dahi, V. Santoni, S. Guénin et al. 2020. Semipurified Rhamnolipid Mixes Protect *Brassica napus* against *Leptosphaeria maculans* early infections. Phytopathology 110(4): 834–842.

Omoboye, O.O., F.E. Oni, H. Batool, H.Z. Yimer, R. DeMot et al. 2019. *Pseudomonas* cyclic lipopeptides suppress the rice blast fungus magnaporthe oryzae by induced resistance and direct antagonism. Front. Plant Sci. 10: 901.

Oni, F.E., O.F. Olorunleke and M. Höfte. 2019. Phenazines and cyclic lipopeptides produced by *Pseudomonas* sp. CMR12a are involved in the biological control of Pythium myriotylum on cocoyam (Xanthosoma sagittifolium). Biological Control 129: 109–114.

Patowary, R., K. Patowary, M.C. Kalita and S. Deka. 2018. Application of biosurfactant for enhancement of bioremediation process of crude oil contaminated soil. International Biodeterioration & Biodegradation 129: 50–60.

Penha, R.O., L.P. Vandenberghe, C. Faulds, V.T. Soccol, C.R. Soccol et al. 2020. *Bacillus* lipopeptides as powerful pest control agents for a more sustainable and healthy agriculture: recent studies and innovations. Planta 251(3): 1–15.

Platel, R., L. Chaveriat, S. Le Guenic, R. Pipeleers, M. Magnin-Robert et al. 2021. Importance of the C12 carbon chain in the biological activity of rhamnolipids conferring protection in wheat against *Zymoseptoria tritici*. Molecules 26(1): 40.

Pradhan, A.K., A. Rath, N. Pradhan, R.K. Hazra, R.R. Nayak et al. 2018. Cyclic lipopeptide biosurfactant from *Bacillus tequilensis* exhibits multifarious activity. 3 Biotech 8(6): 261.

Pršić, J. and M. Ongena. 2020. Elicitors of plant immunity triggered by beneficial bacteria. Frontiers in Plant Science, 11.

Rahman, A., W. Uddin and N.G. Wenner. 2015. Induced systemic resistance responses in perennial ryegrass against *Magnaporthe oryzae* elicited by semipurified surfactin lipopeptides and live cells of *Bacillus amyloliquefaciens*. Molecular Plant Pathology 16: 546–558.

Rai, P.K., M. Singh, K. Anand, S. Saurabh, T. Kaur et al. 2020. Role and potential applications of plant growth-promoting rhizobacteria for sustainable agriculture. In New and Future Developments in Microbial Biotechnology and Bioengineering (pp. 49–60). Elsevier.

Ravindran, A., A. Sajayan, G.B. Priyadharshini, J. Selvin, G.S. Kiran et al. 2020. Revealing the efficacy of thermostable biosurfactant in heavy metal bioremediation and surface treatment in vegetables. Frontiers in Microbiology 11: 222.

Rodríguez, M., A. Marín, M. Torres, V. Béjar, M. Campos et al. 2018. Aphicidal activity of surfactants produced by *Bacillus atrophaeus* L193. Frontiers in Microbiology 9: 3114.

Salamat, N., R. Lamoochi and F. Shahaliyan. 2018. Metabolism and removal of anthracene and lead by a B. subtilis-produced biosurfactant. Toxicology Reports 5: 1120–1123.

Santos, D.K.F., R.D. Rufino, J.M. Luna, V.A. Santos, L.A. Sarubbo et al. 2016. Biosurfactants: multifunctional biomolecules of the 21st century. International Journal of Molecular Sciences 17(3): 401.

Sarubbo, L.A., R.B. Rocha Jr., J.M. Luna, R.D. Rufino, V.A. Santos et al. 2015. Some aspects of heavy metals contamination remediation and role of biosurfactants. Chemistry and Ecology 31(8): 707–723.

Savary, S., L. Willocquet, S.J. Pethybridge, P. Esker, N. McRoberts et al. 2019. The global burden of pathogens and pests on major food crops. Nature Ecology & Evolution 3(3): 430–439.

Schellenberger, R., M. Touchard, C. Clément, F. Baillieul, S. Cordelier et al. 2019. Apoplastic invasion patterns triggering plant immunity: plasma membrane sensing at the frontline. Mol. Plant Pathol. 20: 1602–1616.

Sen, S., S.N. Borah, A. Bora and S. Deka. 2017. Production, characterization, and antifungal activity of a biosurfactant produced by *Rhodotorula babjevae* YS3. Microbial Cell Factories 16(1): 1–14.

Shaikh, S.S., R.Z. Sayyed and M.S. Reddy. 2016. Plant growth-promoting rhizobacteria: an eco-friendly approach for sustainable agroecosystem. In Plant, soil and microbes (pp. 181–201). Springer, Cham.

Shameer, S. and T.N.V.K.V. Prasad. 2018. Plant growth promoting rhizobacteria for sustainable agricultural practices with special reference to biotic and abiotic stresses. Plant Growth Regulation 84(3): 603–615.

Silva, E.J., P.F. Correa, D.G. Almeida, J.M. Luna, R.D. Rufino et al. 2018. Recovery of contaminated marine environments by biosurfactant-enhanced bioremediation. Colloids and Surfaces B: Biointerfaces 172: 127–135.

Singh, R., B.R. Glick and D. Rathore. 2018. Biosurfactants as a biological tool to increase micronutrient availability in soil: A review. Pedosphere 28(2): 170–189.

Soltani-Dashtbozorg, S. 2015. Microbial Rhamnolipids as Environmentally Friendly Biopesticides: Congener Composition Produced, Adsorption in Soil, and Effects on Phytophthora sojae (Doctoral dissertation, University of Akron).

Sur, S., T.D. Romo and A. Grossfield. 2018. Selectivity and mechanism of fengycin, an antimicrobial lipopeptide, from molecular dynamics. The Journal of Physical Chemistry B 122(8): 2219–2226.

Tahseen, R., M. Afzal, S. Iqbal, G. Shabir, Q.M. Khan, Z.M. Khalid and I.M. Banat. 2016. Rhamnolipids and nutrients boost remediation of crude oil-contaminated soil by enhancing bacterial colonization and metabolic activities. International Biodeterioration & Biodegradation 115: 192–198.

Toral, L., M. Rodríguez, V. Béjar and I. Sampedro. 2018. Antifungal activity of lipopeptides from *Bacillus* XT1 CECT 8661 against *Botrytis cinerea*. Frontiers in Microbiology 9: 1315.

Varjani, S.J. and V.N. Upasani. 2017. Critical review on biosurfactant analysis, purification and characterization using rhamnolipid as a model biosurfactant. Bioresource Technology 232: 389–397.

Verma, C., A.J. Das and R. Kumar. 2017. PGPR-assisted phytoremediation of cadmium: an advancement towards clean environment. Current Science 113(4): 715.

Vijayakumar, S. and V. Saravanan. 2015. Biosurfactants-types, sources and applications. Research Journal of Microbiology 10(5): 181.

Wang, H., W. Chu, C. Ye, B. Gaeta, H. Tao et al. 2019. Chlorogenic acid attenuates virulence factors and pathogenicity of *Pseudomonas aeruginosa* by regulating quorum sensing. Applied microbiology and Biotechnology 103(2): 903–915.

Wang, X., L. Sun, H. Wang, H. Wu, S. Chen et al. 2018. Surfactant-enhanced bioremediation of DDTs and PAHs in contaminated farmland soil. Environmental Technology 39(13): 1733–1744.

Wang, Y., J. Liang, C. Zhang, L. Wang, W. Gao et al. 2020. *Bacillus megaterium* WL-3 Lipopeptides collaborate against *Phytophthora infestans* to control potato late blight and promote potato plant growth. Frontiers in Microbiology 11: 1602.

Yamamoto, S., S. Shiraishi and S. Suzuki. 2015. Are cyclic lipopeptides produced by *Bacillus amyloliquefaciens* S13-3 responsible for the plant defence response in strawberry against Colletotrichum gloeosporioides? Letters in Applied Microbiology 60(4): 379–386.

Yan, F., S. Xu, J. Guo, Q. Chen, Q. Meng et al. 2015. Biocontrol of post-harvest Alternaria alternata decay of cherry tomatoes with rhamnolipids and possible mechanisms of action. Journal of the Science of Food and Agriculture 95: 1469–1474.

Yang, Z., W. Shi, W. Yang, L. Liang, W. Yao et al. 2018. Combination of bioleaching by gross bacterial biosurfactants and flocculation: A potential remediation for the heavy metal contaminated soils. Chemosphere 206: 83–91.

Yoshida, S., M. Koitabashi, J. Nakamura, T. Fukuoka, H. Sakai et al. 2015. Effects of biosurfactants, mannosylerythritol lipids, on the hydrophobicity of solid surfaces and infection behaviours of plant pathogenic fungi. Journal of Applied Microbiology 119(1): 215–224.

Zihalirwa Kulimushi, P., A. Argüelles Arias, L. Franzil, S. Steels, M. Ongena et al. 2017. Stimulation of fengycin-type antifungal lipopeptides in *Bacillus amyloliquefaciens* in the presence of the maize fungal pathogen Rhizomucor variabilis. Frontiers in Microbiology 8: 850.

13

Sustainable Farming with Biosurfactants from Fungi

Ahmed Abdul Haleem Khan[1,*] and
Jelena Popović-Djordjević[2]

1. Introduction

The practices of agriculture serve the sources of food and feeds for living organisms, i.e., humans and animals. The increasing world population and limited availability of fertile land has enhanced demand for extensive cropping systems. Cereal crops are highly prone to abiotic and biotic factors (pathogens, pests and weeds) that reduce the yield. The impacts of climate change have resulted in extensive damage to cropping environments due to resistance among the pathogens and pests. Agriculture has turned into a business for profit due to the green revolution and that in turn has increased the risk of hazardous chemicals to ecosystems. The synthetic agrochemicals (fertilizers, herbicides, fungicides and pesticides) tentatively increased crop yields (Khan et al. 2017, Khan 2019). The long term application of these chemical agents has influenced non-target beneficial organisms and developed the resistant forms of harmful target organisms. The synthetic agrochemicals are derivatives of non-renewable fossil energy refineries, i.e., petrochemicals that possess complex polymeric structures in nature that persist in soil for a long time as recalcitrants or xenobiotics. The prolonged persistence of chemicals interfere with the environment and create hazards to microscopic and macroscopic living forms. The fossil sources are limited and need to be replenished on a timely basis to maintain requisite levels for future needs. The search for alternative approaches has highlighted the need for sustainable practices for the next generation of the green revolution. Its time to search for alternative eco-safe agrochemicals is an ongoing concern around the planet (Khan et al. 2016). The

[1] Dept. of Botany, Telangana University, Nizamabad – 503322, Telangana, India.
[2] Department for Chemistry and Biochemistry University of Belgrade, Faculty of Agriculture, Nemanjina 6, 11080 Belgrade, Serbia.
* Corresponding author: aahaleemkhan@gmail.com

use of biosurfactants in farming systems that have served a variety of purposes in safe practices have been reported.

Microorganisms are exploited for sustainable and eco-friendly crop plant growth promotion. The entomopathogenic and antagonistic microbes have proved to be an alternative to agrochemicals that control pests and diseases of crop plants with no harm to non-target organisms (Khan et al. 2010, 2011, 2014). The plant growth promoting organisms produce secondary metabolites during idiophase of growth. Among the secondary metabolites, biosurfactants are important amphiphilic compounds that possess numerous functional applications (cleaning, wetting, dispersing, emulsifying, foaming and anti-foaming agents) (Table 1). This chapter is intended to review the role of fungi in the production of biosurfactants and their significance.

Table 1. Biosurfactant producing fungi and their applications.

Organism	Biosurfactant	Application	Reference
Candida albicans	Lipopeptide	Microbial enhanced oil recovery (MEOR)	Ashish and Debnath 2018
Mucor hiemalis	Glycolipid	Biosurfactant and bioemulsifier from waste soybean oil	Ferreira et al. 2020
Candida sphaerica		Remediation of wastewater and soil contaminated with heavy metals	Luna et al. 2016
Candida utilis	Carboxylic acid	Formulation of cookies	Ribeiro et al. 2020
Aureobasidium pullulans	Polyol lipid	Production, growing and morphological aspects	Brumano et al. 2017
Trametes versicolor	Lipoprotein	Biosurfactant from olive mill waste	Lourenco et al. 2018
Candida lipolytica	Carboxylic acid	Cleaning oil spills	Santos et al. 2016
Leaf and stem endophyte- *Aspergillus niger* and *Glomerell acingulata* from *Piper hispidum* (Piperaceae)		Efficient endophyte for biosurfactant production	Silva et al. 2019
Cutaneotrichosporon mucoides	Sophorolipid	Sustainable production of biosurfactant by yeasts in media with hemicellulosic hydrolysate	Marcelino et al. 2019

2. Biosurfactants

Surfactants and emulsifiers are common terms used in fluid mechanics. The surfactants are soluble compounds that reduce surface tension or interfacial tension that exists between two immiscible liquids with formation aggregate structures, i.e., micelles. The term emulsifier is used for compounds that disperse droplets from one immiscible liquid into another without decrease in surface tension and prevention of

Table 2. Comparison of synthetic and natural surfactants.

	Synthetic Surfactants	**Natural Surfactants**
Source	Petrochemical or Oleochemical	Plant, Animal
Hydrophobic Charged Uncharged	 Anionic or Cationic Non-ionic	Head group with amino acids or saccharide groups
Hydrophobic	Alkyl chain	Alkyl chain Hydrophobic amino acids, triterpenoid, steroidal and steroid alkaloid group

coalescence. The synthetic surfactants originate from petrochemical sources, with low degradability, and unsafe for the environment (Table 2).

The microbial surfactants (biosurfactants) are substitutes for chemical surfactants. In comparison to synthetic surfactants, biosurfactants exhibit biodegradability, low toxicity and stability in conditions, i.e., extreme temperature, salinity, and pH. Biosurfactants are more effective in reducing surface and interfacial tensions, *stabilizing oil-in-water and water-in-oil emulsions than synthetic surfactants.* The biosurfactants are used as detergents, solubilizers, lubricants, food, pharmaceuticals and petrochemicals. Though biosurfactants possess potential applications, their production process is expensive compared to that of synthetic surfactants. 60% of the production costs of biosurfactants need to be spent on culture media. Therefore, numerous attempts have been made to use alternative substrates for the production of biosurfactants (Hafeez et al. 2019). The production process is expected to impact sustainability. The use of substrates from agro-industrial wastes for the production of different biomolecules has been reported in the past. These wastes serve as rich sources of carbohydrates, lipids, and nitrogen (Shekar et al. 2015).

Production of biosurfactants with residues has been reported, i.e., effluents from potato processing and cassava, distilleries, whey, animal fat, molasses, glycerol, corn steep liquor (CSL), wastes from olive oil production and several vegetable oils extractives.

The global market of surfactants was USD 30.64 billion in 2016 and expected to be USD 39.86 billion by 2021. The alternative to synthetic surfactants was biosurfactants. The biosurfactant market accounts for approximately 461991.67 tons and earnings from sales are expected to reach USD 2.6 billion by 2023. Biosurfactants are mainly classified according to their microbial origin and chemical composition (Amaral et al. 2010). In general, biosurfactants possess a hydrophilic portion (carbohydrates, amino acids, cyclic peptides, phosphate groups, carboxylic acids or alcohols) and a hydrophobic portion (long chain fatty acids, hydroxyl fatty acids, α-alkyl-β-hydroxy fatty acids) (Nalini et al. 2020). Based on their chemical structure biosurfactants are classified as: lipopolysaccharides, lipoproteins, complex biopolymers (high molecular weight)—phospholipids, lipopeptides and glycolipids (low molecular weight). The best studied among extracellular glycolipids include—mannosylerythritol lipids, sophorolipids, rhamnolipids, trehalose, lipids and cellobiose lipids (Fig. 1).

The biosurfactants are known to be produced by microbes such as bacteria, yeasts, molds and mushrooms. Biosurfactant producing bacteria includes: *Pseudomonas*

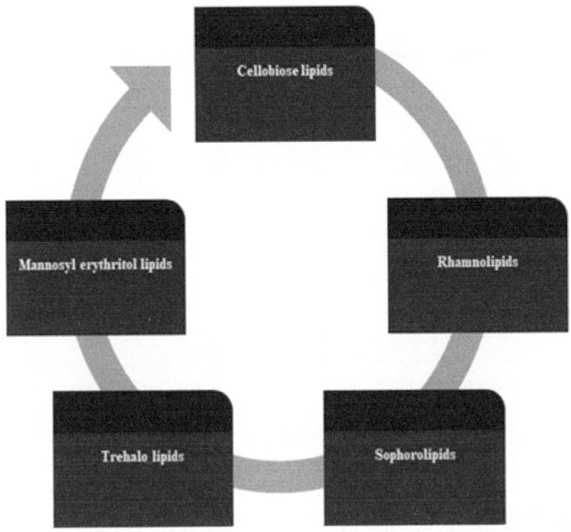

Figure 1. Different classes of biosurfactants.

aeruginosa, *P. putida*, *P. stutzeri*, *Bacillus subtilis*, *B. pumilus*, *B. licheniformis*, *Lactococcuslactis*, *Lactobacillus* sp., *Lachancea thermotolerans*, *Planococcus jake*, *Brevibacterium luteolum*, *Sphingobacterium detergens* and *Paenibacillus* sp., *Streptococcus thermophilus*, and *Nocardioides* sp. from oil reservoirs and marine environments (Bee et al. 2019). The important hurdle in the use of bacteria to produce biosurfactants is their pathogenicity. Though several bacteria are reported to produce biosurfactants, fungi are gaining popularity as potential surfactant producers from different organic wastes of plant and animal origin.

2.1 *Biosurfactants from Fungi*

Fungi are among the predominant heterotrophic eukaryotes that degrade and decompose a variety of substrates to clean the environment. These organisms inhabit the biosphere in a wide range of habitats with microscopic to macroscopic sizes. They act as free living or symbiotic forms that benefit and devour the host by their activities. The biosurfactant producing fungi are shown in Table 3.

Yeasts are known to produce several metabolites with a variety of natural bioactivities, i.e., antimicrobial molecules, biosurfactants, exopolysaccharides, enzymes, lipids, pigments and organic acids. Being unicellular fungi, yeasts received attention from researchers in recent years as they serve as a major source of value added products.

They are easily cultivable in fermentation processes with less labor and low land resources and easy to scale up. Among the yeasts, oleaginous yeasts possess potential in lipid production, commonly referred to as single cell oils (SCO). The SCO cells serve as precursors for biofuels, oleochemicals and food products. Triglycerides (TAGs) are the predominant lipids of SCO (80–90%). TAGs consist of long-chain

Table 3. Biosurfactant producing fungi reported.

Type of Fungi	Biosurfactant producing representatives
Filamentous Fungi	*Absidia corybifera, Aspergillus flavus, A. terreus, A. ustus, A. niger, Beauveria bassiana, Cordyceps javanica, Fusarium* sp., *F. sambucinum, F. fujikuroi, Glomerella cingulata, Penicillium* sp., *P. chrysogenum, Trichoderma harzianum, T. camerunense, Rhizopus arrhizus, Ustilago maydis, U. scitaminea, Xylaria regalis*
Yeasts	*Aureobasidium pullulans, A. thailandense, A. melanogenum, Candida batistae, C. antartica, C. bogorienses, C. bombicola, C. lipolytica, C. ishiwadae, C. ingens, C. floricola, C. utilis, C. sphaerica, C. parapsilosis, C. tropicalis, Cutaneotrichosporon mucoides, Cyberlindnera saturnus, Cryptococcus* sp., *Debaryomyces polymorphus, D. hansenii, Galactomyces pseudocandidum, Geotrichum candidum, Hanseniaspora valbyensis, Kluyveromyces marxianus, Meyerozyma guilliermondii, Moesziomyces antarticus, Pseudozyma rugulosa, P. aphidis, P. fusifornata, P. siamenensis, P. parantartica, P. tsukubabaensis, Rhodotorula glutinis, R. babjevae, Saccharomyces cerevisiae, Sporidiobolus pararoseus, Starmerella bombicola, Torulopsis bombicola, T. petrophilum, T. apicola, Wickerhamomyces anomalus, Wickerhamiella domercqiae, Yarrowia lipolytica*
Mushrooms	*Agaricus bisporus, Trametes versicolor, Pleurotus djamor*

fatty acids that resemble vegetable oil fatty acids. The nitrogen limiting conditions accumulate lipids by yeasts to about 70% of cell dry weight. The process by yeast cells produces lipids in specific conditions and is called "*de novo* lipid accumulation". The *de novo* process is economical and profitable. Yeast cells are grown on inexpensive raw materials (renewable), i.e., industrial wastewater, municipal organic wastes and lignocellulose hydrolysates (Amaral et al. 2010). Yeasts are eukaryotic in cell organization, with rigid cell walls and less prone to damage by biosurfactants and therefore, produce higher concentrations of biosurfactants (Kaur et al. 2017). The different classes of biosurfactant producing yeasts are shown in Table 4.

Table 4. Types of biosurfactants produced by representative fungi.

Chemical composition of biosurfactant	Representative fungi
Sophoroplipids	*Candida bogorienses, Torulopsis bombicola, T. apicola, T. petrophilum*
Mannosyl erythritol lipids	*C.antartica, Pseudozyma aphidis, P. fusifornata, P. siamensis, P. parantartica, P. tsukubabaensis, P. rugulos*
Carbohydrate - Protein - Lipid Complex	*C.lipolytica, C. tropicalis, Debaryomyces polymorphus, Yarrowia lipolytica*
Carbohydrate - Protein - Complex Lipids	*C. lipolytica, Y. lipolytica, Rhodotorula glutinis*
Mannanoprotein Lipids	*S.cerevisiae, Kluveromyces marxianus*

2.1.1 *Rhodotorula babjevae*

Rhodotorula babjevae (oleaginous yeast) was investigated for nitrogen limitation and high carbon content and increased accumulation of intracellular oil content rich in triacylglycerols. The glycolipids from test strains showed a cytotoxic effect against different cancer cell lines. The test strain serves in triacylglycerols production and biosurfactant secretion. The biosurfactants from test strains proved to have therapeutic value and environmental interest (Guerfali et al. 2019). *R. babjevae* was reported to secrete the biosurfactant—sophorolipid (SL). The lactonic and acidic form of sophorolipid was found to exhibit effective antifungal activity against *Colletotrichum gloeosporioides*, *Fusarium verticilliodes*, *Fusarium oxysporum* f. sp. *pisi*, *Corynespora cassiicola*, and *Trichophyton rubrum* (Sen et al. 2017).

2.1.2 *Aureobasidium thailandense*

Aureobasidium thailandense was explored for biosurfactant production. The substrates used were corn steep liquor, sugarcane molasses, and olive oil mill wastewater. Among the low-cost substrates used were olive oil mill wastewater along with yeast extract, and glucose increased biosurfactant production. The molecular structures of biosurfactants secreted from test strains were found similar to lauric acid ester. The oil dispersion assays of biosurfactants from test strains exhibited a better performance than sodium dodecyl sulfate (SDS) (chemical surfactants). The biosurfactants proved to have potential application in the process of bioremediation (Meneses et al. 2017).

2.1.3 *Moesziomyces antarcticus*

The basidiomycetous yeast, *Moesziomyces bullatus* (smut fungus) (Ustilaginales) isolated from the ovary of barnyard grass (*Echinochloa crus-galli*) was reported to degrade certain biodegradable plastics and produce mannosylerythritol lipids (MELs, Fig. 2). The smut fungus in the study was found with similar qualities as the *Moesziomyces antarcticus* (also known as *Pseudozyma antarctica*) (Tanaka et al. 2019).

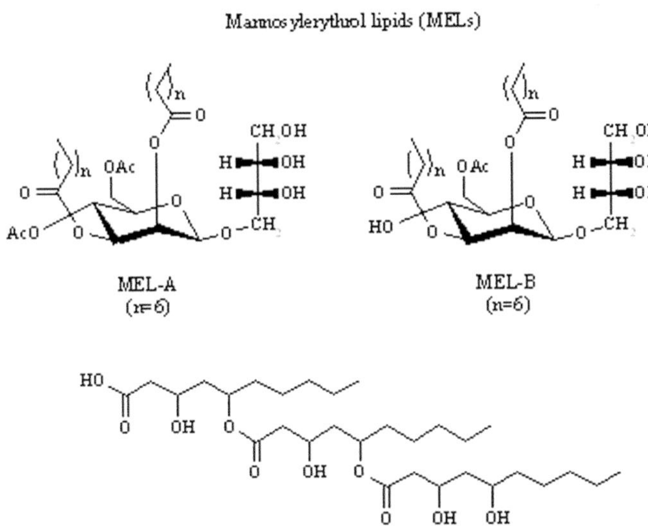

Figure 2. Structures of some glycolipid type biosurfactants (Morita et al. 2007, Bischoff et al. 2015).

2.1.4 *Aureobasidium melanogenum*

Aureobasidium melanogenum (black yeasts) strains were found to produce a wide variety of liamocin structures from xylose, glucose and sucrose. The MALDI-TOF MS analysis showed surfactants as di-, tri-, tetra-, penta- and hexa-acylated C6-liamocins and di-, tri- and tetra-acylated C5-liamocins, Fig. 3 (Saika et al. 2020).

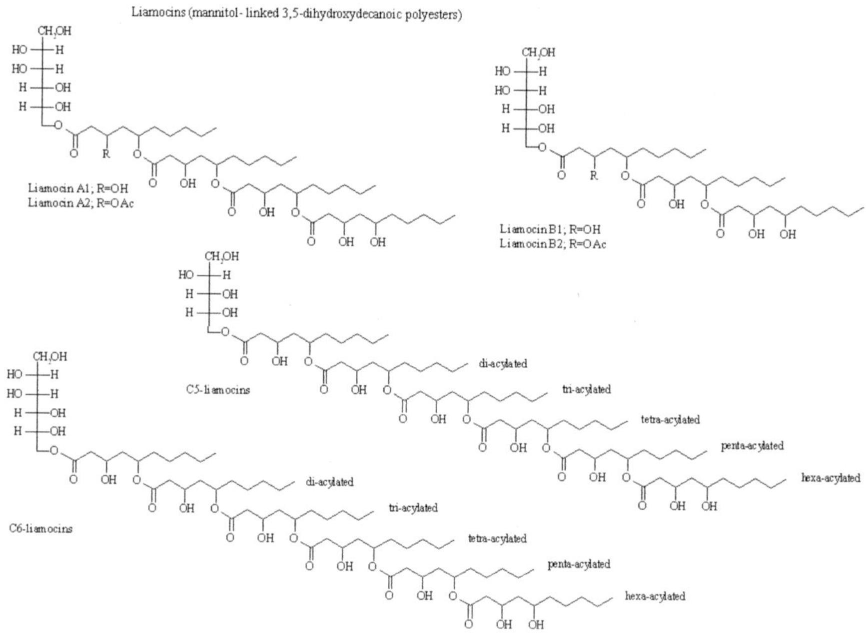

Figure 3. Structures of liamocins (Bischoff et al. 2015, Saika et al. 2020).

2.1.5 *Cryptococcus* sp.

The biosurfactant-producing yeasts were isolated from different foods, waste and sand samples collected from areas contaminated with hydrocarbons in Algeria. The screening tests carried out for isolates were high salinity using emulsification index (E24), drop-collapse test, spread oil technique and hemolytic activity for biosurfactant/bioemulsifier production in mineral salt medium. Among the one hundred isolates, one high-salt-tolerant yeast isolate was the most potent biosurfactants producer, i.e., *Cryptococcus* sp. The partially purified glycolipid biosurfactant from the test yeast isolate showed potential for application in enhanced oil recovery (EOR) (Derguine-Mecheri et al. 2018).

2.1.6 *Candida lipolytica*

The study was investigated on a yeast biosurfactant, Rufisan produced by *Candida lipolytica* to treat contaminants from soil slurries with municipal solid waste. Rufisan was reported to have surfactant properties as it reduces surface tension to 25.29 mN/m and is non-toxic and biodegradable. The findings demonstrated Rufisan as a potential adjuvant to decontaminate petroleum derivatives and free the oil polluted environment (Rufino et al. 2013).

The study investigated the biosurfactant from *Candida lipolytica* as material for remediation of hydrophobic pollutants and heavy metals from the oil industry. The crude biosurfactant removed Cu and Pb (30–40%) from sand and the isolated biosurfactant removed heavy metals (~ 30%) (Santos et al. 2017a,b).

The biosurfactant production was explored by cultivation of *Candida lipolytica* on animal fat and corn steep liquor. The results indicated that the biosurfactant produced had strong potential to clean-up of oil spills at sea and on shorelines (Santos et al. 2014).

The biosurfactant produced by *Candida lipolytica* was characterized as an anionic lipopeptide composed of proteins (50%), lipids (20%), and carbohydrates (8%). The surface tension was reduced from 55 to 25 mN/m in cell-free broth and the yield was 8.0 g/l with CMC (0.03%) (Rufino et al. 2014).

2.1.7 *Smut Fungus*

Pseudozyma antarctica, Ustilago maydis, U. scitaminea, and *P. siamensis* were known to produce Glycolipid biosurfactants, mannosylerythritol lipids (MELs), i.e., MEL-A, MEL-B (Fig. 2), and MEL-C respectively. The biosurfactants were secreted by these strains by using glucose and sucrose as substrate. Among the surfactants, sucrose-derived MELs were found to show interfacial properties, i.e., low critical micelle concentration similar to oil-derived MELs (Morita et al. 2007, Morita et al. 2009a).

A smut fungus *Ustilago scitaminea* from sugar cane (*Saccharum*) was found to accumulate a glycolipid identified as mannosylerythritol lipid-B (MEL-B, 4-O-β-D-(2',3'-di-O-alka(e)noyl-6'-O-acetyl-D-mannopyranosyl)-erythritol). The surfactant MEL-B was found to be produced from a variety of sugars (sucrose, glucose, fructose, and mannose), olive oil and methyl oleate. The MEL secreted by smut fungus using residual oils was complicated for surfactant recovery. Among substrates, sucrose yielded the maximum MEL yield (12.8 g/l) (Morita et al. 2009b).

2.1.8 *Penicillium* sp.

The production of biosurfactants by soil fungi isolated from the Amazon forest was investigated in submerged fermentation. Among the isolates, sixty one strains were found to produce biosurfactants. Among the isolates, *Penicillium* sp. showed the highest emulsification index (54.2%) and the biosurfactant showed stability at 100ºC for 60 min (Sena et al. 2018).

2.1.9 *Candida albicans*

In the study, *Bacillus licheniformis* and *Candida albicans* were investigated for biosurfactant production by determining the emulsification power and surface tension. The extract of crude biosurfactants from supernatant culture growth and the chemical structure was confirmed by FTIR analysis. The bacterial and yeast strains found application in sand pack columns to stimulate oil recovery (El-Sheshtawy et al. 2016).

2.1.10 *Meyerozyma guilliermondii*

The biosurfactant production was investigated by using *M. guilliermondii* through tests, i.e., oil displacement and emulsification (E24) index. The culture medium with yeast extract 1.5% (w/v) and NaCl 3% supported maximum biosurfactant yield (Sharma et al. 2019).

2.1.11 *Yeast consortium*

The synthesis of biosurfactant by yeast consortium (*Rhodotorula* sp., *Debaryomyces hansenii* and *Hanseniaspora valbyensis*) was confirmed by tests, i.e., drop collapse test, methylene blue agar plate method and emulsification test (E24). The test strain was found efficient to remediate Benzo[ghi]perylene (BghiP), a polycyclic aromatic hydrocarbon (PAH) in the presence of nanoparticles (ZnO) (Mandal et al. 2018).

The three yeast cultures, i.e., *Yarrowia* sp., *Barnettozyma californica and Sterigmatomyces halophilus* were used to develop an oleaginous consortium as OYC-Y.BC.SH. The oleaginous consortium accumulated high lipid production and decolorization of Red HE3B dye (above 82%) in the presence of sorghum husk (Ali et al. 2021).

2.1.12 *Xylaria regalis*

In the study on endophytic fungi *Xylaria regalis* isolated from the cones of *Thuja plicata* was reported for biosurfactant production, plant growth-promotion and antagonistic activity against phytopathogens, i.e., *Fusarium oxysporum* and *Aspergillus niger* (Adnan et al. 2018).

2.1.13 *Cyberlindnera saturnus*

The study screened for biosurfactants from (136) marine yeasts with the basic tensio active properties, emulsification and non-hemolytic activities. Among the test strains' isolates, *Cyberlindne rasaturnus* exhibited production of biosurfactant that was characterized as Gal-Gal-Gal-Heptadecanoic acid (named as Cybersan). The biosurfactant was found safe for antimicrobial therapy and other biomedical applications (Balan et al. 2019).

2.1.14 *Candida tropicalis*

The biosurfactant production was investigated by growing the yeast strain *Candida tropicalis* in distilled water (molasses, frying oil and corn steep liquor). The biosurfactant was characterized as an anionic molecule that reduced the surface tension of water (from 70 to 30 mN/m) with no toxic effects on plant seeds and brine shrimp. The biosurfactant showed S removal of heavy metals—Zn, Cu (30 to 80%) and Pb (15%) from contaminated sand (da Rocha et al. 2019).

2.1.15 *Pleurotus djamor*

The study evaluated substrates like sunflower seed shell, grape wastes, potato peels for biosurfactant production by *Pleurotus djamor* in solid state fermentation. The cost of biosurfactant produced was low and proved its use in several applications, i.e., bioremediation, oil recovery and biodegradation (toxic chemicals) (Velioglu and Urek 2015).

2.1.16 *Aspergillus ustus*

The endosymbiotic fungi *Aspergillus ustus* (MSF3) isolated from the marine sponge *Fasciospongia cavernosa* was found to produce a biosurfactant. The biosurfactant produced was characterized as glycolipoprotein and it was found to have a broad spectrum of antimicrobial action. The test strain MSF3 was proved effective for microbial enhanced oil recovery (Kiran et al. 2009).

2.1.17 *Aspergillus* sp.

A marine sponge *Dendrillanigra* associated with *Aspergillus* sp. was isolated and found to have an ability to emulsify hydrocarbons. The biosurfactant production was confirmed with different screening methods (drop collapsing test, emulsification index, hemolytic activity, and oil displacement test). The biosurfactant was characterized as a rhamnolipid that proved its potential against yeast (pathogen: *Candida albicans*) and Gram negative bacteria (Kiran et al. 2010).

2.1.18 *Candida parapsilosis*

In the study on *Candida parapsilosis* the strain was used for different biosurfactant screening tests, i.e., drop collapse, oil spreading, emulsification index and hemolytic activity. The biosurfactant from the test strain showed antibacterial activity against pathogenic *Escherichia coli* and *Staphylococcus aureus*. The biosurfactant produced by *C. parapsilosis* proved to be a broad spectrum antibacterial agent (Garg et al. 2018).

2.1.19 *Starmerella bombicola and Candida floricola*

The isolates from environmental samples were tested for biosurfactant-producing strains (11) using drop-collapse assay and thin-layer chromatography (TLC). The separation patterns of the glycolipid spots corresponded to sophorolipids. The sophorolipids were identified as diacetylatedlactonic and acidic sophorolipids (Fig. 4). The sophorolipids producing strains were identified as *Starmerella bombicola* and *Candida floricola* (Konishi et al. 2016).

2.1.20 *Geotrichum candidum, Galactomyces pseudocandidum and Candida tropicalis*

The study investigated the application of yeast (25) isolates for extracellular biosurfactant production. The efficient strains that produced biosurfactant among the test isolates were *Geotrichum candidum, Galactomyces pseudocandidum* and *Candida tropicalis* (Eldin et al. 2019).

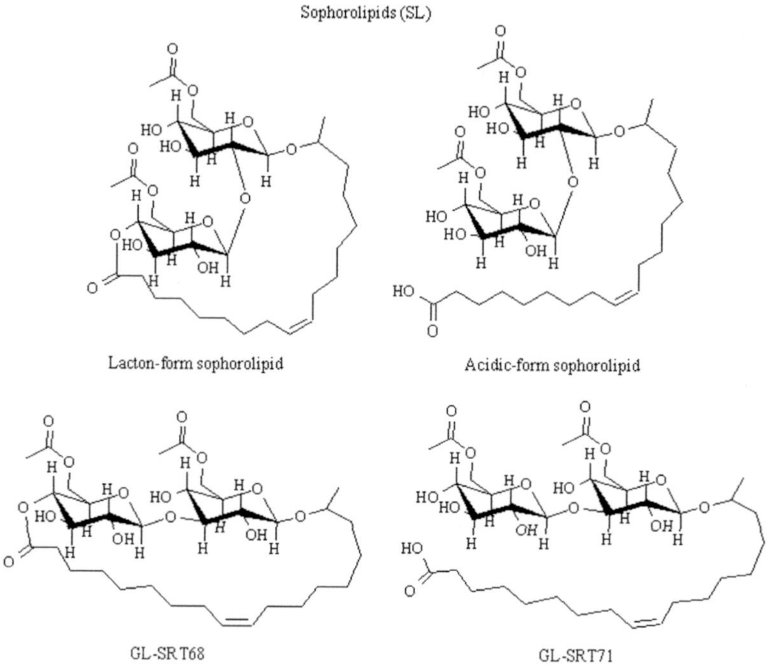

Figure 4. Structures of some sophorolipid biosurfactants (Konishi et al. 2015, 2016).

2.1.21 *Saccharomyces cerevisiae*

The production of extracellular biosurfactants was investigated from *Saccharomyces cerevisiae* by use of cellular stress conditions, i.e., ethanol, hydrogen peroxide, pH, temperature, ultraviolet radiation, and freezing/thawing. The exposure to stress showed higher yeast resistance and yields were high with ultraviolet radiation and hydrogen peroxide stress (Zaparoli et al. 2020).

2.1.22 *Aspergillus niger*

The sequential mutagenesis was induced in *Aspergillus niger* and evaluated for biosurfactant production potential in solid state fermentation of banana stalks powder. The findings proved it to be a cost-effective approach for production of biosurfactant with novel structural and multifunctional characteristics (Asgher et al. 2020).

2.1.23 *Yarrowia lipolytica*

Yarrowia lipolytica (lipolytic oleaginous yeast) and *Bacillus subtilis* (biosurfactant-producing bacteria) were reported for secretion of lipases and biosurfactant with biodegradation of palm oil industrial wastes. The biosurfactant affected saturated fatty acid contents in the yeast lipids that proved suitable for biodiesel feed stocks (Louhasakul et al. 2020).

The biosurfactants (amphisin and viscosinamide) were tested on *Y. lipolytica* (YlLip2) for improved enzyme (lipase) activity. The test strain YlLip2 proved additive in the pharmaceutical, food, cosmetic, and detergent industries (Janek et al. 2020).

2.1.24 *Agaricus bisporus*

Agaricus bisporus (brown mushroom) was investigated for fungal chitosan extraction. The results proved that the fungal chitosan coated liposomes modified with biosurfactant mannosylerythritol lipid A (MEL-A) had an efficient delivery system with increased antioxidant effects (Wu et al. 2019).

2.1.25 *Fusarium fujikuroi*

Fusarium fujikuroi isolate from soil contaminated with hydrocarbons was investigated for biosurfactant production, growth in a low cost medium, and characterized the structure of the biosurfactant. *F. fujikuroi* was reported to reduce surface tension from 72 to 20 mN m^{-1} and the biosurfactant was an α,β-trehalose that belongs to the class of trehalolipids (dos Reis et al. 2018).

2.1.26 *Fusarium sambucinum*

The endophytic strain *Fusarium sambucinum*, isolated from a mangrove forest was exploited for production of biosurfactant and bio emulsifier. The test fungal strain showed biosurfactant ability by decreasing superficial tension to 38 mN/m and exhibited bio emulsifier activity (Martinho et al. 2019).

2.1.27 *Candida sphaerica*

The biosurfactant production was investigated from *Candida sphaerica* by using ground-nut oil refinery residue and corn steep liquor. The crude biosurfactant was found to disperse oil droplets in seawater. The findings indicated the biosurfactant produced by *C. sphaerica* had potential to be used for remediation of oil contaminated water (Luna et al. 2015).

2.1.28 *Sporidiobolus pararoseus*

The bioconversion of biodiesel-derived crude glycerol into carotenoids and lipids was investigated by using *Sporidiobolu spararoseus* (oleaginous red yeast). The study proved the test strain as a source of renewable biodiesel feedstock and natural carotenoids (Chaiyso et al. 2018).

2.1.29 *Starmerella bombicola*

In the study on sophorolipids (SLs) production was investigated using *Starmerella bombicola* in a corncob hydrolysate (CCH) medium derived from lignocellulosic feed stocks as substrates. The SL production was cost-effective (49.2 g/l), with volumetric productivity (12.3 g/l/day) (Konishi et al. 2015).

2.1.30 *Wickerhamomyces anomalus*

The production of biosurfactant was investigated by using *Wickerhamomyces anomalus* with kitchen waste oil and tested against Aedes aegypti larvae. The surfactant from the test strain showed 100% mortality of larvae at a low concentration of 6.25% in 24 hours. The compound was found to be antibacterial (*Bacillus cereus, Salmonella enteritidis, Staphylococcus aureus*, and *Escherichia coli*), anti-adhesive, and antifungal (*Aspergillus, Cercospora, Colletotrichum,* and *Fusarium*). The biosurfactant (50%) inhibited the growth against the test strains (de Andrade et al. 2020).

2.1.31 *Sporidiobolus pararoseus*

Sporidiobolus pararoseus (amylolytic oleaginous red yeast), showed conversion of rice residue from canteen waste into biomass and lipids under shaking flask and upscaling levels. The fatty acids of lipids were found to possess high oleic acid content (60–62%) similar to vegetable oils. The lipids produced were alternative biodiesel feedstock (Chaiyaso et al. 2018).

2.1.32 *Rhizopus arrhizus*

The study was aimed to produce biosurfactants using crude glycerol (3% CG) and corn steep liquor (5% CSL) substrates by inoculating *Rhizopus arrhizus*. The biosurfactant (proteins-38.0%, carbohydrates-35.4%, and lipids-5.5%) produced from *R. arrhizus* reduced the surface tension of water from 72 to 28.8 mN/m and yield was 1.74 g/L. The efficiency of the biosurfactant was demonstrated in the cleaning of the pollutant diesel oil from marine soil (79.4%) (Pele et al. 2019).

2.1.33 *Aureobasidium pullulans*

Aureobasidium pullulans biosurfactants, i.e., liamocin and exophilins were analyzed by hyphenation of high-performance liquid chromatography (HPLC) with high-resolution mass spectrometry (HRMS). The identified biosurfactants were detected with: three liamocins with arabitol as the head group, three mannitol liamocins, and four exophilins (Scholz et al. 2020).

2.1.34 *Candida utilis*

The study was investigated to characterize biosurfactant from *Candida utilis* for use in preparing salad dressings. The mineral medium with 6% glucose and 6% waste frying canola oil was inoculated with *C. utilis*. The crude biosurfactant was found to be resistant to extreme conditions. The biosurfactant was characterized as a carbohydrate-lipid-protein complex that showed application in food emulsions (Campos et al. 2019).

2.1.35 *Aspergillus terreus and Trichoderma harzianum*

The isolates of fungal strains (*Aspergillus terreus* and *Trichoderma harzianum*) from Mediterranean marine sites pervaded by oil spills were investigated for biosurfactant production. The isolated fungi reported to secrete low molecular weight proteins (cerato-platanins), and hydrophobic proteins (fungal surface-active proteins). These proteins were found to stabilize emulsions, and the cerato-platanin from *T. harzianum* lowered the surface tension value compared to a similar protein from *A. terreus* (Pitocchi et al. 2020).

3. Application of Biosurfactants in Agriculture and Farming

Extensive farming of a single crop in all the seasons by the peasant community with the application of synthetic agrochemicals, deteriorates the soil fertility and decreases the yield and becomes an important reason for turning land fertile in nature to barren. To overcome the difficulty of loss of soil fertility, crop rotation as a sustainable approach is suggested. The application of renewable sources in agriculture brought microbial products like fertilizers and pesticides are promoted in the plantation practices. Biosurfactants are reported to control different plant pathogens as biocontrol agents, plant growth promoting agents, insecticidal, mosquitocidal activities, stimulate plant defense and remediation of variety of soil contaminants and enhance the bioavailability of plant nutrients (Singh et al. 2018). The most common phytopathogens controlled by using biosurfactants were *Colletotrichum orbiculare* (cucumber), *C. falcatum* (sugarcane), *Pythium, Phytophthora* sp. (chilli, tomato and strawberry), *Botrytis cinerea* (rapeseed), *Alternaria alternata* (cherry tomato), *Myzus persicae* (tobacco, tomato, spinach, pepper and lettuce) (Singh et al. 2018).

Among the biosurfactants, sophorolipids, cellobiose lipids, mannosyl-erythriol lipids were reported for plant protection against phytopathogenic fungi. The phytopathogenic fungal genera, i.e., *Saccharomyces, Alternaria* (*A. solani, A. alternata, A. tomatophila*), *Aspergillus* (*A. niger*), *Aureobasidium* (*A. pullulans*), *Botrytis* (*B. cinerea*), *Chaetomium* (*C. globosum*), *Penicillium* (*P. digitatum, P. chrysogenum, P. funiculosum*), *Fusarium* (*F. asiaticum, F. austroamericana, F. cerealis, F. oxysporum, F. graminearum*), *Phytophthora* (*P. infestans, P. capsici*), *Gloeophyllum, Cladosporium* and *Schizophyllum* were inhibited by sophorolipids. The cellobiose lipids were active against *Cryptococcus humicola, Pseudozyma fusiformata, P. flocculosa, Sclerotinia sclerotiorum, Phomopsis helianthi, Sphaerotheca fuliginea, Pythium aphanidermatum, Phytophthora infestans* and *Ustilagomaydis*. The rhamnolipid were antifungal against pathogen causing blight—*Phytophthora* sp., anthracnose—*P. capsici, Colletotrichum orbiculare*, brown root rot—*P. cryptogea*, pokkahboeng—*Fusarium sacchari*, wilt—*F. oxysporum* f. sp. *pisi*.

The glycolipid and rhamnolipid biosurfactants were reported to control insect pests and mosquitoes that invade the plants. The biosurfactants served as green pesticides and were effective against aphids, arachnids, box-elder bugs and grasshoppers in the egg and larva stage of the life cycle on host plants (Mnif and Ghribi 2016).

The biosurfactants, i.e., lipopeptides, sophorolipids, rhamnolipids and glycolipids were reported to activate metabolism in plants that support germination and growth. The seed germination accelerated with compounds that increase metabolic rate; these are termed as enhancers or stimulants. The biosurfactants were found effective in pre-sowing seed stimulation which enhanced resistance against pathogens and reduced fertilizer dosage as biostimulants (de Vasconcelos et al. 2020).

4. Conclusion

The time has arrived to reduce the burden of synthetic chemicals in all walks of life and prepare the biosphere suitable for sustaining all organisms in their habitats. Threats to plants, animals and man is increasing due to climate change, anthropogenic activities, industrialization, deforestation and contamination of air, water and soil. Around the globe awareness on organic food and agrochemical free products are widely publicized. The practices for application of renewable sources with lesser harm to ecosystems is the need of the day. The microbial based products are secondary metabolites with a variety of applied values to improve the green environment. The biosurfactants are substitutes for synthetic ones, though expensive, attempts are made to utilize production strategies with cultivation of microbes on cheap substrates. The cheap substrates are inexpensive but the downstream processing costs are 60% of the total expenditure. An attempt is made in the chapter to focus on and review the research carried out on the production of biosurfactants using fungi application in sustainable farming. The near future will be green and eco-safe practices on the planet with regulations by authorities are expected to certainly place biosurfactants in farming systems.

References

Adnan, M., E. Alshammari, S.A. Ashraf, K. Patel, K. Lad et al. 2018. Physiological and molecular characterization of biosurfactant producing endophytic fungi *Xylariaregalis* from the cones of *Thujaplicata* as a potent plant growth promoter with its potential application. BioMed. Research International. Article ID 7362148.

Ali, S.S., R. Al-Tohamy, E. Koutra, A.H. El-Naggar, M. Kornaros et al. 2021. Valorizing lignin-like dyes and textile dyeing wastewater by a newly constructed lipid-producing and lignin modifying oleaginous yeast consortium valued for biodiesel and bioremediation. J. Hazard. Mater. 403: 123575.

Amaral, P.F.F., M.A.Z. Coelho, I.M.J. Marrucho and J.A.P. Coutinho. 2010. Biosurfactants from Yeasts: characteristics, production and application. pp. 236–249. *In*: Sen, R. (eds.). Biosurfactants. Advances in Experimental Medicine and Biology. 672. Springer, New York, NY.

Asgher, M., S. Arshad, S.A. Qamar and N. Khalid. 2020. Improved biosurfactant production from *Aspergillus niger* through chemical mutagenesis: characterization and RSM optimization. SN Applied Sciences 2: 966.

Ashish and M. Debnath (Das). 2018. Application of biosurfactant produced by an adaptive strain of *C.tropicalis* MTCC230 in microbial enhanced oil recovery (MEOR) and removal of motor oil from contaminated sand and water. J. Petr. Sci. Eng. 170: 40–48.

Balan, S.S., C.G. Kumar and S. Jayalakshmi. 2019. Physicochemical, structural and biological evaluation of Cybersan (trigalactomargarate), a new glycolipid biosurfactant produced by amarine yeast, *Cyberlindnera saturnus* strain SBPN-27. Process Biochem. 80: 171–180.

Bee H., M.Y. Khan and R.Z. Sayyed. 2019. Microbial surfactants and their significance in agriculture. pp. 205–215. *In*: Sayyed, R., M. Reddy and S. Antonius (eds.). Plant Growth Promoting Rhizobacteria (PGPR): Prospects for Sustainable Agriculture. Springer, Singapore.

Bischoff, K.M., Leathers, T.D., Price, N.P.J. and Manitchotpisit, P. 2015. Liamocin oil from *Aureobasidium pullulans* has antibacterial activity with specificity for species of *Streptococcus*. J. Antibiot. 68: 642–645.

Brumano, L.P., F.A.F. Antunes, S.G. Souto, J.C. dos Santos, J. Venus et al. 2017. Biosurfactant production by *Aureobasidium pullulans* in stirred tank bioreactor: new approach to understand the influence of important variables in the process. Bioresour. Technol. 243: 264–272.

Campos, J.M., T.L.M. Stamford and L.A. Sarubbo. 2019. Characterization and application of a biosurfactant isolated from *Candida utilis* in salad dressings. Biodegradation 30: 313–324.

Chaiyaso, T. and A. Manowattana. 2018. Enhancement of carotenoids and lipids production by oleaginous red yeast *sporidiobolu spararoseus* KM281507. Prep. Biochem. Biotechnol. 48(1): 13–23.

Chaiyaso, T., W. Srisuwan, C. Techapun, M. Watanabe, S. Takenaka et al. 2018. Direct bioconversion of rice residue from canteen waste into lipids by new amylolytic oleaginous yeast *Sporidiobolus pararoseus* KX709872. Prep. Biochem. Biotechnol. 48(4): 361–371.

da Rocha Junior, R.B., H.M. Meira, D.G. Almeida, R.D. Rufino, J.M. Luna et al. 2019. Application of a low-cost biosurfactant in heavy metal remediation processes. Biodegradation 30(4): 215–233.

de Andrade Teixeira Fernandes, N., A.C. de Souza, L.A. Simoes, G.M. Ferreira Dos Reis, K.T. Souza et al. 2020. Eco-friendly biosurfactant from *Wickerhamomyces anomalus* CCMA 0358 as larvicidal and antimicrobial. Microbiol. Res. 241: 126571.

de Vasconcelos, G.M.D., J. Mulinari, V.K. de Oliveira Schmidt, R.D. Matosinhos, J.V. de Oliveira et al. 2020. Biosurfactants as green biostimulants for seed germination and growth. I.J.R.S.M.B. 6(1): 1–13.

Derguine-Mecheri, L., S. Kebbouche-Gana, S. Khemili-Talbi and D. Djenane. 2018. Screening and biosurfactant/bioemulsifier production from a high-salt-tolerant halophilic *Cryptococcus* strain YLF isolated from crude oil. J. Petr. Sci. Eng. 162: 712–724.

dos Reisa, C.B.L., L.M.B. Morandini, C.B. Bevilacqua, F. Bublitz, G. Ugalde et al. 2018. First report of the production of a potent biosurfactant with α,β-trehalose by *Fusarium fujikuroi* under optimized conditions of submerged fermentation. Brazil J. Microbiol. 49S: 185–192.

Eldin, A.M., Z. Kamel and N. Hossam. 2019. Isolation and genetic identification of yeast producing biosurfactants, evaluated by different screening methods. Microchemical Journal 146: 309–314.

El-Sheshtawy, H.S., I. Aiad, M.E. Osman, A.A. Abo-ELnasr, A.S. Kobisy et al. 2016. Production of biosurfactants by *Bacillus licheniformis* and *Candida albicans* for application in microbial enhanced oil recovery. Egypt. J. Petrol. 25: 293–298.

Ferreira, I.N.S., D.M. Rodriguez, G.M. Campos-Takaki and R.F.d.S. Andrade. 2020. Biosurfactant and bioemulsifier as promising molecules produced by *Mucorhiemalis* isolated from Caatinga soil. Electron. J. Biotechnol. 47: 51–58.

Garg, M., Priyanka and M. Chatterjee. 2018. Isolation, characterization and antibacterial effect of biosurfactant from *Candida parapsilosis*. Biotechnol. Rep. 18: e00251.

Guerfali, M., I. Ayadi, N. Mohamed, W. Ayadi, H. Belghith et al. 2019. Triacylglycerols accumulation and glycolipids secretion by the oleaginous yeast *Rhodotorula babjevae* Y-SL7: Structural identification and biotechnological applications. Bioresour. Technol. 273: 326–334.

Hafeez, F.Y., Z. Naureen and A. Sarwar. 2019. Surfactin: an emerging biocontrol tool for agriculture sustainability. pp. 203–213. *In*: Kumar, A. and V.S. Meena (eds.). Plant Growth Promoting Rhizobacteria for Agricultural Sustainability. Springer, Singapore.

Janek, T., A.M. Mirończuk, W. Rymowicz and A. Dobrowolski. 2020. High-yield expression of extracellular lipase from *Yarrowia lipolytica* and its interactions with lipopeptide biosurfactants: A biophysical approach. Archives of Biochemistry and Biophysics 689: 108475.

Khan, A.A.H., Naseem and S. Samreen. 2010. Antagonistic ability of *Streptomyces* species from composite soil against pathogenic bacteria. BioTechnol. Indian J. 4(4): 222–225.

Khan, A.A.H., Naseem and B. Prathibha. 2011. Screening and potency evaluation of antifungal from soil isolates of *Bacillus subtilis* on selected fungi. Adv. Biotech. 10(7): 35–37.

Khan, A.A.H., A. Sadguna, V. Divya, S. Begum, Naseem et al. 2014. Potential of microorganisms in clean-up the environment. International Journal of Multidisciplinary and Current Research 2: 271–285.

Khan, A.A.H., Naseem and B.V. Vardhini. 2016. Microorganisms and their role in sustainable environment. pp. 60–88. *In*: Prasad, R. (eds.). Environmental Microbiology. IK International Publishing House, New Delhi.

Khan, A.A.H., Naseem and B.V. Vardhini. 2017. Resource-conserving agriculture and role of microbes. pp. 117–152. *In*: Prasad, R. and N. Kumar (eds.). Microbes & Sustainable Agriculture. IK International Publishing House, New Delhi.

Khan, A.A.H. 2019. Plant-bacterial association and their role as growth promoters and biocontrol agents. pp. 389–419. *In*: Sayyed, R. (eds.). Plant growth promoting rhizobacteria for sustainable stress management. Microorganisms for sustainability. 13 Springer, Singapore.

Kiran, G.S., T.A. Hema, R. Gandhimathi, J. Selvin, T.A. Thomas et al. 2009. Optimization and production of a biosurfactant from the sponge-associated marine fungus *Aspergillus ustus* MSF3. Coll. Surf. B Biointerfaces 73: 250–256.

Kiran, G.S., N. Thajuddin, T.A. Hema, A. Idhayadhulla, R.S. Kumar et al. 2010. Optimization and characterization of rhamnolipid biosurfactant from sponge associated marine fungi *Aspergillus* sp. MSF1. Desalination and Water Treatment. 24(1-3): 257–265.

Konishi, M., Y. Yoshida and J. Horiuchi. 2015. Efficient production of sophorolipids by *Starmerella bombicola* using a corncob hydrolysate medium. J. Biosci. Bioeng. 119(3): 317–322.

Konishi, M., M. Fujita, Y. Ishibane, Y. Shimizu, Y. Tsukiyama et al. 2016. Isolation of yeast candidates for efficient sophorolipids production: their production potentials associate to their lineage. Biosci. Biotechnol. Biochem. 80(10): 2058–2064.

Louhasakul, Y., B. Cheirsilp, R. Intasit, S. Maneerat, A. Saimmai et al. 2020. Enhanced valorization of industrial wastes for biodiesel feedstocks and biocatalyst by lipolytic oleaginous yeast and biosurfactant-producing bacteria. Int. J. Biodeterior. Biodegrad. 148: 104911.

Lourenco, L.A., M.D.A. Magina, L.B.B. Tavares, S.M.A.G.U. de Souza, M.G. Roman et al. 2018. Biosurfactant production by *Trametes versicolor* grown on two-phase olive mill waste in solid state fermentation. Environ. Technol. 39(23): 3066–3076.

Luna, J.M., R.D. Rufino, A.M.T. Jara, P.P.F. Brasileiro, L.A. Sarubbo et al. 2015. Environmental applications of the biosurfactant produced by *Candida sphaerica* cultivated in low-cost substrates. Colloids Surf. A Physicochem. Eng. Asp. 480: 413–418.

Luna, J.M., R.D. Rufino and L.A. Sarubbo. 2016. Biosurfactant from *Candida sphaerica* UCP0995 exhibiting heavy metal remediation properties. Process Safety and Environment Protection 102: 558–566.

Mandal, S.K., N. Ojha and N. Das. 2018. Process optimization of benzo[ghi]perylene biodegradation by yeast consortium in presence of ZnO nanoparticles and produced biosurfactant using Box-Behnken design. Front. Biol. 13(6): 418–424.

Marcelino, P.R.F., G.F.D. Peres, R. Teran-Hilares, F.C. Pagnocca, C.A. Rosa et al. 2019. Biosurfactants production by yeasts using sugarcane bagasse hemicellulosic hydrolysate as new sustainable alternative for lignocellulosic biorefineries. Industrial Crops & Products 129: 212–223.

Martinho, V., L.M. Dos Santos Lima, C.A. Barros, V.B. Ferrari, M.R.Z. Passarini et al. 2019. Enzymatic potential and biosurfactant production by endophytic fungi from mangrove forest in Southeastern Brazil. AMB Express 9(1): 130.

Meneses, D.P., E.J. Gudina, F. Fernandes, L.R.B. Gonçalves, L.R. Rodrigues et al. 2017. The yeast-like fungus *Aureobasidium thailandense* LB01 produces a new biosurfactant using olive oil mill wastewater as an inducer. Microbiol. Res. 204: 40–47.

Mnif, I. and D. Ghribi. 2016. Glycolipid biosurfactants: main properties and potential applications in agriculture and food industry. J. Sci. Food Agric. 96(13): 4310–4320.

Morita, T., M. Konishi, T. Fukuoka, T. Imura, H.K. Kitamoto et al. 2007. Characterization of the genus *Pseudozyma* by the formation of glycolipid biosurfactants, mannosylerythritol lipids. FEMS Yeast Res. 7: 286–292.

Morita, T., Y. Ishibashi, T. Fukuoka, T. Imura, H. Sakai et al. 2009a. Production of glycolipid biosurfactants, mannosylerythritol lipids, using sucrose by fungal and yeast strains, and their interfacial properties. Biosci. Biotechn. Biochem. 73(10): 2352–2355.

Morita, T., Y. Ishibashi, T. Fukuoka, T. Imura, H. Sakai et al. 2009b. Production of glycolipid biosurfactants, mannosylerythritol lipids, by a smut fungus, *Ustilago scitaminea* NBRC 32730. Biosci. Biotechn. Biochem. 73(3): 788–792.

Nalini, S., R. Parthasarathi and D. Inbakanadan. 2020. Biosurfactant in food and agricultural application. pp. 75–94. *In*: Gothandam, K.M., S. Ranjan, N. Dasgupta and E. Lichtfouse (eds.). Environmental Biotechnology. Vol. 2. Environmental Chemistry for a Sustainable World. Vol 45. Springer, Cham.

Pele, M.A., D.R. Ribeaux, E.R. Vieira, A.F. Souza, M.A.C. Luna et al. 2019. Conversion of renewable substrates for biosurfactant production by *Rhizopus arrhizus* UCP 1607 and enhancing the removal of diesel oil from marine soil. Electron. J. Biotechnol. 38: 40–48.

Pitocchi, R., P. Cicatiello, L. Birolo, A. Piscitelli, E. Bovio et al. 2020. Cerato-platanins from marine fungi as effective protein biosurfactants and bioemulsifiers. Int. J. Mol. Sci. 21(8): 2913.

Ribeiro, B.G., B.O. de Veras, J.d.S. Aguiar, J.M.C. Guerra, L.A. Sarubbo et al. 2020. Biosurfactant produced by *Candida utilis* UFPEDA1009 with potential application in cookie formulation. Electron. J. Biotechnol. 46: 14–21.

Rufino, R.D., J.M. Luna, P.H.C. Marinho, C.B.B. Farias, S.R.M. Ferreira et al. 2013. Removal of petroleum derivative adsorbed to soil by biosurfactant rufisan produced by *Candida lipolytica*. J. Petr. Sci. Eng. 109: 117–122.

Rufino, R.D., J.M. Luna, G.M. de Campos Takaki and L.A. Sarubbo. 2014. Characterization and properties of the biosurfactant produced by *Candida lipolytica* UCP 0988. Electron. J. Biotechnol. 17: 34–38.

Santos, D.K.F., R.D. Rufino, J.M. Luna, V.A. Santos, A.A. Salgueiro et al. 2013. Synthesis and evaluation of biosurfactant produced by *Candida lipolytica* using animal fat and corn steep liquor. J. Petr. Sci. Eng. 105: 43–50.

Santos, D.K.F., Y.B. Brandao, R.D. Rufino, J.M. Luna, A.A. Salgueiro et al. 2014. Optimization of cultural conditions for biosurfactant production from *Candida lipolytica*. Biocatal. Agric. Biotechnol. 3(3): 48–57.

Santos, D.K.F., A.H.M. Resende, D.G. de Almeida, R. de C.F. S. da Silva, R.D. Rufino et al. 2017a. *Candida lipolytica* UCP0988 biosurfactant: potential as a bioremediation agent and in formulating a commercial related product. Front Microbiol. 8: 767.

Santos, D.K.F., H.M. Meira, R.D. Rufino, J.M. Luna, L.A. Sarubbo et al. 2017b. Biosurfactant production from *Candida lipolytica* in bioreactor and evaluation of its toxicity for application as a bioremediation agent. Process Biochem. 54: 20–27.

Saika, A., T. Fukuoka, S. Mikome, Y. Kondo, H. Habe et al. 2020. Screening and isolation of the liamocin-producing yeast *Aureobasidium melanogenum* using xylose as the sole carbon source. Journal of Bioscience and Bioengineering 129(4): 428–434.

Scholz, K., M. Seyfried, O. Brumhard, L.M. Blank et al. 2020. Comprehensive liamocin biosurfactants analysis by reversed phase liquid chromatography coupled to mass spectrometric and charged-aerosol detection. J. Chromatogr. A. 1627: 461404.

Sen, S., S.N. Borah, A. Bora and S. Deka. 2017. Production, characterization, and antifungal activity of a biosurfactant produced by *Rhodotorula babjevae* YS3. Microb. Cell Fact 16: 95.

Sena, H.H., M.A. Sanches, D.F.S. Rocha, W.O.P.F. Segundo, E.S. de Souza et al. 2018. Production of biosurfactants by soil fungi isolated from the Amazon forest. International Journal of Microbiology 2018, Article ID 5684261.

Sharma, P., S. Sangwan and H. Kaur. 2019. Process parameters for biosurfactant production using yeast *Meyerozyma guilliermondii* YK32. Environ. Monit. Assess. 191: 531.

Shekhar, S., A. Sundaramanickam and T. Balasubramanian. 2015. Biosurfactant producing microbes and their potential applications: a review. Crit. Rev. Environ. Sci. Technol. 45(14): 1522–1554.

Silva, M.E.T., S. Duvoisin, Jr., R.L. Oliveira, E.F. Banhos, A.Q.L. Souza et al. 2019. Biosurfactant production of *Piper hispidum* endophytic fungi. J. Appl. Microbiol. doi: 10.1111/jam.14398.

Singh, P., Y. Patil and V. Rale. 2018. Biosurfactant production: emerging trends and promising strategies. J. Appl. Microbiol. 126: 2–13.

Singh, R., B.R. Glick and D. Rathore. 2018. Biosurfactants as biological tool to increase micronutrient availability in soil: a review. Pedosphere. 28(2): 170–189.

Tanaka, E., M. Koitabashi and H. Kitamoto. 2019. A teleomorph of the Ustilaginalean yeast *Moesziomyces antarcticus* on barnyardgrass in Japan provides bioresources that degrade biodegradable plastics. Antonie Van Leeuwenhoek 112(4): 599–614.

Velioglu, Z. and R.O. Urek. 2015. Optimization of cultural conditions for biosurfactant production by *Pleurotus djamor* in solid state fermentation. Journal of Bioscience and Bioengineering 120(5): 526–531.

Wu, J., Y. Niu, Y. Jiao and Q. Chen. 2019. Fungal chitosan from *Agaricusbisporus* (Lange) Sing. Chaidam increased the stability and antioxidant activity of liposomes modified with biosurfactants and loading betulinic acid. Int. J. Biol. Macromol. 123: 291–299.

Zaparoli, M., N.E. Kreling and L.M. Colla. 2020. Increasing the production of extracellular biosurfactants from *Saccharomyces cerevisiae*. Environ. Qual. Manage 29(4): 51–58.

14

Biosurfactants of Nitrogen Fixers and their Potential Applications

Mohammad Imran Mir,[1] *Humera Quadriya,*[2]
B Kiran Kumar,[1] *Shireen Adeeb Mujtaba Ali,*[2]
Mohamed Yahya Khan[3] and *Bee Hameeda*[2,*]

1. Introduction

The word Surfactant originally is a combination of three terms, 'SURF from surface, ACT from active and ANT of agent'. These compounds decrease the surface and interfacial tension between liquids or a liquid and a solid substance. Petroleum is often used for the chemical synthesis of surfactants. These surfactants are commercially used or exploited in various fields as they possess functionally diversified properties. Agriculture sector uses surfactants in different ways for example, approximately 0.2 million tons reportedly utilized in crop protection and agricultural formulations (Deleu and Paquot 2004). According to Jakia et al. 2020, the global biosurfactant market was evaluated to be 3.99 billion USD during 2016 and would reach 5.52 billion USD by 2022. When applied along with fungicides, insecticides and herbicides, they abet and emphasize surface modifying properties like emulsification, dispersion, spreading, and wetting (Rostas and Blassmann 2009). Commercial production of pesticides currently employs anionic, cationic, amphoteric and nonionic surfactant formulations (Mulqueen 2003). An excessive use of surfactants in formulations may affect the environment and can also be seen as remnants on fruit and vegetable surfaces (Blackwell 2000, Street 1969, Petrovic and Barcelo 2004). Contemplating the above adversities, it is advisable to look for environmentally safe or less harmful surfactants in the pesticide industries

[1] Department of Botany, UCS, Osmania University, Hyderabad, TS, India, 500007.
[2] Department of Microbiology, UCS, Osmania University, Hyderabad, TS, India, 500007.
[3] Kalam Biotech Pvt Ltd, Hyderabad, TS, India, 502032.
* Corresponding author: drhami2009@gmail.com, drhami2009@osmania.ac.in

(Hopkinson et al. 1997). Biosurfactants, as the name suggests are surfactants of biological origin, i.e., from bacteria, yeast or fungi. They are being prioritized due to their excellent properties like low toxicity, biodegradability, higher selectivity, magnificent surface activity, specific activity in extreme conditions (pH, temperature, salinity and others) (Lima et al. 2011b). Biosurfactants are amphiphiles, i.e., they are made up of a polar (hydrophilic) moiety and a non-polar (hydrophobic) group. Because of their amphiphillic structure, biosurfactants not only increase the surface area but also maximize water bio-availability of hydrophobic water-insoluble substances, thereby changing bacterial cell surface properties. Their surface active property outnumbers chemical surfactants as emulsifiers, foaming and dispersing agents (Desai and Banat 1997). Bacterial genera like *Pseudomonas* and *Bacillus* are appreciably studied for the synthesis of biosurfactants and recently nitrogen fixers with biosurfactant features are gaining importance. For several decades, rhizobia were thought to be the only N_2 fixing inhabitants of legume nodules. However, currently a number of α- β- and γ-Proteobacteria have been reported from nodules of legumes and non-leguminous plants like wheat, sugarcane, rice (Saidi et al. 2013, Martinez-Hidalgo and Hirsch 2017, Gopalakrishnan et al. 2017). These non-rhizobial diazotrophic bacteria include *Burkholderia* spp., *Enterobacter* spp., *Pantoea* spp., *Stenotrophomonas* spp., *Paenibacillus* spp., *Ochrobactrum* spp., *Arthrobacter* spp., *Pseudomonas* spp. and more. Biosurfactants synthesized from Cyanobacteria like Dolichospermum, Moorea, Lyngbya and Frankia are *Acidothermus cellulolyticus, Blastococcus saxoobsidens, Geodermatophilus obscurus* and *Modestobacter marinus* among others. Less economic or costly synthesis of biosurfactants is the constraint for its commercial success, which can be endured by valorization of agro-industrial wastes. Because of their potential advantages, biosurfactants are widely used in many industries such as agriculture, soil and water remediation, microbial enhanced oil recovery (MEOR), food industry, cosmetics industry, biomedical science, and nanotechnology (Naughton et al. 2019, Singh et al. 2019). In this chapter, we have provided current information and recent advances on the role of biosurfactants produced by nitrogen fixing bacteria with their application in agriculture.

1.1 Nitrogen Fixation

Nitrogen, is the lightest element of the fifteenth group of the periodic table. Two atoms of nitrogen have triple bond forming colorless, odorless dinitrogen gas which occupies the highest volume, i.e., 78% of dry air/atmosphere, abundant uncombined or unavailable element. This molecular form of nitrogen is made available to living beings by a process known as nitrogen fixation. It can be natural via lightening and fire; biological nitrogen fixation (BNF) by nitrogen fixing microorganisms and chemically/industrially by economically expensive Haber-Bosch process in the ratio 1:46:24 (approximately). Presence or absence of factors like Iron (Fe), Molybdenum (Mo), Phosphorous (P) and Cobalt (Co) may accelerate or limit nitrogen fixation. Nitrogenase enzyme complex can be of three types based on the metal cofactor Mo, V or Fe that bridges Fe and S, and named accordingly: Mo-nitrogenase (Fe-Mo-S), V-nitrogenase (Fe-V-S) and Fe-only nitrogenase (Fe-Fe-S) respectively (Geddes and Oresnik 2016, Humera et al. 2020).

1.2 N2 Fixation ways

a) Chemical process

Haber-Bosch process involves two steps, (i) Synthesis of dihydrogen and carbon dioxide from methane and steam (H_2O) in the presence of nickel catalyst and (ii) The resulting molecules of dihydrogen, three in number, react with a molecule of dinitrogen to synthesize an ammonia molecule, in the presence of high pressure, temperature/heat and an iron catalyst.

b) Biological process

BNF is carried out by microorganisms called nitrogen fixers either (a) Symbiotic (i) aerobic and anaerobic (ii) phototrophic and chemotrophic based on their oxygen requirement, light or chemical requirement respectively or (b) Free-living/Non-symbiotic. Most of the diazotrophs are strict anaerobes or microaerophiles, while few cyanobacterial groups contain heterocysts. Strict aerobes like Azotobacter either maintain a higher respiratory rate in order to tolerate lower concentrations of oxygen inside the cell or modify nitrogenase conformation into its inactive form at greater oxygen levels or hence protect themselves from irreversible damage due to oxygen exposure (Geddes and Oresnik 2016, Abdoulaye et al. 2020).

1.2.1 Mechanism of Nitrogen Fixation

Nitrogenase, a complex protein which comprises Component-I or dinitrogenase-containing Mo, V or Fe and a smaller Component-II or dinitrogenase reductase. Dinitrogenase is a resultant gene product of *nifDK* while homodimer dinitrogenase reductase is of *nifH* gene. Symbiotic nitrogen fixation involves special enzyme complexes and mechanisms. It is an oxygen sensitive enzyme that converts molecular nitrogen to ammonia. It is commonly seen in prokaryotic nitrogen fixers that are physiologically diverse. Along with nitrogenase enzyme, ATP, protons and ferredoxin (Fd), reducing equivalents are essential. Mo-nitrogenase is comparatively less sensitive to oxygen than the other two, i.e., Fe and V-nitrogenase. A gradual progression is observed in the way N_2 fixers protect nitrogenase from oxygen via heterocysts in cyanobacteria, dormant photosystems with zero oxygen synthesis in nocturnal nitrogen fixers (Humera et al. 2020).

1.3 Different ways to Fix Nitrogen Biologically

Unlike legumes, members of family Poaceae like rice, maize and wheat carryout either endophytic or associative nitrogen fixation by colonizing tissue interiors or develop root associations with PGPR or diazotrophic bacteria. Non-symbiotic diazotrophs utilize carbon from the environment, i.e., carbohydrates generated/produced from root exudation and soil organic matter degradation in the rhizospheric soils (Santi carole et al. 2013). The plant-microbe interactions in these regions, plant root interiors produce nitrogen-deficient and energy-sufficient conditions required for N_2 fixation. The root-associated asymbiotic fixation may occur by (a) N_2 fixation independent infection, here bacteria/rhizobia exploit 'crack entry' through cortical cells, invading xylem parenchyma. (b) Rhizobia carryout free-living N_2 fixation in a micro-aerobic environment. (c) Rhizobial hormones carry out plant growth promotion and N_2 fixation simultaneously (Mahmud et al. 2020).

i) By Rhizobia

Symbiotic nitrogen fixation involves. (i) Signal exchange in the rhizosphere. (ii) Root hair invasion. (iii) Bacteroid differentiation. (iv) Nodule development. Legume-rhizobial interaction initiates with chemical signals/inducing molecules like flavonones, betains released from plant root exudates in the rhizospheric region. Transcription of *nod* genes starts as soon as *NodD* recognizes flavonoids by transcriptional activator protein NodD. Other two NodD variants include $NodD_2$—that reacts with the unknown host compound and $NodD_3$—which activates *nod* gene expression in the absence of chemical signals. Approximately 25 proteins encoded by *nod* genes synthesize and export Nod factor (NF), a lipochitooligosaccharide that sends signals to initiate a symbiotic relationship.

Calcium spiking recognizing NF in the nuclear region by calmodulin-dependent protein kinase (CCaMK) induces nodule formation. Phosphorylation and association of IPD3 with CCaMK activates NSP_1 and NSP_2 transcription factors, followed by transcription of early nodulation genes along with NIN and ERN_1 transcription factors (Hoffman et al. 2014). Nodule primordium is formed by cell division in pericycle and cortical cells, as a result of calcium spiking, early nodule expression, cytokine signaling and bacterial infection. The plasma membrane inverts and forms symbiosome membrane which is micro-aerobic, encasing the mature nodule. Bacteria here make use of the nitrogenase enzyme complex and fix dinitrogen, differentiating into bacteroids (Mahmud et al. 2020).

ii) By Frankia

Microorganisms of genus Frankia, from Actinobacteria phylum are reportedly capable of producing biosurfactant besides being diazotrophs. They are developmentally complex and form three cell types: vegetative hyphae, spores located in sporangia, and vesicles. Vesicles are produced under nitrogen-limited, N_2 and NH_4^+ (ammonium) rich conditions and consist of unique lipid-enveloped cellular structures that contain the enzymes responsible for nitrogen fixation. Fatty acids with higher monounsaturated and cyclopropane content than their isoforms in vesicles are greater than in vegetative cells of Frankia, suggesting a similarity in terms of nitrogenase protection against O_2 (Oxygen). Actinorhizal plants are ecologically important as pioneer community plants and have economic value in land reclamation, reforestation, and soil stabilization (Richards et al. 2002, Anders et al. 1989).

Frankia fixes nitrogen in both ways, i.e., via symbiotic species like *EANpec, ACN14a* and *CcI3* and non-symbiotic ones like *Acidothermus cellulolyticus, Blastococcus saxoobsidens, Geodermatophilus obscurus* and *Modestobacter marinus*. According to Santi carole et al. 2013, Frankia associates usually associates with actinorhizal roots. Endophytic association also contributes to nitrogen fixation. Hopanoid lipids envelop these N_2 fixing vesicles thereby protecting inactivation of nitrogenase due to oxygen. Isoflavonoid is reported as one of the signal molecules for the process of nodule formation in actinorhizal plants. Kinase from gene *SymRK* is essential in legumes, actinorhizal plants and arbuscular mycorrhizal fungi in order to synthesize nodules. They are considered to be the less exploited but unique assets to unravel or trace novel and influential biosurfactants (Anita et al. 2013, Medhat et al. 2016).

iii) By Cyanobacteria

Cyanobacteria are directly used as biofertilizers in agriculture fields due to their individual nitrogen fixing ability and also with diverse adsorbents like charcoal and rice straw powder in order to improve yield and soil fertility. They provide available forms of nitrogen, nutrients or growth promoting molecules by exudation and decomposition of microbes. They improve soil compaction, particle/aggregate size, physical and chemical properties and organic matter content (Selvaraj and Sadhana 2020).

1.4 Nitrogen Cycle

Nitrogen cycle is the next important geochemical cycle after carbon cycle, through which nitrogen is circulated in various convertible forms in different ecosystems. The process of conversion of dinitrogen from the atmosphere to bioavailable forms like ammonia involves *nif* genes and is known as nitrogen fixation or diazotrophy. Conversion of ammonia either by autotrophs or heterotrophs into nitrites followed by nitrates is known as nitrification. Single step microbial action completes oxidation of ammonia to nitrates and termed as commamox. Process of ammonia oxidation in oxygen limited or anaerobic conditions is anammox. Few ammonia oxidising microbes are also capable of hydrolyzing urea into ammonia and bicarbonate utilizing enzyme urease. Soils with lower pH contain greater amounts of heterotrophic nitrifiers which utilize organic or inorganic nitrogen compounds as an alternative to ammonia. Major genes involved includes *hzo, hh*-anammox; *pmoA/amoA, norB, hao*-comammox; *amoA, amoB*-ammonification by archea and bacteria respectively; *ure*-urea hydrolysis to ammonia; *nar, nap*-nitrate reduction; *nirS, nirK*-nitrite reduction; *norB*-nitric oxide reduction; *nosZ*-nitric oxide reduction. Decomposition of dead, decaying complex organic matter to simple inorganic bioavailable forms like ammonia, ammonium is termed as ammonification. Process of conversion of nitrite to ammonia with the aid of dehydrogenase and gene *nrfA* is termed as dissimilatory nitrate reduction to ammonia. Process of conversion of available nitrocompounds back to inert or dinitrogen anaerobically involving *nap, nor, nir* and *nos* genes is termed as denitrification. Co-synthesis of dinitrogen along with other nitrocompounds microbiologically is known as codenitrification (Olivia et al. 2017, Florence et al. 2018). Free living rhizospheric nitrogen fixers include Rhizobia, Bradyrhizobia, Rhodobacteria of alphaproteobacteria, Burkholderia, Nitrosospira of betaproteobacteria, Pseudomonas, Xanthomonus of gammaproteobacteria, firmicutes, and cyanobacteria. Examples of endophytes include *Azoarcus* spp., *Herbaspirillum seropedicae* and *Gluconobacter* (Santi carole et al. 2013).

2. Biosurfactants

Microorganisms extracellularly or as part of cell membranes synthesize low molecular weight surface active amphiphillic molecules known as biosurfactants.

They decrease surface tension and interfacial tension at air/water interface and oil/ water interface respectively (Satpute et al. 2010a, Banat et al. 2010).

2.1 Classification of Biosurfactants

Biosurfactants are principally classified based on their microbial origin and chemical composition while chemical surfactants are based on the existing type of polar groups. According to Rosenberg et al. 1999 biosurfactants are classified based on their molecular mass-molecules of (i) low molecular mass that aptly reduce surface and interfacial tension include lipopeptides, glycolipids and phospholipids and (ii) high molecular mass polymers that are potent emulsion-stabilizing agents include polymeric and particulate surfactants. Nearly all the biosurfactants are either anionic or neutral, while few may contain amine groups and are cationic biosurfactants. Hydrophilic moieties of the biosurfactants contain carbohydrates, amino acids, cyclic peptides, phosphate carboxyl acids or alcohol while long chain fatty acids and their derivatives are found in hydrophobic moieties (Desai and Banat 1997, Pacwa-Plociniczak et al. 2011, Nitschke and Coast 2007). The most important groups of biosurfactants and some of their classes are described below:

A. Glycolipids

These are the most familiar biosurfactants that contain carbohydrates in hydrophillic heads and linkage may be with either ether or ester groups. Well-known glycolipids include rhamnolipids, trehalolipids and sophorolipids (Rahman and Gakpe 2008).

i. Rhamnolipids: Rhamnose containing glycolipids were first reported in *Pseudomonas aeruginosa* by Jarvis and Johnson 1949. They are the most researched glycolipids due to their efficient surface activity and physicochemical properties. They contain one or two molecules of rhamnose hooked with one or two β-hydroxydecanoic acid molecules. Karanth et al. 1999 suggested that the 'OH' group of the one acid forms a glycosidic linkage with the rhamnose disaccharide's reducing end while the other acid's 'OH' is involved in ester formation (Desai and Banat 1997, Abalos et al. 2001, Pornsunthorntawee et al. 2009, Abdel-Mawgoud et al. 2009).

ii. Sophorolipids: They are majorly synthesized by yeasts *Torulopsis bombicola* and *Torulopsis petrophilum* (Cooper and Paddock 1984, Hommel et al. 1987). According to Desai and Banat 1997 they are composed of a dimeric sophorose sugar linked to a long chain hydroxyl fatty acid. Sophorolipids usually are a blend of macrolactones and free acid form. Lactone form of sophorolipids is usually adapted in many applications (Hu and Ju 2001).

iii. Trehalolipids: These glycolipids characteristically comprise of disaccharide trehalose linked at C_6 to two β-hydroxy-branched fatty acids and are usually derived from several strains of *Arthrobacter*, *Mycobacterium*, *Brevibacterium*, *Corynebacterium* and *Nocardia* spp. Diversification in structure of a fatty acid and its degree of unsaturation is seen based on the microorganisms synthesizing it (Desai and Banat 1997, Cooper and Zajic 1980).

B. Lipopeptides

These biosurfactants comprise of a hydrophobic fatty acid linked to 7–10 amino acids long peptide as a hydrophillic moiety. Some of the reported cyclic lipopeptides synthesized by *Bacillus* spp. include surfactin, iturin and fengycin families (Kakinuma et al. 1969, Lang 2002).

C. Fatty Acids, Phospholipids and Neutral Lipids

Abundant synthesis of fatty acids and phospholipids (especially in the form of extracellular membrane vesicles) by indefinite bacteria and yeasts when provided with n-alkanes is reported (Rahman and Gakpe 2008). Desai and Banat 1997, suggested that the length of the hydrocarbon chain decides the hydrophillic and lipophilic balance (HLB) structurally. Reduction in surface and interfacial tensions are seen when remarkably active saturated fatty acids are of C_{12}-C_{14} range. Their capability to produce optically clear microemulsions of alkanes in water depict these vesicles to be potential surfactants (Cirigliano and Carman 1985, Rosenberg and Ron 1999). Their surface property is preferably influenced by thee pH and ionic strength of the medium where measurement is carried out.

D. Polymeric biosurfactants

Desai and Banat 1997, reported Emulsan, Liposan, mannoprotein and other polysaccharides-protein complexes as the best studied polymeric biosurfactants. Emulsan, a complex acylated polysaccharide of approximately 1000 kD is of Gram negative bacterial origin like *Acinetobacter calcoaceticus* and usually regarded as industrial emulsifiers. It is an emulsifier for hydrocarbons in water at concentrations as low as 0.001% to 0.01% while liposan is an extracellular water-souble emulsifier synthesized by *Candida lipolytica* and is composed of 83% carbohydrate and 17% protein (Kim et al. 1997, Gorkovenko et al. 1999, Zosim et al. 1982, Cirigliano and Carman 1984).

E. Particulate biosurfactants

Desai and Banat 1997 reported that many of the cellular components like lipoteichoic acid, M protein, fimbriae and protein A have surfactant activity. Particulate biosurfactants, i.e., extracellular membrane vesicles form microemulsions which exert an influence on microbial cells for alkane uptake. Microorganisms like Acinetobacter spp., and *Pseudomonas marginalis* also form vesicles to function as surfactants. For example vesicles of *Acinetobacter* spp. with 20–50 nm diameter and 1.158 cubic g/cm buoyant density comprise of proteins, phospholipids and lipopolysaccharides (Kappeli and Finnerty 1979).

2.2 *Properties of Biosurfactants*

Biosurfactants are more advantageous when compared to their chemical counterparts; they are ecologically safe and acceptable (Cameotra and Makkar 2010).

i) Biodegradability

Biosurfactants are easily broken down by bacteria, fungi or other organisms into basic or smaller components which do not pose a threat to the environment. They are even used in bioremediation and dispersion of oil spills (Mohan et al. 2006, Mulligan et al. 2005).

ii) Low toxicity

Biosurfactants reportedly are negligibly toxic, hence their usage in foods, cosmetics, pharmaceuticals and environmental applications is allowed.

iii) Temperature, pH and ionic strength tolerance

Most of them are active at higher ranges of temperature and pH. Biosurfactants have a salt concentration range upto 10% while chemical surfactants get inactivated at only 20%. McInerney et al. 1990 reported lichenysin synthesized by *Bacillus licheniformis* was unaffected by temperatures (up to 50ºC), pH values of 4.5 to 9 and by NaCl and Ca concentrations up to 50 g/l and 25 g/l respectively. A lipopeptide produced by *Bacillus subtilis* was stable even after autoclaving (121ºC/20 min) and after 6 months at –18ºC; unchanged surface activity at pH 5 to 11 and NaCl concentrations up to 20% (Nitschke and Pastore 2006).

iv) Surface and interfacial activity

It is the important feature for considering a surfactant to be efficient and effective. Efficiency is checked by critical micelle concentration (CMC) while surface and interfacial tensions describe effectiveness of the surfactant. A good surfactant can decrease surface tension of water from 72 to 35 mN/m and the interfacial tension of water/hexadecane from 40 to 1 mN/m (Mulligan 2005). Desai and Banat 1997 suggested biosurfactants to be more effective and efficient and their CMC is about 10–40 times lower than that of chemical surfactants, i.e., smaller amounts of surfactant are needed to get a higher decrease in surface tension.

v) Specificity

Biosurfactants are complex molecules with specific action as they contain specialized functional groups which are significantly used in detoxifying pollutants, de-emulsification of industrial emulsions as well as in specific food, pharmaceutical and cosmetic applications.

3. Biosurfactants produced by different Microorganisms

Recent studies reveal that most of the symbiotic, associative and free living nitrogen fixers produce biosurfactants and play a significant role in bioremediation and also aid in control of plant pathogens. Saisa-ard et al. 2014 reported biosurfactant production by *Azorhizobium doebereinerae, Mesorhizobium* spp. and *Sinorhizobium meliloti* isolated from palm oil contaminated sites in the south of Thailand. Ahmed 2016 reported biosurfactant production by *Rhizobium leguminosarum isolated from Vicia faba* root nodules, growing on mineral salt media (MSM) amended with 1% sunflower oil as the sole source of carbon and ammonium nitrate as the source of nitrogen. Below are the different genera of nitrogen fixers known to produce biosurfactants.

3.1 *Biosurfactant production by Diazotrophs*

a) Azorhizobium

Anuradha et al. 2018 reported biosurfactant production by Azo-rhizobium strains on mineral salt media (MSM) with 2% coconut oil as a source of carbon and ammonium nitrate as nitrogen sources. Crude yield of generated biosurfactant was estimated at 2.5 g/L. The Fourier-Transform infrared spectroscopy (FTIR) analysis of purified biosurfactant confirmed that it was a lipopolysaccharide biosurfactant type.

b) Ochrobactrum

Ibrahim 2018 reported biosurfactant production by *Ochrobactrum anthropi* HM-1 isolated from soil contaminated with engine oil. Bacterial inoculum was added to MSM that contained waste frying oil (2%, v/v) as a carbon source for production of biosurfactants. Apparent yield of biosurfactant after 96 hours of incubation was 4.9 g/l. Surface tension of the growth medium was reduced (70 ± 0.9) to 30.8 ± 0.6 mN/m. It was designated as glycolipids (rhamnolipid) based on Thin-layer chromatography (TLC) and FTIR analysis.

c) Arthrobacter

Biosurfactant called arthrofactin produced by *Arthrobacter* spp. MIS38 has been reported by Morikawa et al. 1993. It is a stronger oil remover than synthetic surfactants, such as Triton X-100 and sodium dodecyl sulphate, in the oil displacement assay. It is reported as five to seven times more effective than surfactin. The minimum surface tension value of arthrofactin was 24 mN/m at a concentration higher than the CMC. It was designated as a lipopeptide type of biosurfactant based on TLC, FTIR, Nuclear magnetic resonance (NMR) and Gas chromatography Mass spectrometry (GC/MS) analysis.

d) Pantoea

Vasileva and Gesheva 2007 reported biosurfactant production by *Pantoea* spp. isolated from Dewart Island's ornithogenic soil, Antarctica. For the biosurfactant's production, bacterial inoculum was added to mineral salts medium (MSM) amended with 2% kerosene or n-paraffins as the sole source of carbon. It was designated as rhamnolipids type of biosurfactant based on TLC analysis pattern.

e) Pseudomonas

Sonja et al. 2019 reported rhamnolipid synthesis by *Pseudomonas aeruginosa* and *Pseudomonas nitroreducens* isolated from soil contaminated with petrol. Glucose and sodium nitrate were the best sources of carbon and nitrogen, with rhamnolipid yield of 5.46 g/l and surface tension of water was reduced from 72 to 37 mN/m (Onwosi et al. 2012)

f) Azospirillum

Ojeda-Morales et al. 2015 investigated the biosurfactant production by *Azospirillum lipoferum* and *Azospirillum brasilense*, isolated from soils contaminated with oil. Biosurfactant that *A. brasilense* produced showed 229 min emulsion stability, 0.1375 g L^{-1} yield, 80% emulsion capacity and 38 mNm^{-1} reduction of surface tension while the bio-tensoactive produced by *A. lipoferum* had 260 min emulsion stability, 0.22 g L^{-1} yield, 90% emulsion capacity and 35.5 mN m^{-1} reduction of surface tension.

g) Serratia

Varied carbon sources like 0.2% lactose, 6% cassava flour wastewater, 5% corn waste oil were used for biosurfactant production and oil remediation by *Serratia marcescens*. Cassava flour waste is considered favourable for reducing surface tension up to 25.92 mN/m and CMC to 1.5% (w/v). Infrared spectra revealed anionic polymers with greater stability under wide ranges of temperature, pH and salt concentrations (Helvia et al. 2009).

h) Bacillus

Biosurfactants like surfactin, lichenysin, fengysin and iturin are synthesized by different species of *Bacillus* spp., using varied carbon sources such as glucose, sucrose, glycerol, sodium lactate, crude oil, diesel and olive oil and reduce surface tension upto 24.6 mN/m and CMC to 50 mgL^{-1} (Sonja et al. 2019, Zhou et al. 2015).

i) Paenibacillus

Mesbaiah et al. 2016 documented the production of biosurfactants isolated from Algerian polluted soil by *Paenibacillus popilliae*. Olive oil (1%, v/v) was used as the source of carbon. During exposure to elevated temperatures (70ºC), moderately high salinity (20% NaCl), and at wide range of pH values (2–10), the biosurfactant preserved its properties. Infrared spectroscopy (FTIR) showed that it belonged to the lipopeptide class in its chemical structure. The biosurfactant production by *Paenibacillus* spp. D_9 was reported by Jimoh and Lin 2019. The biosurfactant that *Paenibacillus* spp. D_9 produced, displayed higher hydrophobicity for the long-chain hydrocarbons mixtures tested such as 71.5% diesel fuel, 70.0% engine oil and 76.0% n-paraffin. Suitable conditions for growth and biosurfactant production were provided in a medium containing 10% (v/v) diesel fuel. The apparent biosurfactant yield was 4.7 g/L and decreased surface tension used against the carbon source from 71.4 to 30.1 mN/m. With emulsification efficiencies against a wide range of hydrophobic contaminants, the CMC value of the biosurfactant was 200 mg/L. The study showed that the genus *Paenibacillus* provided a low molecular weight lipopeptide biosurfactant using various physiochemical and analytical methods.

j) Enterobacter

Biosurfactant production by a microbial consortium of *Enterobacter cloacae* and *Pseudomonas* spp. has been reported by Darvishi et al. 2011, isolated from soil contaminated with heavy crude oil in South Iran. A biosurfactant mixture with excessive oil spreading and emulsification properties was produced by the consortium. At temperatures up to 70ºC, pressures up to 6000 psia, salinities up to 15% (w/v), and pH range 4–10, this consortium was able to grow and produce biosurfactant. For biosurfactant production, the optimum temperature and pH value were found to be 40ºC and 7.0 respectively. When the cells were grown on a MSM containing 1.0% (w/v) olive oil, 1.0% (w/v) sodium nitrate amended with 1.39% (w/v) K_2HPO_4 at 40ºC and 150 rpm after 48 h of incubation, these conditions gave the best biosurfactant production of 1.74 g/1. Jadhav et al. 2011 recorded biosurfactant production by *Enterobacter* spp. MS16 isolated from soil contaminated with diesel. For production of biosurfactant, 2% Enterobacter inoculum was added to MSM that contained natural wastes as carbon sources at a concentration of 1% w/v. Groundnut oil cake, sunflower oil cake and molasses were among the wastes used. Analysis of biosurfactant confirmed that it was composed of glucose, galactose and arabinose, and a fatty acid moiety at C_{16} and C_{18}.

k) Stenotrophomonas

Nogueira et al. 2020 documented biosurfactant production by *Stenotrophomonas maltophilia* growing in a saline mineral medium amended with 10% waste soybean oil. An emulsification index of 78.57, 54.07 and 58.62%, using soybean, corn, and diesel oils, respectively, proved the emulsifying property of the biomolecule produced. The yield of bioemulsifier produced was 2.8 g/L, with an anionic and polymeric character (37.6% lipids, 28.2% proteins and 14.7% carbohydrates), which was confirmed by FTIR.

l) Azotobacter

A number of workers have reported the production of biosurfactant by *Azotobacter* spp. Thavasi et al. 2009 reported that biosurfactant can be produced by *Azotobacter chroococcum* on economically cheaper carbon sources such as waste motor lubricant oil or peanut oil cake. The component biosurfactant of *A. chroococcum* is a lipopeptide with a 31.30:68.69 percent mixture of lipid and protein. The biosurfactant's mass spectral analysis showed ionized compounds with molecular weights of m/z = 326.5, 663.4, and 1,347.3 which could represent the molecules of lipids and proteins, respectively. Devianto et al. 2020 recorded the biosurfactant production by *Azotobacter vinelandii* with molasses and glucose as sources of carbon. Thavasi et al. 2006 reported development of biosurfactant by marine nitrogen fixing bacterium *A. chroococcum* growing on MSM with 0.5% crude oil as the sole source of carbon for application in crude oil degradation.

m) Burkholderia

Wattanaphon et al. 2008 documented the production of biosurfactants isolated from fuel-oil contaminated soil by *Burkholderia cenocepacia* BSP3. Suitable conditions for growth and biosurfactant production were provided by glucose containing a medium supplemented with nitrate or sunflower seed oil. Analysis of the purified biosurfactant produced by *B. cenocepacia* BSP3 confirmed that it was a glucolipid biosurfactant type having a molecular mass of 550.4 g mol^{-1}. The apparent biosurfactant yield was 6.5± 0.7 gl^{-1}. Tavares et al. 2013 investigated the biosurfactant (rhamnolipid) production by *Burkholderia kururiensis* $KP23^T$, a N-fixing, plant growth-enhancing and trichloroethylene (TCE)-degrading bacterium. The biosurfactant produced by *B. kururiensis* $KP23^T$ was identified as a rhamnolipid. Suitable conditions for growth and biosurfactant production were provided by MSM medium supplemented with glycerol. The apparent biosurfactant yield was 0.78 g/l. Dubeau et al. 2009 also recorded rhamnolipids type of biosurfactant production by *Burkholderia thailandensis* and *Burkholderia pseudomallei* when grown in nutrient broth amended with 4% canola oil or 4% glycerol. Rhamnolipids produced by *B.*

thailandensis decreased surface tension of water to 42 mN/m at a CMC of around 225 mg/L.

n) Frankia

Frankia spp. is reported to produce neutral, glycolipids and polar lipids. Their biosurfactant molecular structures have not been determined yet (Kugler et al. 2012).

3.2 Biosurfactants produced by Cyanobacteria

Phylum Cyanobacteria and diatoms garner copious promissory exopolysaccharides (EPS). EPS of biological origin being non toxic, and non hazardous impart and facilitate compensation of environmental losses due to industrialization. Structurally diverse secondary metabolites from Cyanobacteria like *Moorea bouillonii* are proposed as novel biosurfactants. They are efficient photoautotrophs even under controlled conditions, and are able to synthesize novel, economically potent EPSs, like emulcyan; somocystinamide A; apratoxin from *Cylindrotheca closterium, Cyanobacterium Phormidium* spp.; *Lyngbya majuscula*; *Moorea producens* respectively. These organisms are explored for their industrial significance to unveil potent EPSs and their improved forms (Jakia et al. 2020, Jose et al. 2014, Eduardo et al. 2016).

3.3 Biosurfactants produced by Nitrifiers/denitrifiers

i. Nitrosomonas

(Hydroxy) fatty acids bound to proteinogenic or non-proteinogenic amino acids form lipoamino acid biosurfactants, for example by *Nitrosomonas europaea*. However, not many reports exist about biosurfactants produced by nitrifying bacteria (Kubicki et al. 2019).

ii. Pseudomonas

Nitrate reducing or denitrifying bacteria like *Pseudomonas stutzeri* are reported to remediate hydrocarbons, both aliphatic and aromatic from varied polluted environments like petroleum reservoirs. Glycolipids are the significant biomolecules reported along with other fatty acids and lipids in the biosurfactant produced when grown on glycerol and glucose media with CMC of 35 mgL^{-1} (Fuqiang et al. 2018).

4. Applications of Biosurfactants produced by Nitrogen Fixers

In many industries such as agriculture, soil and water remediation, microbial enhanced oil recovery (MEOR), food and cosmetic industry, biomedical science and nanotechnology, most biosurfactants have potential applications (Naughton et al. 2019, Singh et al. 2019).

4.1 Applications in Agriculture

In agriculture-related areas, biosurfactants can be widely used to improve the biodegradation of contaminants in order to improve quality of agricultural soils, indirectly promote plant growth, as they have antimicrobial activity, and can also

Table 1. Biosurfactant production by different nitrogen fixing bacteria grown on various carbon sources and their extraction and characterization methods employed.

Name of the isolate	Biosurfactant yield (g/l)	Carbon and Nitrogen source used	Method of extraction	Characterization method employed	Author and year of publication
Azotobacter chrococcum	4.6 g/l	Crude oil, Peanut oil cake	Chloroform: Methanol:2:1	FTIR MS	Thavasi et al. 2006, 2009
Azotobacter vinelandii	9.19 g/l	Glucose, Molasses	Chloroform: Methanol:2:1	Not Mentioned	Devianto et al. 2020
Azo-rhizobium	2.5 g/l	Coconut oil (2%) Ammonium nitrate	Chloroform: Methanol:2:1	FTIR TLC	Anuradha et al. 2018
Azorhizobium doebereinerae, Mesorhizobium spp., *Sinorhizobium meliloti*	2.81 g/l, 1.78 g/l, 1.17 g/l	Palm oil (1%), Rubber serum (1%), Palm oil (1%)	Ethyl acetate	Not Mentioned	Saisa-ard et al. 2014
Rhizobium leguminosarum	NA	Sunflower oil (1%), Sesame oil (1%) Ammonium nitrate	Chloroform: Methanol:1:1	FTIR	Ahmed 2016
Azospirillum brasilense	0.1375 g/l	Crude oil Glucose (2%)	Chloroform: Methanol:2:1	FTIR TLC	Ojeda-Morales et al. 2015
Azospirillum lipoferum	0.22g/l	Crude oil Glucose (2%)	Chloroform: Methanol:2:1	FTIR TLC	Ojeda-Morales et al. 2015
Burkholderia cenocepacia BSP3	6.5 g/l	Glucose, Sunflower seed oil Nitrate	Ethyl acetate	TLC/MS	Wattanaphon et al. 2008
Burkholderia kururiensis KP23^T	0.78 g/l	Glycerol (3%) Yeast extract	Ethyl acetate	LTQ Orbitrap Hybrid Mass Spectrometer	Tavares et al. 2013

Table 1 contd. ...

...Table 1 contd.

Name of the isolate	Biosurfactant yield (g/l)	Carbon and Nitrogen source used	Method of extraction	Characterization method employed	Author and year of publication
Burkholderia thailandensis and *Burkholderia pseudomallei*	1.473	Glycerol, Canola oil (4%)	Ethyl acetate	LC/MS	Dubeau et al. 2009
Enterobacter cloacae and *Pseudomonas* spp. (consortium, ERCPPI-2)	1.74 g/l	Olive oil (1%) Sodium nitrate (1%)	Chloroform: Ethanol:2:1	TLC^{1}H NMR/^{13}C NMR MS	Darvishi et al. 2011
Enterobacter spp. MS16	1.5 g/L	Sunflower oil cake, Groundnut oil cake	Chloroform: Methanol:2:1	TLC, FTIR NMR, GC/MS	Jadhav et al. 2011
Stenotrophomonas maltophilia	2.8 g/L	Waste soybean oil (10%)	Acetone: Ethanol:1:1	FTIR	Nogueira et al. 2020
Paenibacillus popilliae	0.5 g/l	Olive oil	Chloroform:Methanol:2:1	FTIR	Mesbaiah et al. 2016
Paenibacillus spp. D9	4.7 g/L	Diesel fuel (10%)	Silica gel column with chloroform: Methanol	TLC FTIR LC/MS GC/MS	Jimoh and Lin 2019
Ochrobactrum anthropi HM-1	4.9 g/l	Waste frying oil (2%)	Chloroform: Methanol:2:1	TLC FTIR	Ibrahim 2018
Arthrobacter spp.	2.7 g/l	Peptone Yeast extract	Hexane Extraction HPLC	TLC FTIR NMR GC/MS	Morikawa et al. 1993
Pantoea spp.	1.2 g/l	n-Paraffins (2%), Kerosene	Chloroform: Methanol:2:1	TLC	Vasileva and Gesheva 2007

Pseudomonas nitroreducens	5.46 g/l	Palm oil (10%), Diesel (10%), Glucose Sodium nitrate	Acetone precipitation	Not Mentioned	Onwosi et al. 2012
Pseudomonas aeruginosa	2.65 g/l	Sodium citrate 0.5% Yeast extract	Silica gel column with chloroform: Methanol:water 60:20:0.5 Methanol precipitation	TLC FTIR NMR GC/MS	Rajni et al. 2018
Pseudomonas stutzeri	6.1 g/l	Glucose, Glycerol Nitrate	Silica gel column with chloroform: Methanol:water 5:4:1 Methanol precipitation	TLC FTIR GC-FID	Fuqiang et al. 2018
Azotobacter chrococcum	2.7 g/l	Sunflower oil, Heavy oil 150 Yeast extract, Ammonium sulfate	Chloroform: Methanol:2:1	Not mentioned	Auhim and Mohamed 2013
Serretia marcescens	NA	Lactosee, Corn oil Cassava flour waste water	Chloroform: Methanol:1:2	FTIR Gel chromatography	Helvia et al. 2019
Bacillus spp.	NA	Glucose Peptone	Methanol	Sephadex LH-20 chromatography, RP-HPLC ESI-QTOF-MS	Shimei et al. 2019
Frankia spp.	NA	Glucose N2, NH4 containing media	NA	NA	Kugler et al. 2012
Moorea bouillonii	NA	Marine Cyanobacteria	Chloroform	Chiral phase HPLC NMR FTIR	Jakia et al. 2020

increase plant microbe interactions that are beneficial to plants. The existence of organic and inorganic contaminants that impart abiotic stress to a cultivated crop plant, affects the productivity of agriculture land. The bioremediation process is needed to increase the quality of such soils polluted by hydrocarbons and heavy metals (Sun et al. 2006). Since biosurfactants are known to enhance bioavailability and biodegrade hydrophobic compounds, various technologies, such as soil washing technology and clean-up combined technologies, use biosurfactants to effectively extract hydrocarbons and metals, respectively (Sachdev and Swaranjit 2013). Biosurfactants accelerate the very critical phenomenon of desorption of hydrophobic contaminants closely bound to soil particles which is quite important in the process of bioremediation. There are several reports on biosurfactants produced by nitrogen fixing bacteria, to improve the health of agricultural soil through the soil remediation process.

Addition of biosurfactants can be expected to increase biodegradation of hydrocarbons by mobilization, solubilization or emulsification (Nguyen et al. 2008, Nievas et al. 2008). Thavasi et al. 2006 reported biodegradation properties of biosurfactant produced by free living, nitrogen fixing strain of *Azotobacter chroococcum*. They recorded that at the end of 120 hours of incubation, 58% of the crude oil was degraded. Devianto et al. 2020 analyzed application of biosurfactants produced by *Azotobacter vinelandii* and Tween 80 as possible soil washing agents to remove total petroleum hydrocarbons (TPH). Odukkathil and Vasudevan 2013 stated *Arthrobacter* spp. produced the rhamnolipid type of biosurfactant which solubilized endosulfan and its main endosulfate metabolite. Better emulsification of endosulfan was shown by the biosurfactant produced by *Arthrobacter* spp. and 99% of endosulfan and 94% of endosulfate were degraded.

Mesbaiah et al. 2016 also reported that biosurfactants produced by *Paenibacillus popilliae* were capable of solubilization of polyaromatic hydrocarbons (phenanthrene, anthracene, and pyrene) in water which was stated to be higher. Application of biosurfactant at a concentration of 1.5 g/l resulted in a solubilization rate of phenanthrene, anthracene and pyrene of 5.14 ± 0.130, 4.17 ± 0.050 and 2.91 ± 0.017 times higher than their respective controls. Tavares et al. 2013 reported rhamnolipid production by N2-fixing, and plant growth-promoting bacterium, *Burkholderia kururiensis* KP23T that could degrade Trichloroethylene (TCE). Biodegradative properties of biosurfactants produced by *Stenotrophomonas maltophilia* were evaluated under submerged conditions for their pyrene bioremediation ability in minimal media spiked with pyrene at 20, 50, and 100 $\mu g\ mL^1$. They recorded that within 15–20 days, this strain could degrade 87% of the pyrene (Singh et al. 2015).

Pesticide solubilization by *Burkholderia cenocepacia* BSP3 (glucolipid) was tested by Wattanaphon et al. 2008. They recorded that this strain exhibited better solubilization efficiency of pesticides (ethyl parathion, methyl parathion and trifluralin) than chemical surfactants such as Tween 80 and sodium dodecyl sulphate. Ramirez et al. 2020 reported rhamnolipids produced by *Burkholderia* spp. degrade dibenzothiophene (DBT). The lipophilic nature of DBT enables it to concentrate and bioaccumulate in the environment via the food chain, giving it an ecotoxicological and food safety risk. El-Sheekh et al. 2014 suggested that cyanobacteria like *S. platensis* and *N. punctiforme* are able to carry out bioremediation in areas polluted with hydrocarbons. Concentration of *S. platensis* upto 1%, if applied completely treats or removes oils. While Santra S.C. et al. 2015 suggest the presence of cyanobacteria along with other aerobic organotrophs are help to degrade petroleum and other complex organic compounds.

Several biosurfactants produced by nitrogen fixing microbes have demonstrated antimicrobial activity against plant pathogens and are thus considered to be promising bio-control molecules for sustainable agriculture. It is recognized that biosurfactants developed by rhizobacteria have antagonistic properties (Nihorimbere et al. 2011). Also, their antimicrobial mechanism is based on the disruption of the cytoplasmic membrane, decreasing the possibility of microorganism resistance (Haba et al. 2003, Vollenbroich et al. 1997). Jadhav et al. 2011 reported glycolipid type of biosurfactants produced by *Enterobacter* spp. MS16 having antifungal activity against *Penicillium chrysogenum*, *Aspergillus niger* and inhibition of fungal spore germination. Biosurfactants developed by *Stenotrophomonas maltophilia* have a bactericidal property against pathogenic *P. aeruginosa*, reported by Tripathi et al. 2019. Ochrosin, a novel biosurfactant developed by the halophilic *Ochrobactrum* spp. strain BS-206, reported strong bactericidal activity against both Gram-negative and Gram-positive bacteria and various *Candida* strains with minimum inhibitory concentration (MIC) values that ranged from 4.68 to 150 ug mL^{-1} (Kumar et al. 2014). Recently it has been reported that rhamnolipid type of biosurfactants produced by *Pseudomonas* spp. regulate the quorum sensing (QS) mechanism (Dusane et al. 2010). Yalcin and Ergene 2009 have reported antifungal activity of the lipopeptide biosurfactant type produced by *Burkholderia cepacia*. Singh et al. 2020 reported biosurfactants produced by *Stenotrophomonas* spp. BAB-6435 have antifungal and antibacterial activity.

The presence of even low amounts of heavy metals in the soils, due to their highly toxic nature, has been shown to have severe effects on plants such as root tissue necrosis and foliage purpling. The utility of biosurfactants for the bioremediation/removal of soil polluted with heavy metals is mainly based on their ability to form metal complexes. The anionic biosurfactants create complexes with metals in a nonionic form by ionic bonds. These bonds are stronger than the bonds of the metals to the soil, and due to the lowering of the interfacial tension, metal-biosurfactant complexes are desorbed from the soil matrix to the soil solution. Biosurfactant micelles are also able to extract metal ions from soil surfaces. The micelles polar head groups bind metals that mobilize them in water. Mohammed Yahya Khan et al. 2015 suggest another level of utilizing heavy metals after degradation or biosurfactant activity. The reaped metal can be employed for the green-nanoparticle production or plant extract-microbe-metal unification. With the biosurfactant employed, efficiency of removal was 52.2% for Mn, 44.0% for Zn, 37.7% for Cd, 32.5% for Pb, 31.6% for As and 24.1% for Cu (Mulligan and Gibbs 2004, Juwarkar et al. 2007). Patil et al. 2012 reported rhamnolipids produced by *Stenotrophomonas koreensis* removed heavy metals from 200 ppm metal solutions. The availability of micronutrients to plants is further increased by biosurfactants produced by rhizobacteria (Singh et al. 2018, Pacwa-Plociniczak et al. 2011). Jimoh and Lin 2020 reported lipopeptides produced by *Paenibacillus* spp. remove heavy metals from contaminated sites. For Ca, Cu, Fe, Mg, Ni and Zn, the removal efficiency was 85.90%, 98.68%, 99.97%,

63.28%, 99.93% and 94.22% respectively. Yang et al. 2016 reported glycolipid type of biosurfactants produced by *Burkholderia* spp. could remove heavy metals from contaminated soils. *Nostoc linckia, Neptis rivularis, Nostoc muscorum, Anabaena subcylindrica, Gloeocapsa* spp. are some biosurfactants used for heavy metal bioremediation due to their increased interior pH as a result of photosynthesis, which allows them to lock the contaminants in their biofilms (Singh et al. 2016). Cyanobacteria being safe also contribute to a sustainable environment and agriculture. Its application in molecular biology and genetic engineering is paving ways to bio-fuel, biosurfactant production and surplus food supply (Singh et al. 2016).

In order to ensure the beneficial impact of rhizobacteria on plants, it is very important for these microbes to communicate with plant surfaces such as roots (Nihorimbere et al. 2011). In order to create an association with the plant, microbial factors such as motility, the ability to form biofilms on the root surfaces and the release of quorum sensing (QS) molecules are required. For the synthesis of antifungal compounds and for root surface colonization by rhizobacteria, QS molecules such as acyl homoserine lactone (AHL) are necessary. Studies also show that the concentration of these molecules was high in the rhizosphere region compared to that of bulk soil (soil away from the roots of plants) indicating the role of AHL and AHL-like molecules in rhizosphere competence (beneficial microorganism ability to colonize the surface of root). It is also stated that these AHLs contribute to the regulation of biosurfactants and exopolysaccharides, which are important for biofilm formation. Long-chain AHLs act as biosurfactants in *Rhizobium etli*, the symbiotic partner of *Phaseolus vulgaris* that regulates the swarming phenotype and surface movement that induces liquid flow through a superficial tension gradient, and thus facilitates the bacteria's attachment to surfaces (Daniels et al. 2006).

5. Conclusion and Future Perspectives

This chapter presents a description of biosurfactants produced by nitrogen fixing bacteria and their uses in agriculture. Chemical surfactants are familiarly exploited due to different functional properties in many commercial sectors in addition to agriculture. However, due to soil pollution potency of chemical surfactants it is suggested for utilization of biosurfactants that have low toxicity and are biodegradable. They are less successful commercially due to higher synthesis costs. Optimized growth conditions utilizing low cost renewable substrates (agro-industrial wastes) and novel, efficient methods for isolation and purification of biosurfactants would make their synthesis more economically feasible and commercially viable. Most biosurfactant production by nitrogen fixing microorganisms is studied *in vitro* while very less knowledge is available *in situ*. More investigation and studies are needed to explain the significant role of biosurfactants of nitrogen fixers and nitrifying bacteria.

Off the varied environmental pollutants, oil spills are the most calamitous causing colossal damage to ecosystems at all trophic levels. Most of the research in the past was focused on carbon degraders; however the impact of these hydrocarbon pollutants on microbes responsible for nitrification or nitrogen fixing bacteria has been less explored. Nitrifiers, ammonia oxidizing bacteria and archaea are sensitive to hydrocarbon toxicity and hence nitrification process can be lost in different coastal ecosystems. In addition microbial community shifts also may be found after the oil spills which need to be established using different metagenomic studies. Nevertheless, a better understanding of nitrogen fixers and nitrifying bacteria with biosurfactant production will add information to better understand the nitrogen cycle, improved bioremediation processes and agricultural systems to attain sustainability.

Acknowledgements

Authors (HB, HQ, SA) acknowledge the partial financial support under DST-PURSE (C-DST-PURSE-II/43/2018).

References

Abalos, A., A. Pinazo, M.R. Infante, M. Casals, F. Garcia et al. 2001. Physicochemical and antimicrobial properties of new rhamnolipids produced by *Pseudomonas aeruginosa* AT10 from soybean oil refinery wastes. Langmuir. 17(5): 1367–1371.

Abdel-Mawgoud, A.M., M.M. Aboulwafa and N.A.H. Hassouna. 2009. Characterization of rhamnolipid produced by *Pseudomonas aeruginosa* isolates Bs20. Appl. Biochem. Biotechnol. 157(2): 329–345.

Abdoulaye Soumare, Abdala G. Diedhiou, Moses Thuita, Mohamed Hafidi, Yedir Ouhdouch et al. 2020. Exploiting biological nitrogen fixation: a route towards a sustainable agriculture. Plants 9(8): 1011.

Ahmed, S.A. 2016. Determining the optimum conditions for bioemulsifier produced by *Rhizobium leguminosarum* biovar viciae isolated from *Vicia faba* root nodules. Iraqi J. Sci. 57(3C): 2188–2196.

Almatawah, Q. 2017. An indigenous biosurfactant producing *Burkholderia cepacia* with high emulsification potential towards crude oil. J. Environ. Anal. Toxicol. 7(6): 528.

Anders, T. Nancy, A. Schultz, R. David, D. Benson et al. 1989. Differences in fatty acid composition between vegetative cells and N2-fixing vesicles of *Frankia* spp. strain CpI1, 3399–3403. The Proceedings of the National Academy of Sciences, USA. Vol. 86. J. Microbiol.

Anita Sellstedt and Kerstin H. Richau. 2013. Aspects of nitrogen-fixing Actinobacteria, in particular free-living and symbiotic Frankia. FEMS Microbiol. Lett. 342(2): 179–86.

Anuradha, P., M. Rasika and K. Aruna. 2018. Optimization of bio-surfactant production by Azo-rhizobium strain isolated from oil-contaminated soil. GSC Biol. Pharm. Sci. 3(3): 035–046.

Auhim, H.S. and A.I. Mohamed. 2013. Effect of different environmental and nutritional factors on biosurfactant production from *Azotobacter chroococcum*. IJAPBS 2: 477–481.

Bagampriyal Selvaraj and Sadhana Balasubramanian. 2020. Formulations of BGA for Paddy Crop. Agroecosystems. [Online First], IntechOpen, DOI: 10.5772/intechopen.92821.

Banat, I.M. 1995. Biosurfactants production and possible uses in microbial enhanced oil recovery and oil pollution remediation: a review. Bioresour. Technol. 51(1): 1–12.

Banat, I.M., A. Franzetti, I. Gandolfi, G. Bestetti, M.G. Martinotti et al. 2010. Microbial biosurfactants production, applications and future potential. Appl. Microbiol. Biotechnol. 87(2): 427–444.

Blackwell, P.S. 2000. Management of water repellency in Australia and risks associated with preferential flow, pesticide concentration and leaching. J. Hydrol. 231–232: 384–395.

Cameotra, S.S., R.S. Makkar, J. Kaur and S.K. Mehta. 2010. Synthesis of biosurfactants and their advantages to microorganisms and mankind. Adv. Exp. Med. Biol. 672: 261–280.

Cirigliano, M.C. and G.M. Carman. 1984. Isolation of a bioemulsifier from *Candida lipolytica*. Appl. Environ. Microbiol. 48(4): 747–750.

Cirigliano, M.C. and G.M. Carman. 1985. Purification and characterization of liposan, a bioemulsifier from *Candida lipolytica*. Appl. Environ. Microbiol. 50(4): 846–850.

Cooper, D.G. and J.E. Zajic. 1980. Surface-active compounds from microorganisms. Adv. Appl. Microbiol. 26: 229–253.

Cooper, D.G. and D.A. Paddock. 1984. Production of a biosurfactant from *Torulopsis bombicola*. Appl. Environ. Microbiol. 47(1): 173–176.

Daniels, R., S. Reynaert, H. Hoekstra, C. Verreth, J. Janssens et al. 2006. Quorum signal molecules as biosurfactants affecting swarming in *Rhizobium etli*. Proc. Natl. Acad. Sci. USA 103: 14965–14970.

Darvishi, P., S. Ayatollahi, D. Mowla and A. Niazi. 2011. Biosurfactant production under extreme environmental conditions by an efficient microbial consortium, ERCPPI-2. Colloids and Surfaces B: Biointerfaces 84(2): 292–300.

Deleu, M. and M. Paquot. 2004. From renewable vegetables resources to microorganisms. New trends in surfactant. CR Chimie 7: 641–646.

Desai, J.D. and I.M. Banat. 1997. Microbial production of surfactants and their commercial potential. Microbiol. Mol. Biol. Rev. 61(1): 47–64.

Devianto, L.A., C.E.L. Latunussa, Q. Helmy and E. Kardena. 2020. Biosurfactants production using glucose and molasses as carbon sources by *Azotobacter vinelandii* and soil washing application in hydrocarbon-contaminated soil. In IOP Conference Series: Earth and Environmental Science 475(1): 012075.

Déziel, É., G. Paquette, R. Villemur, F. Lépine, J. Bisaillon et al. 1996. Biosurfactant production by a soil *Pseudomonas* strain growing on polycyclic aromatic hydrocarbons. Appl. Environ. Microbiol. 62(6): 1908–1912.

Dhara, P. Sachdev and Swaranjit S. Cameotra. 2013. Biosurfactants in agriculture. Appl. Microbiol. Biotechnol. 97(3): 1005–1016.

Dubeau, D., E. Déziel, D.E. Woods and F. Lépine. 2009. *Burkholderia thailandensis* harbors two identical rhl gene clusters responsible for the biosynthesis of rhamnolipids. BMC Microbiol. 9(1): 263.

Dusane, D., P. Rahman, S. Zinjarde, V. Venugopalan, R. McLean et al. 2010. Quorum sensing; implication on rhamnolipid biosurfactant production. Biotech Genetic Eng. Rev. 27: 159–184.

Eduardo, J. Gudiña, José A. Teixeira and Lígia R. Rodrigues. 2016. Biosurfactants produced by marine microorganisms with therapeutic applications. Mar. Drugs 14(2): 38.

El-Sheekha, M.M. and R.A. Hamouda. 2014. Biodegradation of crude oil by some cyanobacteria under heterotrophic conditions. Desalination and Water Treatment 52: 1448–145.

Florence, M., B.A. Alexander, P. Natasha, C.S. Lance, W.P. John et al. 2018. Exploring the alternatives of biological nitrogen fixation. Metallomics 10: 523–538.

Fuqiang Fan, Baiyu Zhang, Penny L. Morrillb and Tahir Husain et al. 2018. Isolation of nitrate-reducing bacteria from an offshore reservoir and the associated biosurfactant production, RSC Adv. 47(8): 26596–26609.

Geddes, B.A. and I.J. Oresnik. 2016. The mechanism of symbiotic nitrogen fixation. pp. 69–97. *In*: The Mechanistic Benefits of Microbial Symbionts. Springer, Cham.

Gopalakrishnan, S., V. Srinivas and S. Samineni. 2017. Nitrogen fixation, plant growth and yield enhancements by diazotrophic growth-promoting bacteria in two cultivars of chickpea (*Cicer arietinum* L.). Biocatal. Agric. Biotechnol. 11: 116–123.

Gorkovenko, A., J. Zhang, R.A. Gross and D.L. Kaplan. 1999. Control of unsaturated fatty acid substituents in emulsans. Carbohydr. Polym. 39(1): 79–84.

Haba, E., A. Pinazo, O. Jauregui, M.J. Espuny, M.R. Infante et al. 2003. Physicochemical characterization and antimicrobial properties of rhamnolipids produced by *Pseudomonas aeruginosa* 47T2 NCBIM 40044. Biotechnol. Bioeng. 81(3): 316–322.

Hangcheng Zhou, Jixiang Chen, Zhi Yang, Bo Qin, Yanlin Li et al. 2015. Biosurfactant production and characterization of *Bacillus* spp. ZG0427 isolated from oil-contaminated soil. Ann. Microbiol. 65: 2255–2264.

Hélvia, W.C. Araújo, Rosileide F.S. Andrade, Dayana Montero-Rodríguez, Daylin Rubio-Ribeaux et al. 2019. Sustainable biosurfactant produced by *Serratia marcescens* UCP 1549 and its suitability for agricultural and marine bioremediation applications. Microb. Cell Fact. 18: 2.

Hoffman, B.M., D. Lukoyanov, Z.Y. Yang, D.R. Dean, L.C. Seefeldt et al. 2014. Mechanism of nitrogen fixation by nitrogenase: the next stage. Chem. Rev. 114(8): 4041–4062.

Hommel, R., O. Stiiwer, W. Stuber, D. Haferburg, H.P. Kleber et al. 1987. Production of water-soluble surface-active exolipids by *Torulopsis apicola*. Appl. Microbiol. Biotechnol. 26(3): 199–205.

Hopkinson, M.J., H.M. Collins and G.R. Goss. 1997. Pesticide formulations and application systems: ASTM Committee E-35 on Pesticides. ASTM International 17(1328): 1–331.

Hu, Y. and L.K. Ju. 2001. Purification of lactonic sophorolipids by crystallization. J. Biotechnol. 87(3): 263–272.

Humera Quadriya, Mohammed Imran Mir, K. Surekha, Gopal S. Krishnan, Yahya M. khan et al. 2020. Bee Hameeda 2020. Diverse microbial entities in N cycling to acquire environmental equilibrium, In book- Rhizosphere Microbes.

Ibrahim, H.M. 2018. Characterization of biosurfactants produced by novel strains of *Ochrobactrum anthropi* HM-1 and *Citrobacter freundii* HM-2 from used engine oil-contaminated soil. Egypt. J. Pet. 27(1): 21–29.

Jadhav, M., A. Kagalkar, S. Jadhav and S. Govindwar. 2011. Isolation, characterization, and antifungal application of a biosurfactant produced by *Enterobacter* spp. MS16. Eur. J. Lipid Sci. Technol. 113(11): 1347–1356.

Jakia Jerin Mehjabin, Liang Wei, Julie G. Petitbois, Taiki Umezawa, Fuyuhiko Matsuda et al. 2020. Biosurfactants from marine cyanobacteria collected in sabah, Malaysia. J. Nat. Prod. 83(6): 1925–1930.

Jarvis, F.G. and M.J. Johnson. 1949. A glycolipid produced by *Pseudomonas aeruginosa*. J. Am. Chem. Soc. 71(12): 4124–4126.

Jay Shankar Singh, Arun Kumar, Amar N. Rai and Devendra P. Singh. 2016. Cyanobacteria: A precious bio-resource in agriculture, ecosystem, and environmental sustainability. Front Microbiol. 7: 529.

Jimoh, A.A. and J. Lin. 2019. Production and characterization of lipopeptide biosurfactant producing *Paenibacillus* spp. D9 and its biodegradation of diesel fuel. Int. J. Environ. Sci. Technol. 16(8): 4143–4158.

Jimoh, A.A. and J. Lin. 2020. Biotechnological applications of *Paenibacillus* spp. D9 lipopeptide biosurfactant produced in low-cost substrates. Appl. Biochem. Biotechnol. 1–21.

José de Jesús Paniagua-Michel, Jorge Olmos-Soto and Eduardo Roberto Morales-Guerrero. 2014. Algal and Microbial Exopolysaccharides: New Insights as Biosurfactants and Bioemulsifiers. Adv. Food Nutr. Res. 73. Elsevier Inc.

Juwarkar, A.A., A. Nair, K.V. Dubey, S.K. Singh, S. Devotta et al. 2007. Biosurfactant technology for remediation of cadmium and lead contaminated soils. Chemosphere 68. 10: 1996–2002.

Kakinuma, A., A. Ouchida, T. Shima, H. Sugino, M. Isono et al. 1969. Confirmation of the structure of surfactin by mass spectrometry. Agric. Biol. Chem. 33(11): 1669–1671.

Kappeli, O. and W.R. Finnerty. 1979. Partition of alkane by an extracellular vesicle derived from hexadecane-grown Acinetobacter. J. Bacteriol. 140(2): 707–712.

Karanth, N.G.K., P.G. Deo and N.K. Veenanadig. 1999. Microbial production of biosurfactants and their importance. Curr. Sci. 116–126.

Kim, P., D.K. Oh, S.Y. Kim and J.H. Kim. 1997. Relationship between emulsifying activity and carbohydrate backbone structure of emulsan from *Acinetobacter calcoaceticus* RAG-1. Biotechnol. Lett. 19(5): 457–459.

Kumar, C.G., P. Sujitha, S.K. Mamidyala, P. Usharani, B. Das et al. 2014. Ochrosin, a new biosurfactant produced by halophilic *Ochrobactrum* spp. strain BS-206 (MTCC 5720): purification, characterization and its biological evaluation. Process Biochem. 49(10): 1708–1717.

Lang, S. 2002. Biological amphiphiles (microbial biosurfactants). Curr. Opin. Colloid Interface Sci. 7(1): 12–20.

Lima, T.M., L.C. Procópio, F.D. Brandão, A.M. Carvalho, M.R. Tótola et al. 2011b. Biodegradability of bacterial surfactants. Biodegradation 22: 585–592.

Mahmud, K., S. Makaju, R. Ibrahim and A. Missaoui. 2020. Current progress in nitrogen fixing plants and microbiome research. Plants 9(1): 97.

Martinez-Hidalgo, P. and A.M. Hirsch. 2017. The nodule microbiome: N2-fixing rhizobia do not live alone. Phytobiomes 1: 70–82.

McInerney, M.J., M. Javaheri and D.P. Nagle. 1990. Properties of the biosurfactant produced by *Bacillus licheniformis* strain JF-2. J. Ind. Microbiol. 5(2-3): 95–101.

Medhat Rehan, Erik Swanson and Louis S. Tisa. 2016. Frankia as a Biodegrading Agent. In book: Actinobacteria - Basics and Biotechnological Applications. Edition: 1. InTech Editors. Dharumadurai Dhanasekaran and Yi Jiang.

Mesbaiah, F.Z., K. Eddouaouda, A. Badis, A. Chebbi, D. Hentati et al. 2016. Preliminary characterization of biosurfactant produced by a PAH-degrading *Paenibacillus* spp. under thermophilic conditions. Environ. Sci. Pollut. Res. 23(14): 14221–14230.

Mohamed Yahya Khan, T.H., Swapna T.H. Swapna, Bee Hameeda, Gopal Reddy et al. 2015. Bioremediation of heavy metals using biosurfactants. Advances in Biodegradation and Bioremediation of Industrial Waste. CRC Press, Taylor and Francis group.

Mohan, P.K., G. Nakhla and E.K. Yanful. 2006. Biokinetics of biodegradation of surfactants under aerobic, anoxic and anaerobic conditions. Water Res. 40(3): 533–540.

Morikawa, M., H. Daido, T. Takao, S. Murata, Y. Shimonishi et al. 1993. A new lipopeptide biosurfactant produced by *Arthrobacter* spp. strain MIS38. J. Bacteriol. 175(20): 6459–6466.

Müller, M.M., J.H. Kügler, M. Henkel, M. Gerlitzki, B. Hörmann, M. Pöhnlein, C. Syldatk and R. Hausmann 2012. Rhamnolipids—next generation surfactants? J. Biotechnol, 162(4): 366–80. doi: 10.1016/j.jbiotec.2012.05.022.

Mulligan, C.N. and B.F. Gibbs. 2004. Types, production and applications of biosurfactants. Proc-Indian Nat Sci Acad. Part B. 70(1): 31–56.

Mulligan, C.N. 2005. Environmental applications for biosurfactants. Environ. Pollut. 133(2): 183–198.

Mulqueen, P. 2003. Recent advances in agrochemical formulations. Adv. Colloid Interface Sci. 106: 83–107.

Naughton, P.J., R. Marchant, V. Naughton and I.M. Banat. 2019. Microbial biosurfactants: current trends and applications in agricultural and biomedical industries. J. Appl. Microbiol. 127: 12–28. doi: 10.1111/jam.14243.

Nguyen, T.T., N.H. Youssef, M.J. McInerney and D.A. Sabatini. 2008. Rhamnolipid biosurfactant mixtures for environmental remediation. Water Res. 42(6-7): 1735–1743.

Nievas, M.L., M.G. Commendatore, J.L. Esteves and V. Bucalá. 2008. Biodegradation pattern of hydrocarbons from a fuel oil-type complex residue by an emulsifier-producing microbial consortium. J. Hazard. Mater. 154(1-3): 96–104.

Nihorimbere, V., Marc M. Ongena, M. Smargiassi and P. Thonart. 2011. Beneficial effect of the rhizosphere microbial community for plant growth and health. Biotechnol. Agron. Soc. Environ. 15: 327–337.

Nitschke, M. and G.M. Pastore. 2006. Production and properties of a surfactant obtained from *Bacillus subtilis* grown on cassava wastewater. Bioresour. Technol. 97(2): 336–341.

Nitschke, M. and S.G.V.A.O. Costa. 2007. Biosurfactants in food industry. Trends Food Sci. Technol. 18(5): 252–259.

Nogueira, I.B., D.M. Rodríguez, R.F. da Silva Andradade, A.B. Lins, A.P. Bione et al. 2020. Bioconversion of agro industrial waste in the production of bioemulsifier by *Stenotrophomonas maltophilia* UCP 1601 and Application in Bioremediation Process. Int. J. Chem. Eng. DOI: 10.1155/2020/9434059.

Odukkathil, G. and N. Vasudevan. 2013. Enhanced biodegradation of endosulfan and its major metabolite endosulfate by a biosurfactant producing bacterium. J. Environ. Sci. Health B. 48(6): 462–469.

Ojeda-Morales, M.E., M. Domínguez-Domínguez, M.A. Hernández-Rivera and J. Zavala-Cruz. 2015. Biosurfactant production by strains of *Azospirillum* isolated from petroleum-contaminated sites. Water Air Soil Pollut. 226(12): 401.

Olivia, R., J. Schmitt, S.M.J. Mike and L. Claudia. 2017. Metagenomic potential for and diversity of N cycle driving microorganisms in the Bothnian Sea sediment. Microbiology Open 6(4). PMID: 28544522.

Onwosi, C.O. and F.J.C. Odibo. 2012. Effects of carbon and nitrogen sources on rhamnolipid biosurfactant production by *Pseudomonas nitroreducens* isolated from soil. World J. Microbiol. Biotechnol. 28(3): 937–942.

Pacwa-Płociniczak, M., G.A. Płaza, Z. Piotrowska-Seget and S.S. Cameotra. 2011. Environmental applications of biosurfactants: recent advances. Int. J. Mol. Sci. 12(1): 633–654.

Patil, S.N., B.A. Aglave, A.V. Pethkar and V.B. Gaikwad. 2012. *Stenotrophomonas koreensis* a novel biosurfactant producer for abatement of heavy metals from the environment. Afr. J. Microbiol. Res. 6(24): 5173–5178.

Petrovic, M. and D. Barcelo. 2004. Analysis and fate of surfactants in sludge and sludge-amended soil. Trends Analyt. Chem. 23: 10–11.

Pornsunthorntawee, O., S. Chavadej and R. Rujiravanit. 2009. Solution properties and vesicle formation of rhamnolipid biosurfactants produced by *Pseudomonas aeruginosa* SP4. Colloids Surf B Biointerfaces 72(1): 6–15.

Rahman, P.K. and E. Gakpe. 2008. Production, characterization and applications of biosurfactants-review. Biotechnology 7: 360–370.

Ramirez, C.A.O., A. Kwan and Q.X. Li. 2020. Rhamnolipids induced by glycerol enhance dibenzothiophene biodegradation in *Burkholderia* spp. C3. Eng. J. DOI: 10.1016/j.eng.2020.01.006.

Richards, J.W., G.D. Krumholz, M.S. Chval and L.S. Tisa. 2002. Heavy metal resistance patterns of Frankia strains. Appl. Environ. Microbiol. 68: 923–927.

Rosenberg, E. and E.Z. Ron. 1999. High-and low-molecular-mass microbial surfactants. Appl. Microbiol. Biotechnol. 52(2): 154–162.

Rostas, M. and K. Blassmann. 2009. Insects had it first: surfactants as a defense against predators. Proc. R. Soc. B. 276: 633–638.

Saïdi, S., S. Chebil, M. Gtari and R. Mhamdi. 2013. Characterization of root-nodule bacteria isolated from *Vicia faba* and selection of plant growth promoting isolates. World J. Microbiol. Biotechnol. 29(6): 1099–1106.

Saisa-ard, K., A. Saimmai and S. Maneerat. 2014. Characterization and phylogenetic analysis of biosurfactant-producing bacteria isolated from palm oil contaminated soils. Songklanakarin J. Sci. Technol. 36: 2.

Santi Carole, D. Bogusz and C. Franche. 2013. Biological nitrogen fixation in non-legume plants. Ann. Bot. 111(5): 743–767.

Santra, S.C., A. Mallick and A.C. Samal. 2015. Biofertilizer for Bioremediation. pp. 205–234. *In*: Pati, B.R. and S.M. Mandal (eds.). Recent trends in biofertilizers. I K International Publishing House Pvt. Ltd.

Satpute, S.K., A.G. Banpurkar, P.K. Dhakephalkar, I.M. Banat, B.A. Chopade et al. 2010. Methods for investigating biosurfactants and bioemulsifiers: a review. Crit. Rev. Biotechnol. 30(2): 127–144.

Singh, A., K. Kumar, A.K. Pandey, A. Sharma, S.B. Singh et al. 2015. Pyrene degradation by biosurfactant producing bacterium *Stenotrophomonas maltophilia*. Agric. Res. 4(1): 42–47.

Singh, P., Y. Patil and V. Rale. 2019. Biosurfactant production: emerging trends and promising strategies. J. Appl. Microbiol. 126: 2–13. doi: 10.1111/jam.14057.

Singh, R., B.R. Glick and D. Rathore. 2018. Biosurfactants as a biological tool to increase micronutrient availability in soil: A review. Pedosphere 28(2): 170–189.

Singh, R., S.K. Singh and D. Rathore. 2020. Analysis of biosurfactants produced by bacteria growing on textile sludge and their toxicity evaluation for environmental application. J. Dispers. Sci. Technol. 41(4): 510–522.

Sonja, Kubicki, B. Alexander, Nadine Katzke, Karl-Erich Jaeger, Anita Loeschcke et al. 2019. Marine biosurfactants: biosynthesis, structural diversity and biotechnological applications, Mar. Drugs 17. 408. doi:10.3390/md17070408.

Street, J.C. 1969. Methods of removal of pesticides residues. Canad. Med. Ass. J. 100: 154–160.

Sun, X., L. Wu and Y. Luo. 2006. Application of organic agents in remediation of heavy metals-contaminated soil. Ying Yong Sheng Tai Xue Bao. 17: 1123–1128.

Tavares, L.F., P.M. Silva, M. Junqueira, D.C. Mariano, F.C. Nogueira et al. 2013. Characterization of rhamnolipids produced by wild-type and engineered *Burkholderia kururiensis*. Appl. Microbiol. Biotechnol. 97(5): 1909–1921.

Thavasi, R. 2006. Biodegradation of crude oil by nitrogen fixing marine bacteria *Azotobacter chroococcum*, R. Thavasi, S. Jayalakshmi, T. Balasubramanian and Ibrahim M. Banat CAS in Marine Biology. Res. J. Microbiol. 1(5): 401–408.

Thavasi, R., V.S. Nambaru, S. Jayalakshmi, T. Balasubramanian, I.M. Banat et al. 2009. Biosurfactant production by *Azotobacter chroococcum* isolated from the marine environment. Mar. Biotechnol. 11(5): 551.

Tripathi, V., V.K. Gaur, N. Dhiman, K. Gautam, N. Manickam et al. 2019. Characterization and properties of the biosurfactant produced by PAH-degrading bacteria isolated from contaminated oily sludge environment. Environ. Sci. Pollut. Res. 1–11.

Vasileva-Tonkova, E. and V. Gesheva. 2007. Biosurfactant production by antarctic facultative anaerobe *Pantoea* spp. during growth on hydrocarbons. Curr. Microbiol. 54(2): 136–141.

Vollenbroich, D., G. Pauli, M. Ozel and J. Vater. 1997. Antimycoplasma properties and application in cell culture of surfactin, a lipopeptide antibiotic from *Bacillus subtilis*. Appl. Environ. Microbiol. 63(1): 44–49.

Wattanaphon, H.T., A. Kerdsin, C. Thammacharoen, P. Sangvanich, A.S. Vangnai et al. 2008. A biosurfactant from *Burkholderia cenocepacia* BSP3 and its enhancement of pesticide solubilization. J. Appl. Microbiol. 105(2): 416–423.

Yalcin, E. and A. Ergene. 2009. Screening the antimicrobial activity of biosurfactants produced by microorganisms isolated from refinery wastewaters. J. Appl. Biol. Sci. 3(2): 163–168.

Yang, Z., Z. Zhang, L. Chai, Y. Wang, Y. Liu et al. 2016. Bioleaching remediation of heavy metal-contaminated soils using *Burkholderia* spp. Z-90. J. Hazard. Mater. 301: 145–152.

Zosim, Z., D. Gutnick and E. Rosenberg. 1982. Properties of hydrocarbon in water emulsions stabilized by Acinetobacter RAG1 emulsan. Biotechnol. Bioeng. 24(2): 281–292.

Index

S

Y